Numerical Simulation of Mechanical Behavior of Composite Materials

Lecture Notes on Numerical Methods in Engineering and Sciences

Aims and Scope of the Series

This series publishes text books on topics of general interest in the field of computational engineering sciences.

The books will focus on subjects in which numerical methods play a fundamental role for solving problems in engineering and applied sciences. Advances in finite element, finite volume, finite differences, discrete and particle methods and their applications are examples of the topics covered by the series.

The main intended audience is the first year graduate student. Some books define the current state of a field to a highly specialised readership; others are accessible to final year undergraduates, but essentially the emphasis is on accessibility and clarity.

The books will be also useful for practising engineers and scientists interested in state of the art information on the theory and application of numerical methods.

Titles:

Numerical Simulation of Mechanical Behavior of Composite Materials

Sergio Oller

Centre Internacional de Mètodes Numèrics en Enginyeria (CIMNE)
School of Civil Engineering
Universitat Politècnica de Catalunya (UPC)
Barcelona, Spain

CIMNE

 Springer

ISBN: 978-3-319-04932-8 (HB)
ISBN: 978-3-319-04933-5 (e-book)

Depósito legal: B-2223-2014

A C.I.P. Catalogue record for this book is available from the Library of Congress

Lecture Notes Series Manager: **Mª Jesús Samper,** CIMNE, Barcelona, Spain

Cover page: **Pallí Disseny i Comunicació,** www.pallidisseny.com

Printed by: **Artes Gráficas Torres S.L.**
Huelva 9, 08940 Cornellà de Llobregat (Barcelona), España
www.agraficastorres.es

Printed on elemental chlorine-free paper

Numerical Simulation of Mechanical Behavior of Composite Materials
Sergio Oller

First edition, 2014

© International Center for Numerical Methods in Engineering (CIMNE), 2014
Gran Capitán s/n, 08034 Barcelona, Spain
www.cimne.com

This work is dedicated to my wife and son, and also to all my loved ones

Preface

This is a summary of the research work on "composite materials" carried out by the author since 1990. It reflects my personal work with other researchers and also with senior engineering and PhD students whose thesis I have advised on this research line[1].

The subjects included in the present work are oriented mainly towards the constitutive modeling and structural evaluation of structures built with composite materials. Many of these subjects were and are still today original. Therefore, they do not represent a simple continuity of the research line trends of the past.

All the concepts presented here are a summary of a more extensive work carried out by the author and other collaborators whose texts can be found in the references mentioned in each case. It is worthwhile mentioning that all these subjects are still under development and require the additional participation and dedication of more people. This summary shows the subjects are not at the same conceptual and technical level. Some ideas are more developed than others, however, in general, all of them show that the numerical-mechanical study of composite materials is a very promising area for both, research work and industrial potential, and consequently to improve their use for the best interest of society.

This research work started in 1990 and has progressed on two different research lines: the "Mixing Theory" and its modifications and the "Homogenization Theory" and its comprehensive study. Both lines have given way to a significant scientific production.

---

[1] **PhD theses involved in this book:**

- Luccioni B. (1993). Formulación de un Modelo Constitutivo para el Tratamiento de Materiales Ortótropos. Universidad Nacional de Tucumán - Laboratorio de Estructuras Tucumán, Argentina. Advisors: S. Oller, R. Danesi.
- Car E. (2000). Modelo Constitutivo para el Estudio del Comportamiento Mecánico de los Materiales Compuestos. Universidad Politécnica de Cataluña - Departamento de Resistencia de Materiales y Estructuras en la Ingeniería – Barcelona, Spain. Advisors: S. Oller, E. Oñate.
- Zalamea F. (2001). Modelización de Compuestos Mediante la Teoría de Homogeneización. Universidad Politécnica de Cataluña - Departamento de Resistencia de Materiales y Estructuras en la Ingeniería – Barcelona, Spain. Advisors: J. Miquel, S. Oller.
- Nallim L. (2002). Mecánica de placas anisótropas: Un enfoque variacional. Universidad Nacional de Salta - Facultad de Ingeniería – Salta, Argentina. Advisors: R. O. Grossi, S. Oller.
- Rastellini F. (2006). Modelización Numérica de la No-linealidad Constitutiva de Laminados Compuestos. Universidad Politécnica de Cataluña - Departamento de Resistencia de Materiales y Estructuras en la Ingeniería – Barcelona, Spain. Advisors: S. Oller, E. Oñate.
- Martínez X. (2008). Micro-Mechanical Study of Composites by Using the Serial/Parallel Mixing Theory. Departamento de Resistencia de Materiales y Estructuras en la Ingeniería – Barcelona, Spain. Advisor: S. Oller.
- Molina M. (2012). Simulación de los materiales compuestos como refuerzo en estructuras de hormigón armado. Departamento de Resistencia de Materiales y Estructuras en la Ingeniería - Barcelona, Spain. Advisor s: S. Oller, A. Barbat.
- Badillo H. (2012). Numerical modelling based on the multiscale homogenization theory. Application in composite materials and structures. Departamento de Resistencia de Materiales y Estructuras en la Ingeniería – Barcelona, Spain. Advisor: S. Oller.

The original objective of the research project was and is still today to design composite materials from their own components basic properties, without handling the composite as a single material. This "concern" has permanently pushed forward the research development and has opened the path towards the "material design" suitable for each structure. The subject's complexity and difficulty have demanded the intensive implementation of many techniques previously developed by the author in other research lines such as plasticity, anisotropy, introduction of rigid-body movements using the constitutive equation, local instability through a constitutive formulation, large strains, delamination problems, composites fatigue, etc. Simultaneously, a thorough data base and parallelization intensive computational work has been carried out. Moreover, a finite element code −PLCd[2]− has been developed and used as a "development and testing tool" to implement these formulations. All this technology has been transferred to the COMET[3] program, which is a finite element code developed at CIMNE for more general purposes.

Currently, research on composites is still under way and the temperature and humidity effects on the behavior have been included for a more comprehensive study of the matrix-fiber sliding phenomena. We are also trying to take these formulations to the structural finite elements, such as shells and membranes.

This work has been possible thanks to the institutional support of CIMNE (International Center for Numerical method in Engineering), which has financially supported this book since its first edition in Spanish in 2003, and later in its English edition. In this latter task many people have participated, and particularly I thank Ms. Hamdy Briceño and Prof. Miguel Cerrolaza for their careful translation and revision of this text. I also thank all my students who have contributed to the correction of the text during the eleven years that this book has been used as a syllabus of "the Nonlinear Dynamics course" in the Department of Strength of Materials, of the Technical University of Catalonia, Spain.

I hope these notes will contribute to a better understanding of the non-linear dynamics and encourage the reader to study this subject in greater depth.

Sergio Oller

Barcelona, March 2014

---

[2] PLCd-Manual. Non-linear thermomachanic finite element code oriented to PhD student education. Code developed at CIMNE, 1991-to present (https://web.cimne.upc.edu/users/plcd/).
[3] Cervera M., Agelet C. and Chiumenti M. (2002). COMET: Coupled Mechanical and Thermal Analysis. Data input manual. *Technical Report CIMNE N° IT-308*. Barcelona, Spain.

CONTENTS

1 INTRODUCTION

The use of composite materials in structural design has grown significantly over the past few years. This is because the mechanical properties of these materials are better than the traditional ones as a result of their designing process possibilities.

These materials have a suitable weight-strength and weight-stiffness relationship. They are corrosion-resistant, thermally stable and appropriate for structures that depend mainly on the weight variable in the designing process. Composite materials are suitable for structural components requiring high stiffness, impact-resistance, complex shapes and a high production volume. Moreover, the use of these materials has been extended to the spare parts manufacturing for the aeronautic, aerospace, navy and automobile industries during the last years[1,2].

One of the problems regarding the use of composite materials in structures is the lack of technology to guarantee the pieces joints and reliable designs. Conventional analytical techniques are not suitable for composite materials. There are uncertainties about the durability and aging of materials. Further studies are required to guarantee their integrity over long periods of time. Moreover, composite materials represented by a single orthotropic material containing the properties of the whole set did not turn out as expected, particularly when the behavior of at least one of its components exceeds the elasticity limit. As observed in the references below, many attempts have been made to model the behavior of composite materials through the finite element technique for structural design and analysis, where the correlation between the analysis and the experimental results are not entirely satisfactory[1,3].

1.1 Uses of composite materials

1.1.1   The use of composite materials in the automobile industry

Composite materials are not commonly used in the structural and automobile industries[4]. However, they are important in the car racing business due to their high potential, as proved in the tests labs. Special cars, such as the Formula 1 ones, have a sophisticated design, manufacture and operation. A lot of money is invested in technology compared to other sectors of the automotive industry. Due to the great push and development of the aeronautic industry, composite materials have been used since 1980. Car racing designers

[1] Ali R. (1996). Use of finite element technique for the analysis of composite structures. *Computers & Structures*. 58, No. 5. pp. 1015-1023.

[2] O'Rourke B. P. (1989). The use of composite materials in the design and manufacture of formula 1 racing cars. *Proc. Inst. Mech. Engng.* C387/025. pp. 39-48.

[3] Klintworth J. and Macmillian S. (1992). *Effective analysis of laminated composite structures.* Benchmark (NAFEMS). pp. 20-22.

[4] Noguchi M. (1993). Present and Future of Composite Materials for Automotive Application in Japan. *Proceedings of the Ninth International Conference on Composite Materials ICCM 9 Metal Matrix Composites.* Ed. A. Miravete. pp. 97 - 110. University of Zaragoza - Woodhead Publishing Limited

must comply with specific car strength, stiffness and weight regulations. The weight is also important for the definition of the car's hull material in the manufacturing process. This problem is solved with the optimization of geometry and the quality of the materials.

Composite reinforced fiber materials were used for the first time in the automobile industry in the fifties. Many of the polyester-resine matrix composites and randomly-oriented reinforced fiber glass materials were used since then until the end of the eighties.

Carbon-fiber-reinforced composite materials were used in the eighties for the first time. Their use was extended due to their high weight-strength and weight-stiffness relationship. They are also corrosion-resistant, thermally-stable and suitable for structures that depend mainly on weight variables in the designing process. Seventy percent of the Formula 1 cars are currently manufactured with composite materials and this percentage is still growing.

Composite materials have improved both car performance and safety due to their great impact absorption capacity. In addition to the structural parts, composite materials are used in other car parts such as carbon-carbon brake discs, clutches and thermal insulation. The weight of cars is reduced significantly with carbon-carbon brake discs while their braking capacity is increased.

The latest suspension system innovation designed by MacLaren in 1992 involved a 50%-60% weight reduction and therefore an inertia reduction.

1.1.2   The use of composite materials in the aeronautic industry

Composite materials were used in the aeronautic industry before they were used in the automotive industry. This industry is pioneer in the use of composite materials for different applications. High resistance and stiffness along with lower costs and weight are among the main factors for using composite materials in the aerospace industry. Weight is the main variable in the design of aircraft structure materials. The large-scale use of these materials in aviation engines is due to Roll-Royce. Reinforced plastic fiber-glass RB108 was used in the turbine compression system.

In aero engines, the use of polymer-matrix composite materials (PMC) competes with titaniun due to their high weight-strength and weight-stiffness relationship. Metal-matrix composite materials (MMC) are potentially appropriate for high temperatures that may damage the polymer-matrix composite materials. Ceramic-matrix composite materials are used for areas subjected to high temperatures (CMC). CMC are materials conceptually different from PMC and MMC materials. Their fibers modify how crackings spread in the material and do not work as a reinforced matrix[5].

The introduction of "big fan" engines in the early seventies coincides with the wide availability of carbon-fiber-reinforced composite materials. The materials' characteristics were improved and weight was reduced considerably. Despite the aforementioned advantages, these big fan engines did not pass the collision tests against birds and composite materials were replaced by titaniun.

Except for the engines and the combustion area, most of the external part of the turbine structure is made of carbon-fiber-reinforced composite material. Very few parts are made of composite materials in the internal part of the turbine. However, they account for 10% of the turbine weight.

---

[5] Ruffles R. (1993). Applications of advanced composites in gas turbine aero engines. *Proceedings of the Ninth International Conference on Composite Materials ICCM 9 Metal Matrix Composites.* Ed. A. Miravete. pp. 123 - 130. University of Zaragoza - Woodhead Publishing Limited.

In the future, the use of these materials will be extended to the internal part of the turbine. Big fan engines and their supporting structure will be built in carbon-fiber-reinforced composite materials within the engine development programs for civil aviation pushed by Roll-Royce.

The aero engine parts made of composite materials for military purposes are different from those for civil use mainly because these engines are assembled directly on the aircraft structure and do not have an external cover. Several parts made of MMC and CMC materials are currently under development programs.

1.1.3 Composite materials in the naval industry

The use of organic composite materials (OMC) in marine applications started in the forties. Nowadays it has extended to small ships. However, they are not used in large ships and offshore structures because designers and constructors of this type of structures are not familiar with these materials and the application of technologies in the naval industry. Currently there is a strong demand of composite materials for ships, submarines and offshore structures because weight can be reduced to 25-50% in comparison with traditional materials (aluminium and steel). Likewise, they are highly corrosion-resistant (maintenance cost cuts) and have low thermal conductivity. They are not attracted by magnetic fields and repairing is simple.

The use of reinforced fiber-glass composite materials has grown in the naval industry due to the good cost-strength relationship. Fiber-glass materials are very suitable and resistant against marine agents and also very structurally resistant.

The most relevant naval application for these materials is the manufacturing of mine searcher vessels. For this type of ship low-magnetic and high-impact resistant materials are required. These materials are also applied for rapid patrol boats. In the commercial marine, there are also applications for small fishing boats (< 25 m long) due to their low weight and maintenance costs. In offshore structures, composite materials are used because of their low weight and high corrosion-resistance. In boat racing, reinforced carbon-fiber-composite materials are commonly used because they are more resistant than the reinforced fiber-glass composite materials. Moreover, their weight is lighter and boat performance is better. In the boat racing business, designers use edge technology that provides valuable data on materials performance, which can be used in other applications in the naval industry.

Thermal stable resine matrix composites and high performance reinforced fiber materials (aramid, carbon) are suitable for lightweight structures that require good mechanical properties. Although they meet the requirements, further tests and improvements are necessary to evaluate their response under service conditions such as degragation due to long exposure under water, fatigue, impact loads or flame retardant.

Composite materials are particularly suitable for strain, temperature, vibration, stiffness control, etc., as they can carry inside sensors and actuators during the manufacturing process. They provide designers with useful information about structural design maximization and structural safety factors.

CMC and MMC materials are used only when high temperatures and stiffness resistance are required.

Composite materials test methods are based on aeronautical tests and some cannot be applied to the aeronautical industry. Consequently, tests and methodologies for different industry applications must be defined.

1.1.4   The use of composite materials in Civil Engineering

The use of composite materials in civil engineering has become more relevant during the last decade. Its presence in this field was almost negligible. In the construction industry, composite materials are used in structures subjected to aggressive environmental conditions such as parts of offshore platforms, warehouses, ground anchors, non-conductive and non-magnetic constructions, structure reinforcements, both passive and active reinforcements, wires, cross-walking decks, profiles for bars and tunnel coating.

1.1.4.1   Reinforcement structures

One of the most widely used and efficient systems in reinforced structures is the bonding of steel sheets using structural adhesive. Despite the efficiency of the bonding of steel sheets, they have some problems:

Steel sheets are heavy elements and their handling and assembly are difficult. Auxiliary resources are required for their assembly for long periods of time, causing nuisances in traffic when level crossings are reinforced.

- Steel corrosion affecting their resistance and adherence between steel sheets and concrete.

- Concrete plane surfaces are needed for perfect sheet bonding.

Due to the high number of structures requiring reinforcement, materials tests (EMPA) for the reinforced structure system with composite materials have been developed in Swiss Federal labs since the eighties in order to solve the aforementioned problems.

Carbon-fiber and epoxy-resine composite sheets for concrete, metal and wood structures, among others, are alternatives to conventiomal reinforced systems through the bonding of steel sheets. The growing use of these composite materials for reinforcement is owed mainly to their mechanical performance, good corrosion resistance, light weight which facilitates transportation, handling and assembly. The first large-scale use of this type of reinforcement was in 1991 in the Ibach bridge in Lucern.

This system is used in different countries, but only in Switzerland and Germany the number of structural reinforcements with the bonding of sheets of composite material goes up to 250 and there are more than 1000 in the world, mainly in countries such as Switzerland, Germany, U.S.A, Japan and Canada. Thanks to lighter auxiliary resources for shorter periods of time, savings can go up to 25% for the total structure reinforcement to compensate the higher cost of composite material as compared to steel[6].

1.1.4.2   Reinforcement in Concrete

Another use of fiber-reinforced composites is replacing the reinforcement in concrete or in pre-stressed tendons. It is applied mainly when steel reinforcement in concrete poses some corrosion or magnetism problems. The alternative is fiber-glass composite bars and thermosetting matrix (polyester or epoxi) coating .The adherence between the concrete and this type of reinforcement is very high and the textured surfaces obtained are comparable to steel bars.

---

[6] G. Pulido M. D. y Sobrino J. (1996). Los materiales compuestos en el refuerzo de puentes. *Revista Internacional de Ingeniería de Estructuras.* Vol. 3, pp. 75-95. No.1.

Nowadays there are various applications such as non-magnetic slabs, thin layers of concrete (reinforcement can be very close to the outer surface), reinforced earth and bracing bars.

1.1.4.3   Bars

Fiber bars are used mainly when there are corrosion problems or in communication or transmission buildings where metalic bars would be susceptible to heating by energy absorption or electromagnetic waves transmission interferences.

Fiber-glasses are used for medium to high tensile stress (up to 2700 MPa.) while carbon or aramid fibers are used for higher stresses. The 268 meters Colserolla Telecommunication Tower in Barcelona city is an example of this application where aramid fibers were used and each one of the bars consists of seven 50 mm cords placed within a polyethylene sheath.

1.1.4.4   "Pultruded " profiles

Profiles manufactured by the pultruded process are used in industrial buildings with corrosion risk because of the presence of acids, for example in aluminium production plants. Sodium hydroxide and hydrochloric acid vapors produce serious corrosion problems in these plants. In industrial buildings for computer and electronic product tests, these materials are also used due to their insulating properties against electromagnetic waves.

1.2 Composite properties

Composite materials are obtained by searching alternative materials having one or more of the following characteristics:

- Low cost.

- Good structural behavior.

- Low weight.

- Feasible massive production.

These premises can be satisfied in different ways. Thus, the characteristics of the synthetic polymers can be increased, up to a large extent, by mixing different components that produce a material with more structural capacities.

Composite materials are used in different industrial sectors due to economic factors: the substantial volume reduction of high cost material and their exclusive allocation where they are really necessary.

The materials obtained have the following characteristics:

- Good thermal and acoustic isolation.

- Resistance to some chemical agents.

- Energy dissipation caused by micro cracking in the interfaces of its components so that the general behavior of the structure does not show a sharp drop of its limit resistance.

- Low weight, reduction of easy-to-transport dead weights.

- Excellent behavior against corrosion in aggressive environments.

- High mechanical properties.

- High fire retardant.

- Sandwich configuration which enables them to obtain low coefficients of thermal conductivity

- Waterproof and resistant to various chemical agents.

The most widely used component is in particles or fiber form. In the first case, the material particles or specific materials are attached to each other by a continuous matrix with low modulus of elasticity. In fibrous composites, the reinforcement can be oriented towards any direction required to provide optimum resistance and stiffness. As a result of the material moldability, the most effective structural forms can be chosen.

In the construction industry, fiber-glass, carbon or aramid fiber or the combination of both are used to obtain hybrid fibers. Fiberglass and polyester polymer or epoxy is used to make a fiber composite material called polyester. The main disadvantages of using composite materials are:

- The materials generally show a non-linear behavior even at very low stresses.

- When the composite materials manufacture process is finished it usually shows residual stresses that are very hard to quantify due to temperature variations.

- Reinforced materials in a composite (fibers, particles, etc.) normally change their mechanical properties due to environmental effects.

The use of composite materials for structural components requires a particular design. This is mainly due to the high anisotropy and strength relationship between the fiber and the matrix. The composite designing process is based on empirical methods. In the related literature there is a lack of behavior analysis or simulations of these materials subjected to stress levels above the elastic limit. Conventional techniques are not suitable for the analysis of composite materials. Likewise, the use of the finite element method to study composites which are represented by a single material having properties of the whole set has not been satisfactory. The main difficulty when using the finite element method in conventional constitutive models is that the behavior of highly anisotropic materials subjected to loads above the elasticity limit of at least one of the materials of the composite cannot be modeled. Consequently, it is necessary to model composite materials through theories leading to the behavior simulation of highly anisotropic materials and permanent directioned strains; with differentiated behavior for each substance conforming the composite, good handling of the reinforcement phase with fibers; ability to represent the relative displacement between the matrix and the fiber (debonding); capacity to represent the local buckling of the reinforcement; capacity to simulate large displacements and strains and other collateral phenomena. All these phenomena lead to a strength and stiffness global loss of the composite with linearity loss in the set response.

1.3 Classification of composite materials

Composite materials are very hard to define due to their qualities, composition, properties, manufacture, etc. There are different ways to classify them and they are all probably correct. However, and in order to be consistant with the development of the present work, the composite materials can be classified as follows:

1.3.1 Classification by topology

Among the different classifications, there is one based on the topology configuration, which refers to what the components are and how they are distributed: matrix composite materials, short and/or long fiber composite materials, laminates and a combination of each aforementioned type. Figure 1.1 shows the classification of composite materials according to their topology configuration.

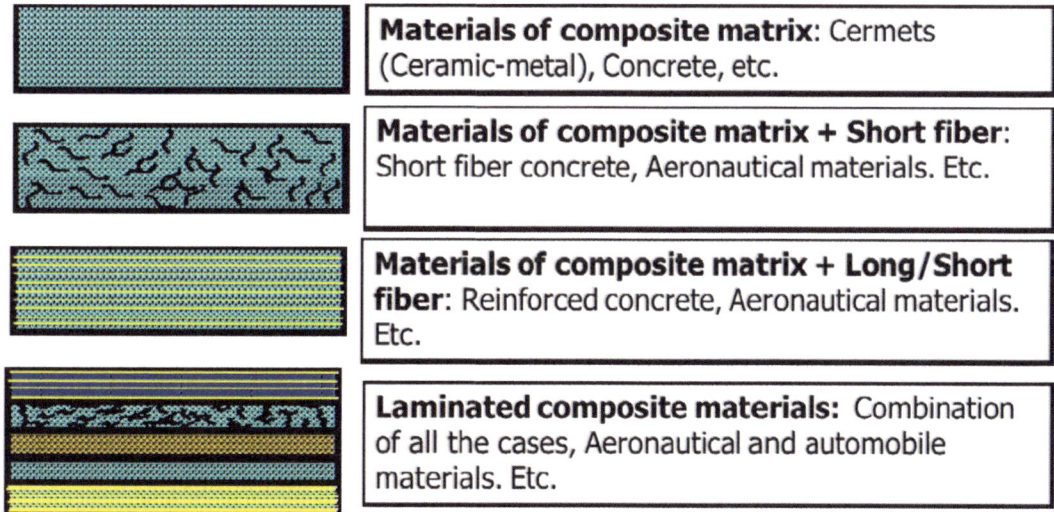

Figure 1.1 – Classification of composite materials according to their topology.

1.3.2 Classification by components

Composite materials can be classified according to their type and form as follows:

- **Fibers**: Made up by short or long continuous fibers in one or three directions or randomly distributed and bound by a matrix. This matrix can be formed by two or more materials.

- **Particles**: Made up by particles working agglutinated by a matrix.

- **Laminates**: Made up by layers or laminate constituents with different strengths depending on their magnitude and directions.

- **Flaked**: Made up by flat flakes embedded in a matrix.

- **Skeleton fill**: Made up for one skeleton filled with another material.

The most widely used materials are the fibrous type. Fibers are in charge of resisting mechanical actions and the matrix works as a binder and protector of the environment. The fiber's mechanical strength is about 25 to 50 times greater than the matrix´s strength. In concrete in traction this relation is of order 100, which leads to a strongly anisotropic behavior.

When loading is applied to a composite material, stresses are produced in the interior and in order to achieve a good transition of these between the fiber and the matrix a resine is placed on top.

The matrix's function is to distribute and transmit the loads to the fibers. For composite laminates, shear strength is very important. The matrix's function is to guarantee the displacement continuity among the laminates throughout the laminates thicknes and influences the cracking mode.

In laminates subjected to compression, the matrix influences the buckling length of the fibers. The most commom matrices are made up for polymers, metal and ceramic.

The most widely used polymers are: thermoplastic polymers, themostable polymers and foamed polymers.

The mechanical properties in solid state for thermoplastic materials are generally non-linear with a very low modulus of elasticity and a ductile behavior. The thermosetting materials are generally linear with a high elasticity modulus and a very brittle behavior.

1.3.3   Classification by structure

From the mechanical behavior point of view, composite materials can be classified as follows:

- **Basic structure**: It is studied at single molecule level or crystalline grids.

- **Microscopic structure**: It is taken into consideration for the fiber-matrix interaction, its influence on stress distribution and failure occurrences, discontinuities or crackings under elementary loading conditions.

- **Macroscopic structure**: It is considered in the composite material classification from a macroscopic point of view, as a combination of different substances that contribute to the equilibrium state of the set.

In this work, a structural study will be presented from the macroscopic point of view considering the following hypotheses:

- Fibers are uniformly distributed in the matrix.

- There is perfect adherence between the matrix and the reinforcement.

- The matrix contains neither voids nor defects.

- There are no residual stresses in composite materials with manufacturing defects. However, they might be included as initial conditions.

The literature papers by Miravete (2000)[7] and Car (2000)[8] are recommended for an extensive description of composites types and components, manufacture and industrial application.

---

[7] Miravete A. (2000). *Materiales Compuestos. Vol. 1 y Vol. 2*. Director de la obra: Antonio Miravete.
[8] Car E. (2000*). Modelo constitutivo continuo para el estudio del comportamiento mecánico de los materiales compuestos*. Tesis Doctoral, Universidad Politécnica de Cataluña.

2 MECHANICAL ANISOTROPY

2.1 Introduction

The constitutive models developed for the behavior simulation of simple isotropic materials are not suitable for the analysis of composite materials due to the strong anisotropy of these latters. There are different reasons and degrees of importance. The composite representation by a single orthotropic material having properties of the whole set has not been satisfactory either. Therefore, the mixing theory will also be presented in this chapter.

There exist different formulations for anisotropic materials presenting a non-linear constitutive response (Hill (1971)[1]), (Bassani (1977)[2], (Barlat and Lian (1989)[3], (Barlat et al. (1991)[4]). These theories are based on threshold functions of discontinuity (yield functions) and anisotropic plastic potentials. Thus, new procedures must be developed to integrate the constitutive equation.

The anisotropic formulation presented in this chapter is a generalization of any classic linear or non-linear isotropic formulation such as plasticity, viscoelasticity, damage, etc. This is based on the translation of all the material constitutive parameters, stress and strain states from a *real anisotropic space* to another *fictitious isotropic* space. Once they are there, an isotropic constitutive model is used along with other techniques and procedures for the isotropic constitutive equations.

The anisotropic formulation presented in this chapter is the generalization of the general properties of the anisotropic formulation.

---

[1] Hill .R (1971). *The Mathematical Theory of Plasticity.* Oxford University Press.

[2] Bassani J. L. (1977). Yield characterization of metals with transversely isotropic plastic properties. *Int. J. Mech. Sci.* vol 19, pp. 651.

[3] Barlat F. and Lian J. (1989). Plastic behavior and stretchability of sheet metals. Part I: A yield function for othotropic sheet under plane stress conditions. *Int. Journal of Plasticity*, vol. 5, pp. 51.

[4] Barlat F. and Lege D. J. and Brem J. C. (1991). A six-component yield function for anisotropic materials. *Int. Journal of Plasticity*, vol. 7, pp. 693.

2.2 Generalities on the anisotropic formulation

A suitable formulation of a constitutive law for orthotropic solids or non-proportional anisotropic elastoplastics [*] is a very complex problem and it gets even more complicated when a behavior occurs in large displacements and strains. Reinforced fibers composite materials are simplified forms of anisotropic non-proportional materials made up by two substances. The elastic behavior description of an anisotropic solid is not very complicated. The elasticity theory general forms[5] (Matthews and Rawlings (1994)[6]), (Pendleton and Tuttle (1989)[7]) can be used. However, for the inelastic field there are no general and satisfactory experiences.

The formulation of the threshold function of discontinuity or yield function and all the concepts deriving from it entails a very hard problem to solve from the mechanics point of view. The general formulation for anisotropic yield functions must tend to the behavior of isotropic materials as a particular case of the anisotropic materials and must have the properties of the isotropic functions as described by Malvern (1969)[8], Gurtin (1981)[9] and Chen and Han (1988)[10]. These properties refer to the symmetry[**] and convexity of the yield function.

The first attempts to formulate yield functions for non-proportional orthotropic materials were made by Hill[11], who in 1948 managed to extend von Mises's isotropic model to the orthotropic case and modified it in subsequent publications (Hill (1965)[12]), (Hill (1979)[13]), (Hill (1990)[14]). The principal limitation of this formulation and its subsequent modifications is that it is imposible to represent the composite materials behavior, which is sensitive to pressure such as geomaterials or composite materials. Several authors have proposed yield functions in the stress state for anisotropic materials; some of them are Bassani (1977)[2] and Barlat et al. (1989, 1991)[3,4]. Barlat et al. (1991) used a linear transformation of the anisotropic material in stress state multiplying all the stress tensor components by different constants. Karafillis and Boyce (1993)[15] proposed a general expresion for policrystal materials yield surfaces which can describe isotropic and anisotropic materials. The material anisotropy is described by introducing a set of irreducible tensor variables. This set of tensor variables allows a linear tranformation from

---

[*] NOTE: A non proportional material is the material in which the relationship among the material elastic modules in any two directions is not equal to the relation among the strength in the same directions.

[5] Hull D. (1987). *An Introduction to Composite Materials*. Cambridge University Press.

[6] Matthews F. L. and Rawlings R. D. (1994). *Composite Materials: Engineering and Science*. Chapman and Hall.

[7] Pendleton R.L. and Tuttle M.E. (1989). *Manual on Experimental Methods for Mechanical Testing of Composites*. Elsevier Applied Science Publishers.

[8] Malvern L.E. (1969). *Introduction to the Mechanics of a Continuous Medium*. Prentice-Hall.

[9] Gurtin M. E. (1981). *An introduction to continuum mechanics*. Academic Press. New York.

[10] Chen W. F. and Han D. J. (1988). *Plasticity for structural engineers*. Springer-Verlag. New York.

[**] NOTA: Dado un estado de tensiones principales σ_1, σ_2, σ_3, la función de fluencia resulta simétrica si se cumple $f(\sigma_1, \sigma_2, \sigma_3) \equiv f(\sigma_3, \sigma_1, \sigma_2)$.

[11] Hill R. (1948). *A theory of the yielding and plastic flow of anisotropic metals*. Proc. Roy. Soc. London. Ser. A. Vol. 193, pp. 281 - 297.

[12] Hill R. (1965). Micro mechanics of elastoplastic materials. *J. Mech. Phis. Solids*. Vol. 13, pp. 89-101.

[13] Hill R. (1979). Theoretical plasticity of textured aggregates. *Math. Proc. Cambridge Philos. Soc.*. Vol. 85, pp. 179 - 191, No 1.

[14] Hill R. (1990). Constitutive modelling of orthotropic plasticity in sheet metals. *J. Mech. Phys. Solids*. Vol. 38, pp. 405 - 417, No. 3.

[15] Karafillis A. P. and Boyce M. C. (1993). A general anisotropic yield criterion using bounds and a transformation weighting tensor. *J. Mech. Phys. Solids*. Vol. 41, pp. 1859 - 1886.

number = 12

a stress state of the anisotropic material to a state called IPE (Isotropic Plasticity Equivalent Material).

Dvorak and Bahei-El-Din (1982)[16] also used stress operators along with von Mises's yield criterion for the analysis of composite materials. Several authors have used fourth-order tensors in the formulation of yield criteria for anisotropic materials; some of these authors are Shih and Lee (1978)[17], Eisenberg and Yen (1984)[18] and Voyiadjis and Foroozesh(1990)[19]. Voyiadjis and Thiagarajan (1995)[20] carried out a study of unidirectionally reinforced composite materials formulating a general yield surface depending on a fourth-order tensor.

A generalization of the classic isotropy theory can be carried out through the anisotropic formulation developed in this chapter. This formulation is based on a linear transformation of both the stress and strain tensors through the fourth-order tensors containing the material anisotropic information. This linear transformation guarantees the yield function convexity and plastic potential (Eggleston (1969))[21]. The yield function convexity guarantees that the second law of thermodynamics is met and that after applying a plastic loading, growing monotonous, any unloading leads to an elastic state. This theory is called space transformation and is based on the ideas proposed by Betten (1981)[22] (1988)[23] and uses the concept of *"mapped stress tensor"* which allows the stress tensor to be transported from *a real space to* a *fictitious stress space*. The generalized transformation of the stress tensor of the anisotropic material is presented schematiacally in Figure 2.1. The algorithms for isotropic materials can be used and consequently a computational model can be implemented.

In previous studies, several authors have developed a generalization of the isotropic plasticity theory into the anisotropic one (Oller et. al. (1993)[24]), (Oller et. al. (1993)[25]). The basic idea was to model the behavior of a solid in the *real anisotropic space* through an ideal solid in the *fictitious isotropic space* (see Figure 2.1). The model is based on a linear transformation of the stress tensor assuming the elastic strains are identical in both spaces, which introduces a limitation in the mapped anisotropy theory. This limitation requires respecting the proportionality between the strength limit and the elasticity module for each direction of the material. To avoid this simplified hypothesis, Oller et. al. (1995)[26] proposed applying the linear transformation on both the stress and the strain spaces.

[16] Dvorak G. J. and Bahei-El-Din Y. A. (1982). Plasticity analysis of fibrous composites. *J. App. Mech.* Vol. 49, pp. 327 - 335.

[17] Shih C. F. and Lee D. (1978). Further developments in anisotropic plasticity. *J. Engng Mater. Technol.* Vol. 105, pp. 242

[18] Eisenberg M. A. and Yen C. F. (1984). The anisotropic deformation of yield surfaces. *J. Engng. Mater. Technol.* Vol. 106, pp. 355

[19] Voyiadjis G. Z. and Foroozesh M. (1990). Anisotropic distortional yield model. *J. Appl. Mech.* Vol. 57, pp. 537.

[20] Voyiadjis G.Z. and Thiagarajan G. (1995). A damage ciclic plasticity model for metal matrix composites. *Constitutive Laws, Experiments and Numerical Implementation.*

[21] Eggleston H. G. (1969). *Convexity.* Cambridge University Press.

[22] Betten J. (1981). Creep theory of anisotropic solids. *J. Rheol.* Vol. 25, pp. 565-581.

[23] Betten J. (1988). Application of tensor functions to the formulation of yield criteria for anisotropic materials. *Int. J. Plasticity,* Vol. 4, pp. 29-46.

[24] Oller S., Oñate E., Miquel J. and Botello S. (1993). A finite element model for analysis of multiphase composite materials. *Ninth International Conferences on Composite Materials.* Ed. A. Miravete. Zaragoza - Spain.

[25] Oller S., Oñate E. and Miquel J. (1993). Simulation of anisotropic elastic-plastic behaviour of materials by means of an isotropic formulation. *2nd. US Nat. Congr. Comput. Mech.* Washington DC.

[26] Oller S., Botello S., Miquel J. and Oñate E. (1995). An anisotropic elastoplastic model based on an isotropic formulation. *Engineering Computations,* Vol. 12, No. 3, pp. 245-262.

The constitutive model detailed in this chapter can be applied to materials presenting a strong anisotropy such as fiber-reinforced composite materials and it results from a generalization of the classic plasticity theory. The anisotropic behavior of the material is formulated through a fictitious isotropic space of stresses and strains resulting from a linear tensor transformation of the real spaces of the anisotropic stresses and strains. The parameters involved in the transformation tensor definition are obtained from experimental tests. The advantage of using this type of model is that the same yield function can be used, as well as the plastic potential and integration methods of the constitutive equation developed for isotropic materials.

All the information of the material anisotropy is contained in the fourth-order tensors of transformation of the stress spaces ($\mathbf{A}^S$, o $\boldsymbol{a}^\tau = \vec{\phi}(\mathbf{A}^S)$) in the referential configuration or updated, respectively) and strains ($\mathbf{A}^E$, o $\boldsymbol{a}^e = \vec{\phi}(\mathbf{A}^E)$) in the referential or updated configurations, respectively). A methodology can be obtained for definition of the anisotropic models through the space transformation theory.

The formulation resulting from this procedure is quite general and can be used to carry out the analysis of composite materials with high anisotropy level.

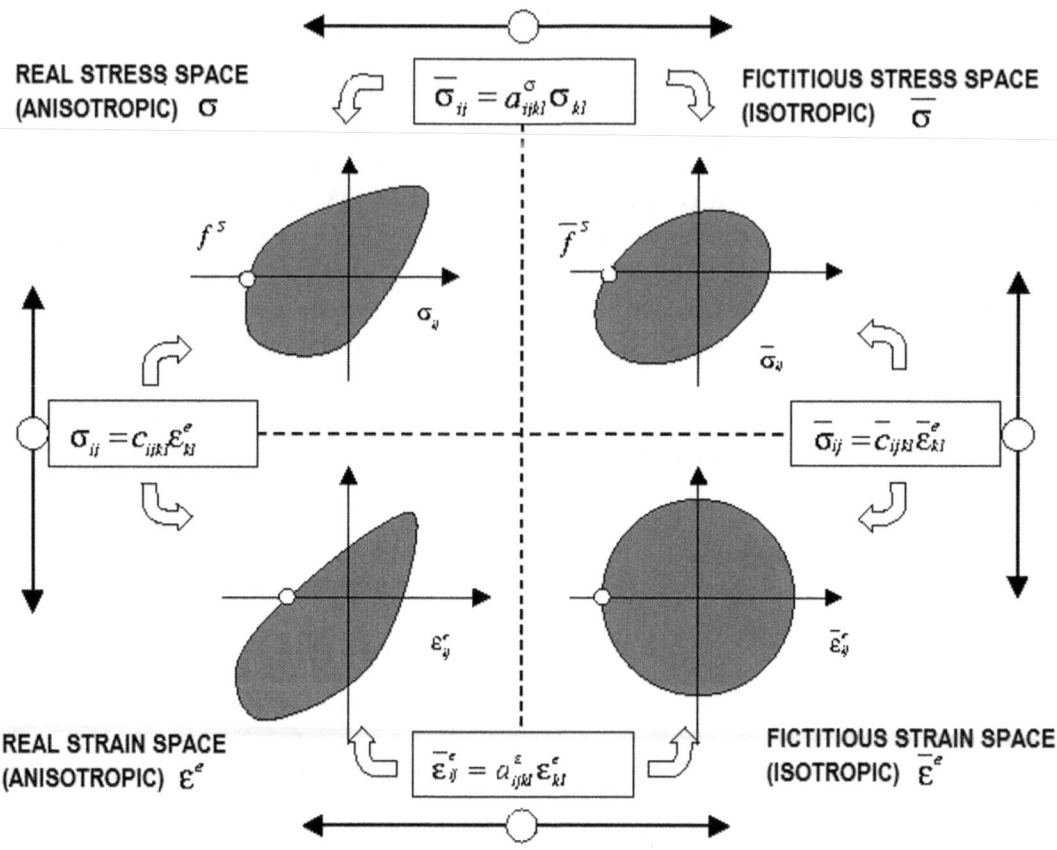

Figure 2.1 – Space Transformation. Real and fictitious stresses and strain spaces in small strains.

2.3 Yield function and plastic potential for isotropic materials

In the following sections an elastoplastic model will be presented. It can be formulated either by the referential or the updated configuration using the total or updated Langrangean kinematics (Green and Naghdi (1971)[27], (Lubliner (1990)[28]). The non-linear behavior of materials subjected to large plastic strains and small elastic strains can be simulated with this constitutive model as a cuadratic potential (García Garino and Oliver (1992)[29], (Lubliner (1990)[28]) which is used for the definition of elastic free energy.

The yield and plastic potential functions can be defined in the Kirchhoff stress space in the updated configuration as follows:

$$\begin{cases} \text{Yield function} & : \quad \mathbb{F}^{\tau}(\boldsymbol{\tau}, g, \boldsymbol{\alpha}, \boldsymbol{f}^{\tau}) = 0 \\ \text{Potential function:} & \mathbb{G}^{\tau}(\boldsymbol{\tau}, g) = \text{cte} \end{cases} \tag{2.1}$$

where $\mathbb{F}^{\tau}(\boldsymbol{\tau}, g, \boldsymbol{\alpha}, \boldsymbol{f}^{\tau}) = 0$ and $\mathbb{G}^{\tau}(\boldsymbol{\tau}, g) = \text{cte}$ represent the yield and plastic potential functions in the updated configuration, $\boldsymbol{\tau}$ is the Kirchhoff stress tensor, g the metric tensor of the space configuration, $\boldsymbol{\alpha}$ the internal variables group and $\boldsymbol{f}^{\tau}$ the strength tensor in the elastic limit threshold of the material in the updated configuration. These functions can also be defined in the referential configuration as follows,

$$\begin{cases} \text{Yield function} & : \quad \mathbb{F}^{S}(\mathbf{S}, \boldsymbol{G}, \boldsymbol{\alpha}, \boldsymbol{f}^{S}) = 0 \\ \text{Potential function:} & \mathbb{G}^{S}(\mathbf{S}, \boldsymbol{G}) = \text{cte} \end{cases} \tag{2.2}$$

where $\mathbb{F}^{S}(\mathbf{S}, \boldsymbol{G}, \boldsymbol{\alpha}, \boldsymbol{f}^{S}) = 0$ and $\mathbb{G}^{S}(\mathbf{S}, \boldsymbol{G}) = \text{cte}$ represent the yield and plastic potential functions in the referential configuration, $\mathbf{S}$ is the Piola Kirchhoff stress tensor, $\boldsymbol{G}$ is the metric tensor in the referential configuration for an orthogonal coordinated system $\boldsymbol{G} = \mathbf{I}$, $\boldsymbol{\alpha}$ is the internal variables group and $\boldsymbol{f}^{S}$ is the strength tensor in the elastic limit threshold of the material in the referential configuration. It is important to highlight that the $\mathbb{F}^{\tau}(\boldsymbol{\tau}, g, \boldsymbol{\alpha}, \boldsymbol{f}^{\tau})$, $\mathbb{F}^{S}(\mathbf{S}, \boldsymbol{G}, \boldsymbol{\alpha}, \boldsymbol{f}^{S})$, $\mathbb{G}^{\tau}(\boldsymbol{\tau}, g)$, and $\mathbb{G}^{S}(\mathbf{S}, \boldsymbol{G})$ functions are isotropic, symmetric and convex.

The yield and plastic potential functions can be expressed in the most general case as a function of six independent variables, which are the stress tensor components.

$$\begin{cases} \text{Yield function} & : \quad \mathbb{F}^{S}(S_{11}, S_{22}, S_{33}, S_{12}, S_{13}, S_{23}, \boldsymbol{\alpha}, \boldsymbol{f}^{S}) = 0 \\ \text{Potential function:} & \mathbb{G}^{S}(S_{11}, S_{22}, S_{33}, S_{12}, S_{13}, S_{23}) = \text{cte} \end{cases} \tag{2.3}$$

or as a function of the main stresses in the isotropic case:

---

[27] Green A. E. and Nagdhi P. M. (1971). Some remarks on elastic-plastic deformation at finite strains. *Int. J. of Eng. Sc.*, Vol 9, pp. 1219-1229.

[28] Lubliner J. (1990). *Plasticity Theory*. Macmillan Publishing, U.S.A.

[29] Garcia Garino G. and Oliver J. (1992). A numerical model for elastoplastic large strain problems. Fundamentals and applications. *Computational Plasticity III*. Ed. D.R.J. Owen and E. Oñate and E. Hinton. Vol. 1, pp. 117-129, CIMNE, Barcelona, Spain.

$$\begin{cases} \text{Yield function} & : \quad \mathbb{F}^S(S_1, S_2, S_3, \boldsymbol{\alpha}, \boldsymbol{f}^S) = 0 \\ \text{Potential function:} & \mathbb{G}^S(S_1, S_2, S_3) = \text{cte} \end{cases} \qquad (2.4)$$

as a result of a transformation of the stress state. The yield and plastic potential functions are isotropic provided that for any orthogonal transformation the invariance condition of the yield or plastic potential functions is verified. (Malvern(1969)[8]), that is,

$$\begin{cases} \text{Yield function} & : \quad \mathbb{F}^S(a_{ip}\,a_{jq}\,S_{pq}, \boldsymbol{\alpha}, \boldsymbol{f}^S) = \mathbb{F}^S(S_{ij}, \boldsymbol{\alpha}, \boldsymbol{f}^S) = 0 \\ \text{Potential function:} & \mathbb{G}^S(a_{ip}\,a_{jq}\,S_{pq}) = \mathbb{G}^S(S_{ij}) = \text{cte} \end{cases} \qquad (2.5)$$

where $a_{ik}\,a_{jk} = \delta_{ij}$ represents an orthogonal transformation and δ_{ij} are the Kronecker tensor components. The isotropic materials satisfy the invariance condition and the yield and plastic potential functions can be written as a function of the stress tensor invariants, that is,

$$\begin{cases} \text{Yield function} & : \quad \mathbb{F}^S(I_1, I_2, I_3, \boldsymbol{\alpha}, \boldsymbol{f}^S) = 0 \\ \text{Potential function:} & \mathbb{G}^S(I_1, I_2, I_3) = \text{cte} \end{cases} \qquad (2.6)$$

In the case of materials satisfying the plastic incompressibility condition [30], the yield and plastic potential functions are,

$$\begin{cases} \text{Yield function} & : \quad \mathbb{F}^S(J_2, J_3, \boldsymbol{\alpha}, \boldsymbol{f}^S) = 0 \\ \text{Potential function:} & \mathbb{G}^S(J_2, J_3) = \text{cte} \end{cases} \qquad (2.7)$$

where J_2 and J_3 represent the second and third invariant of the deviatory part of the second tensor of the Piola-Kirchhoff (referential configuration). Similarly, in the updated configuration, the yield and plastic potential functions are isotropic if the invariance condition of the yield or plastic potential is verified, and for materials that satisfy the plastic incomprehensibility condition are written as,

$$\begin{cases} \text{Yield function} & : \quad \mathbb{F}^\tau(j_2, j_3, \boldsymbol{g}, \boldsymbol{\alpha}, \boldsymbol{f}^\tau) = 0 \\ \text{Potential function:} & \mathbb{G}^\tau(j_2, j_3, \boldsymbol{g}) = \text{cte} \end{cases} \qquad (2.8)$$

where j_2 and j_3 represent the second and third invariant of the deviatory part of the Kirchhoff stress tensor in the updated configuration.

2.4 General explicit definition of the isotropic yield criterion in the referential configuration

A general definition of the isotropic yield criterion is established in this section. The strength in the traction yield stress f_t is different from the threshold in compression f_c. That is:

---

[30] NOTE: The condition of plastic incompressibility is verified when the yield function is not affected by the spheric part of the stresses tensor.

$$\mathbb{F}^S(\mathbf{S},f) = \frac{1}{f^2}\left(S_x^2 + S_y^2 + S_z^2\right) - \frac{\hat{\lambda}}{f^2}\cdot\left(S_y S_z + S_z S_x + S_x S_y\right) +$$
$$+\frac{(2+\hat{\lambda})}{f^2}\cdot\left(S_{yz}^2 + S_{xz}^2 + S_{xy}^2\right) + \frac{2\cdot\hat{\alpha}}{f^2}\left(S_x + S_y + S_z\right) - 1 = 0 \tag{2.9}$$

being $f = \sqrt{f_t f_c}$ the equivalent strength threshold, $\hat{\alpha} = \frac{1}{2}\left[(f_c/f_t) - (f_t/f_c)\right] = \frac{1}{2f}\left(f_c^2 - f_t^2\right)$ an adjustment coefficient and $\hat{\lambda}$ a parameter to be determined for the yield criteria specification. The general isotropic yield criteria can also be represented by equation (2.1) as,

$$\mathbb{F}^S(\mathbf{S},f) = \mathbf{S}^T\cdot\mathbf{P}\cdot\mathbf{S} - 1 = 0$$

or by developing the form,

$$\{S_x,S_y,S_z,S_{xy},S_{yz},S_{xz}\}\cdot\frac{1}{f^2}\cdot
\begin{bmatrix}
1+\dfrac{2\hat{\alpha}f}{S_x} & -\dfrac{\hat{\lambda}}{2} & -\dfrac{\hat{\lambda}}{2} & 0 & 0 & 0 \\[2ex]
-\dfrac{\hat{\lambda}}{2} & 1+\dfrac{2\hat{\alpha}f}{S_y} & -\dfrac{\hat{\lambda}}{2} & 0 & 0 & 0 \\[2ex]
-\dfrac{\hat{\lambda}}{2} & -\dfrac{\hat{\lambda}}{2} & 1+\dfrac{2\hat{\alpha}f}{S_z} & 0 & 0 & 0 \\[2ex]
0 & 0 & 0 & 2+\hat{\lambda} & 0 & 0 \\[1ex]
0 & 0 & 0 & 0 & 2+\hat{\lambda} & 0 \\[1ex]
0 & 0 & 0 & 0 & 0 & 2+\hat{\lambda}
\end{bmatrix}
\cdot
\begin{Bmatrix} S_x \\ S_y \\ S_z \\ S_{xy} \\ S_{yz} \\ S_{xz} \end{Bmatrix} - 1 = 0 \tag{2.10}$$

where $\mathbf{S}^T = \{S_x,S_y,S_z,S_{xy},S_{yz},S_{xz}\}$ is a column matrix that represents the symmetric part of the stress tensor and $\mathbf{P}$ square matrix to recover the canonical form as expressed in equation (2.1).

This criterion defines the elastic behavior threshold for different types of materials ranging from geomaterials to metals by defining appropriately their coefficients $\hat{\alpha}$ and $\hat{\lambda}$. Assuming the hypothetical case of a metal in which the following strength relationshiship is met $f_c \equiv f_t \equiv f \Rightarrow \hat{\alpha} = 0$ y $\hat{\lambda} = 1$, then the classic *von Mises criterion* (Lubliner (1990)[28]) is recovered, and it can be expressed as follows,

$$\mathbb{F}^S(\mathbf{S},f) = \frac{1}{f^2}\left(S_x^2 + S_y^2 + S_z^2\right) - \frac{1}{f^2}\left(S_y S_z + S_z S_x + S_x S_y\right) +$$
$$+\frac{3}{f^2}\cdot\left(S_{yz}^2 + S_{xz}^2 + S_{xy}^2\right) - 1 = 0 \tag{2.11}$$

And its matrix representation as a function of the von Mises matrix $\mathbf{P}^{\text{Mises}}$ takes the following form,

$$\mathbb{F}^S(\mathbf{S}, f) = \mathbf{S}^T \cdot \mathbf{P}^{\text{Mises}} \cdot \mathbf{S} - 1 = 0$$

$$\{S_x, S_y, S_z, S_{xy}, S_{yz}, S_{xz}\} \cdot \frac{1}{f^2}
\begin{bmatrix}
1 & -\frac{1}{2} & -\frac{1}{2} & 0 & 0 & 0 \\
-\frac{1}{2} & 1 & -\frac{1}{2} & 0 & 0 & 0 \\
-\frac{1}{2} & -\frac{1}{2} & 1 & 0 & 0 & 0 \\
0 & 0 & 0 & 3 & 0 & 0 \\
0 & 0 & 0 & 0 & 3 & 0 \\
0 & 0 & 0 & 0 & 0 & 3
\end{bmatrix}
\cdot
\begin{Bmatrix}
S_x \\
S_y \\
S_z \\
S_{xy} \\
S_{yz} \\
S_{xz}
\end{Bmatrix}
- 1 = 0 \tag{2.12}$$

The form, according to the *Mises-Schleicher criterion* (Lubliner (1990)[28]), is recovered assuming that $f_c > f_t \Rightarrow \hat{\alpha} > 0$ and $\hat{\lambda} = 1$, and the *Drucker-Prager criterion* (Lubliner (1990)[28]) is obtained by reordering equation (2.1) and establishing the equivalent resistance $f = \sqrt{f_t f_c}$, $\hat{\alpha} = \frac{1}{2}[(f_c/f_t) - (f_t/f_c)] = \frac{1}{2f}(f_c^2 - f_t^2)$, and $\hat{\lambda} = 1 + 3\hat{\alpha}^2$. Then,

$$\mathbb{F}^s(\mathbf{S}, f) = \sqrt{\left(S_x^2 + S_y^2 + S_z^2\right) - \left(S_y S_z + S_z S_x + S_x S_y\right) + 3 \cdot \left(S_{yz}^2 + S_{xz}^2 + S_{xy}^2\right)} +$$
$$+ \frac{\hat{\alpha}}{1 + \hat{\alpha}^2}\left(S_x + S_y + S_z\right) - \frac{f}{\sqrt{1 + \hat{\alpha}^2}} = 0 \tag{2.13}$$

These general definitions of yield criteria will add another degree of generality which will lead to the definition of orthotropic yield criteria.

2.5 General explicit definition of the orthotropic yield criterion in the referential configuration

Just as Hill (1948)[11], Hill (1990)[14] had formulated the yield criterion for orthotropic materials from a generalization of the von Mises criterion, equation (2.1) can be generalized to an explicit orthotropic formulation as follows,

$$\mathbb{F}^S(\mathbf{S}, \boldsymbol{f}^S) = \left(\frac{S_x^2}{f_x^2} + \frac{S_y^2}{f_y^2} + \frac{S_z^2}{f_z^2}\right) - \hat{\lambda}_1\left(\frac{S_y S_z}{f_y f_z}\right) - \hat{\lambda}_2\left(\frac{S_z S_x}{f_z f_x}\right) - \hat{\lambda}_3\left(\frac{S_x S_y}{f_x f_y}\right) +$$
$$+ \left(\frac{S_{yz}^2}{f_{yz}^2} + \frac{S_{xz}^2}{f_{xz}^2} + \frac{S_{xy}^2}{f_{xy}^2}\right) + 2 \cdot \left(\hat{\alpha}_x \frac{S_x}{f_x} + \hat{\alpha}_y \frac{S_y}{f_y} + \hat{\alpha}_z \frac{S_z}{f_z}\right) - 1 = 0 \tag{2.14}$$

in which x, y, z are the orthotropic axes directions, and $\boldsymbol{f}^S = \{f_x, f_y, f_z, f_{xy}, f_{yz}, f_{xz}\}$ are the strength thresholds in the elastic limit expressed in such referential system and $\hat{\alpha}_i = \frac{1}{2}[(f_c/f_t)_i - (f_t/f_c)_i]$, with $i = x, y, z$. As observed, equation (2.14) is transformed into an isotropic form, equation (2.1), when the following conditions are met:

$$\hat{\lambda}_1 = \hat{\lambda}_2 = \hat{\lambda}_3 \equiv \hat{\lambda} \qquad ; \qquad \hat{\alpha}_x = \hat{\alpha}_y = \hat{\alpha}_z = \hat{\alpha} \equiv \frac{1}{2}\left[\left(\frac{f_c}{f_t}\right) - \left(\frac{f_t}{f_c}\right)\right]$$

$$f_x = f_y = f_z \equiv f = \sqrt{f_c f_t} \quad ; \qquad f_{yz} = f_{xz} = f_{xy} \equiv f_\tau = \frac{f}{\sqrt{2+\hat{\lambda}}} = \frac{\sqrt{f_c f_t}}{\sqrt{2+\hat{\lambda}}}$$

(2.15)

In these last expressions a dependency is observed between the isotropic and orthotropic magnitudes.

The Mises-Hill orthotropic yield criterion is obtained from equation (2.14) forcing equality between the tension and compression strengths in each orthotropy direction $\hat{\alpha}_x = \hat{\alpha}_y = \hat{\alpha}_z = 0$, and also subjecting $\hat{\lambda}_i$ to the following restriction,

$$\hat{\lambda}_1\hat{\lambda}_2\hat{\lambda}_3 + \hat{\lambda}_1^2 + \hat{\lambda}_2^2 + \hat{\lambda}_3^2 = 4 \qquad (2.16)$$

As a result of this particularization, the Mises-Hill yield criterion is obtained,

$$\mathbb{F}^S(\mathbf{S}, f^S) = (G+H)S_x^2 + (F+H)S_y^2 + (F+G)S_z^2 - 2FS_yS_z - 2GS_zS_x -$$
$$- 2HS_xS_y + 2LS_{yz}^2 + 2MS_{xz}^2 + 2NS_{xy}^2 - 1 = 0 \qquad (2.17)$$

Hill's criterion can be written in the following matrix form,

$$\mathbb{F}^S(\mathbf{S}, f^S) = \mathbf{S}^T \cdot \mathbf{P}^{\text{Mises}} \cdot \mathbf{S} - 1 = 0$$

$$\mathbb{F}^\sigma(\mathbf{S}, f^S) = \{S_x, S_y, S_z, S_{xy}, S_{yz}, S_{xz}\} \cdot \begin{bmatrix} (H+G) & -H & -G & 0 & 0 & 0 \\ -H & (F+H) & -F & 0 & 0 & 0 \\ -G & -F & (G+H) & 0 & 0 & 0 \\ 0 & 0 & 0 & 2N & 0 & 0 \\ 0 & 0 & 0 & 0 & 2L & 0 \\ 0 & 0 & 0 & 0 & 0 & 2M \end{bmatrix} \cdot \begin{Bmatrix} S_x \\ S_y \\ S_z \\ S_{xy} \\ S_{yz} \\ S_{xz} \end{Bmatrix} - 1 = 0 \quad (2.18)$$

where $\mathbf{S}^T = \{S_x, S_y, S_z, S_{xy}, S_{yz}, S_{xz}\}$ is the symmetric part of the stress tensor and $\mathbf{P}^{\text{Hill}}$ a square matrix to recover the canonic form expressed in equation (2.17).

By comparing equation (2.17) with equation (2.14), the following relationship among Mises-Hill parameters and the strength thresholds is obtained,

$$\begin{cases} \dfrac{1}{f_x^2} = (G+H) & ; & \dfrac{\hat{\lambda}_1}{f_y f_z} = 2F & ; & \dfrac{1}{f_{yz}^2} = 2L \\[2mm] \dfrac{1}{f_y^2} = (F+H) & ; & \dfrac{\hat{\lambda}_2}{f_x f_z} = 2G & ; & \dfrac{1}{f_{xz}^2} = 2M \\[2mm] \dfrac{1}{f_z^2} = (F+G) & ; & \dfrac{\hat{\lambda}_3}{f_x f_y} = 2H & ; & \dfrac{1}{f_{xy}^2} = 2N \end{cases} \qquad (2.19)$$

By clearing F, G and H from the second colunm of equation (2.19) and replacing it into the first one a system of equations in $\hat{\lambda}_i$ is obtained and its roots are,

$$\begin{cases} \hat{\lambda}_1 = \dfrac{f_z^2 f_x^2 - f_z^2 f_y^2 + f_y^2 f_x^2}{f_y f_z f_x^2} \\[2ex] \hat{\lambda}_2 = \dfrac{-f_z^2 f_x^2 + f_z^2 f_y^2 + f_y^2 f_x^2}{f_x f_z f_y^2} \\[2ex] \hat{\lambda}_3 = \dfrac{f_z^2 f_x^2 + f_z^2 f_y^2 - f_y^2 f_x^2}{f_y f_x f_z^2} \end{cases} \qquad (2.20)$$

from where it follows that the condition required by equation (2.16) is satisfied. By replacing equation (2.20) into equation (2.19) Hill's constants are obtained as a function of the strength thresholds,

$$\begin{cases} F = \dfrac{\hat{\lambda}_1}{2 f_y f_z} = \dfrac{f_z^2 f_x^2 - f_z^2 f_y^2 + f_y^2 f_x^2}{2 f_y^2 f_z^2 f_x^2} & ; & L = \dfrac{1}{2 f_{yz}^2} \\[2ex] G = \dfrac{\hat{\lambda}_2}{2 f_x f_z} = \dfrac{-f_z^2 f_x^2 + f_z^2 f_y^2 + f_y^2 f_x^2}{2 f_y^2 f_z^2 f_x^2} & ; & M = \dfrac{1}{2 f_{xz}^2} \\[2ex] H = \dfrac{\hat{\lambda}_3}{2 f_x f_y} = \dfrac{f_z^2 f_x^2 + f_z^2 f_y^2 - f_y^2 f_x^2}{2 f_y^2 f_z^2 f_x^2} & ; & N = \dfrac{1}{2 f_{xy}^2} \end{cases} \qquad (2.21)$$

As shown by Hill's parameter definition, the yield criterion, equation (2.17), recovers von Mises isotropic form, equation (2.11), if the following conditions are met,

$$\hat{\lambda}_1 = \hat{\lambda}_2 = \hat{\lambda}_3 \equiv \hat{\lambda} = 1 \;\; ; \;\; f_x = f_y = f_z \equiv f \;\; ; \;\; f_{yz} = f_{xz} = f_{xy} \equiv f_\tau = \dfrac{f}{\sqrt{2 + \hat{\lambda}}} = \dfrac{f}{\sqrt{3}} \qquad (2.22)$$

Equation (2.14) is transformed into Hill-Schleicher orthotropic criterion (Lubliner (1990)[28]), when in addition to equation (2.16) the orthotropic ratio is also met $\hat{\alpha}_x + \hat{\alpha}_y + \hat{\alpha}_z > 0$.

2.6 General implicit definition of the orthotropic criterion in the referential configuration

As shown in the previous section, the traditional procedures to obtain the constitutive equations for elastoplastic anisotropic materials are based on the description of yield surface and plastic potential surface as a function of the material's own characteristics. It is hard to satisfy the invariance conditions of these cases and sometimes it is simply imposible.

The *general implicit definition* of an orthotropic yield criterion is based on an "*isotropic formulation in the fictitious space*" to transform it into an "*implicit orthotropic formulation in the real space*". This means that it is not necessary to explicitly express the mathematical form of this orthotropic criterion, but simply to express mathematically its isotropic form and recognize the existance of a numeric transformation to change the isotropic criterion to another implicit orthotropic one.

This implicit orthotropy formulation qualititatively mentioned is known as "*space mapping theory*" (Sobotka (1969)[31], Boehler and Sawczuk (1970)[32], Beten (1981)[22], Oller et al. (1995)[26], (1996)[33,34], (1998)[35], Car et al. (1999)[36,37], (2000)[41]), and it involves accepting the existance of two stress spaces: one expressed as $\boldsymbol{\sigma}$ defined as "stresses real space in the referential configuration" Ω^S, where the "implicit orthotropy yield" $\mathbb{F}^S(\mathbf{S}, f^S) = 0$ lies, and another one expressed as $\overline{\mathbf{S}}$ known as "stresses fictitious space in the referential configuration" $\overline{\Omega}^S$, where the "explicit isotropy yield" $\overline{\mathbb{F}}^S(\overline{\mathbf{S}}, f^{\overline{S}}) = 0$ lies. Moreover, between both spaces there is a symmetric fourth-order tensor of space transformation $\mathbf{A}^S$, that allows transforming in a biunivocal manner a tensor image of the stresses defined in one space to other and vice versa. This procedure guarantees the invariance condition (Oller et al. (1995)[26]), (Casas et al. (1998)[38]) (see Figure 2.1).

2.6.1   Stress space transformation

The second tensor transformation of Piola-Kirchhoff stresses from the anisotropic real space to the fictitious isotropic one, which guarantees the invariance previously defined, is carried out through the following form (Oller et al. (1995)[26])

$$\overline{S}_{IJ} \overset{def}{=} A^S_{IJKL}\, S_{KL} \tag{2.23}$$

where $\mathbf{A}^S$ is a fourth-order tensor defined as the linear operator $\mathbf{A}^S : S_2 \to S_2$ which establishes the relationshipship between the symmetric tensors space of the real and fictitious stresses, respectively, in the referential configuration, $\overline{\mathbf{S}}$ is the second tensor of Piola-Kirchhoff stresses in the fictitious space and $\mathbf{S}$ is the second tensor of Piola-Kirchhoff stresses in the anisotropic real space. The fourth-order tensor $\mathbf{A}^S$ is defined in the referential configuration and remains constant in this configuration. The yield condition of the anisotropic material in the isotropic fictitious space is expressed as:

$$\overline{\mathbb{F}}^S(\overline{\mathbf{S}}, \mathcal{C}, \boldsymbol{\alpha}, f^{\overline{S}}) = 0 \tag{2.24}$$

where $\overline{\mathbb{F}}^S$ is the yield function in *the referential configuration* and *in the isotropic fictitious space* and is different from the yield function in the *anisotropic space* $\mathbb{F}^S$ in the arguments that both functions contain.

[31] Sobotka, Z. (1969). Theorie des plastischen Fließens von ainsotropen Körpern. Z. *Angew. Math. Mech., vol. 49, pp. 25-32.*

[32] Boehler, J. P., and Sawczuk, A. (1970). Équilibre limite des sols anisotropes. *J. Méc., Vol. 9, pp. 5-32.*

[33] Oller, S., Oñate, E., Miquel Canet, J., and Botello, S. (1996-a). A plastic damage constitutive model for composite materials. *Int. Jour. Solids Struct., Vol. 33, No. 17, pp. 2501-2518.*

[34] Oller, S., Oñate, E., and Miquel Canet, J. (1996-b). A mixing anisotropic formulation for analysis of composites. *Communications in Numerical Methods in Engineering, Vol. 12, 471-482.*

[35] Oller, S., Rubert, J., Las Casas, E., Oñate, E., and Proença, S. (1998). Large Strains Elastoplastic Formulation For Anisotropic Materials. *First Esaform Conference on Material Forming,ed. by J. Chenot, J. Agassant, P. Montnitinnet, B. Bergnes, N. Billon. pp. 191194. Sophia Antipolis (France) March. 1998.*

[36] Car, E., Oller, S. and Oñate, E. (1999-a). Numerical Constitutive Model For Laminated Composite Materials. *Second ESAFORM Conference on Material Forming. Minho, Portugal, Ed. J. A. Covas, pp. 147-150.*

[37] Car, E., Oller, S., and Oñate, E. (1999-b). A Large Strain Plasticity Model for Anisotropic Material - Composite Material Application. *Int. J. Plasticity, Vol.17, No. 11, pp. 1437-1463.*

[38] Las Casas E., Oller S., Rubert J., Proença and Oñate E. (1998). A Large Strain Explicit Formulation for Composites. Proceedings of the Fourth World Congress on Computational Mechanics. Ed. S. Idelsohn, E. Oñate and E. Dvorkin. CIMNE, Barcelona, Spain.

The definition of the *"implicit orthotropic yield criterion"* is achieved by admitting that in the isotropic fictitious space there exists an isotropic yield criterion of the type $\overline{\mathbf{S}}^T \cdot \mathbf{P} \cdot \overline{\mathbf{S}} - 1 = 0$, which is equivalent to the one defined in equation (2.12), and that is an image of the orthotropic criterion expected to be approximated. As shown below, by substituting the stress space transformation into the orthotropic yield criterion, *"the orthotropic yield criterion in the isotropic fictitious space"* is obtained,

$$
\left\{
\begin{array}{l}
\text{a) Orthotropic yield criterion in the real stresses space:} \quad \mathbb{F}^S(\mathbf{S}, \boldsymbol{f}^S) = \mathbf{S}^T \cdot \mathbf{P}^{\mathrm{Ort}} \cdot \mathbf{S} - 1 = 0 \\[2mm]
\qquad \mathbb{F}^S(\mathbf{S}, \boldsymbol{f}^S) \equiv \mathbb{F}^S(\overline{\mathbf{S}}, \mathbf{A}^S, \boldsymbol{f}^{\overline{S}}) = \left[\left(\mathbf{A}^S \right)^{-1} \cdot \overline{\mathbf{S}} \right]^T \cdot \mathbf{P}^{\mathrm{Ort}} \cdot \left[\left(\mathbf{A}^S \right)^{-1} \overline{\mathbf{S}} \right] - 1 = 0 \\[2mm]
\qquad \mathbb{F}^S(\overline{\mathbf{S}}, \mathbf{A}^S, \boldsymbol{f}^{\overline{S}}) = \overline{\mathbf{S}}^T \cdot \underbrace{\left[(\mathbf{A}^S)^{-T} \cdot \mathbf{P}^{\mathrm{Ort}} \cdot (\mathbf{A}^S)^{-1} \right]}_{\mathbf{P}} \cdot \overline{\mathbf{S}} - 1 = 0 \\[4mm]
\text{b) Isotropic yield criterion in the fictitious stresses space:} \quad \overline{\mathbb{F}}^S(\overline{\mathbf{S}}, \boldsymbol{f}^{\overline{S}}) = \overline{\mathbf{S}}^T \cdot \mathbf{P} \cdot \overline{\mathbf{S}} - 1 = 0
\end{array}
\right. \tag{2.25}
$$

from where the tensor $\mathbf{A}^S$ can be deducted to obtain the relationship $\mathbf{P}^{\mathrm{Ort}} = (\mathbf{A}^S)^T \cdot \mathbf{P} \cdot (\mathbf{A}^S)$ between matrices $\mathbf{P}^{\mathrm{Ort}}$ (defined by the implicit orthotropic criterion, equation (2.25) to be approximated, and $\mathbf{P}$ (defined by the adopted isotropic fictitious criterion) can be obtained. Once the *isotropic yield criterion* is chosen $\overline{\mathbb{F}}^S(\overline{\mathbf{S}}, \boldsymbol{f}^{\overline{S}}) = \overline{\mathbf{S}}^T \cdot \mathbf{P} \cdot \overline{\mathbf{S}} - 1 = 0$, it will be used for the definition of the *implicit orthotropic criterion* $\mathbb{F}^S(\mathbf{S}, \boldsymbol{f}^S) = \mathbf{S}^T \cdot \mathbf{P}^{\mathrm{Ort}} \cdot \mathbf{S} - 1 = 0$, only the definition of the *space transformation tensor* $\mathbf{A}^S$ remains, which will be shown below.

In the definition of the introduced tensor operator form and properties $\mathbf{A}^S$ it is necessary to take into account the following symmetry:

$$
A_{IJKL}^S = A_{JIKL}^S = A_{IJLK}^S \tag{2.26}
$$

It is assumed that the fourth-order tensor of the stress space trasformation $\mathbf{A}^S$ is also symmetric.

$$
A_{IJKL}^S = A_{KLIJ}^S \tag{2.27}
$$

As a first approximation, the linear operator $\mathbf{A}^S$ can be defined in the following simplefied form (Oller *et al* (1995)[26]), leading to a fourth-order tensor and a diagonal square matrix representation due to its symmetries.

$$
A_{IJKL}^S = f_{IK}^{\overline{S}} \cdot (f_{JL}^S)^{-1} \tag{2.28}
$$

where $f_{IK}^{\overline{S}}$ y f_{JL}^S are the stress second order tensors representing the corresponding isotropic fictitious space and real anisotropic strengths. The linear operator definition given in equation (2.28) is not univocal as it is generally not possible to obtain a fourth-order tensor (81 components) from the second order tensors information (9 components each). Consequently, equation (2.28) is only verified for diagonal tensors.

More generally, the stress transformation A_{IJKL}^S can be adjusted to any anisotropic behavior in the stress space from the material properties and the plastic yield criterion form established in the isotropic and anisotropic spaces. This is,

$$A^S_{IJKL} = \left(W_{IJRS}\, w_{RSKL}\right)^{-1} \tag{2.29}$$

where W_{IJRS} contains the strength threshold information in each orthotropic direction and w_{RSKL} is a tensor to adjust the proposed orthotropic criteria. Both tensors are defined further[39].

The elastic limit determination or strength threshold is based on real experimental tests and therefore the structure is in a deformed state (updated configuration). Consequently, this obtained limit value is found in the updated configuration.

The necessary strength threshold to find tensor $\mathbf{A}^S$ is obtained by the strength limit value in the referential configuration and therefore the pull back transport of the stress limit obtained in the uniaxial test must be carried out, thus

$$\boldsymbol{f}^S = \bar{\bar{\phi}}(\boldsymbol{f}^\tau) \tag{2.30}$$

where $\boldsymbol{f}^S$ and $\boldsymbol{f}^\tau$ are strength tensors in the referential and updated configurations, respectively, and $\bar{\bar{\phi}}(\bullet)$ is the transport operator between the two *pull-back* configurations, the strain gradient tensor function . This tensor is calculated component by component from the gradient tensor of the strains, which for one dimensional case is

$$\mathbf{F} = \frac{\partial \boldsymbol{x}}{\partial \mathbf{X}} = \frac{l_f}{l_0} \tag{2.31}$$

where l_0 is the length of the specimen tested in an uniaxial form, and l_f is the length of the specimen on the updated configuration. Taking into account that $\boldsymbol{\tau} = J\boldsymbol{\sigma}$ and $\mathbf{S} = \mathbf{F}^{-1} : (J\,\boldsymbol{\sigma}) : \mathbf{F}^{-T}$, the material strength value is:

$$\boldsymbol{f}^S = \frac{l_0}{l_f}(\boldsymbol{f}^\sigma) \tag{2.32}$$

where $\boldsymbol{f}^S$ and $\boldsymbol{f}^\sigma$ are the material resistances in the referential and updated configuration, respectively.

2.6.2   Strain space transformation

The space transformation defined in equation (2.23) can only be applied to proportional materials, which are materials that have a constant relationship between the yield stress and the Young module for each space direction,

$$f_{11}/C_{11} = f_{22}/C_{22} = \cdots = f_{23}/C_{23} \tag{2.33}$$

where f_{ij} and C_{ij} represent the material strength and the elastic module, respectively, in the referential configuration for the space $i-j$. This means that the elastic strain tensor should be the same in the real and fictitious spaces (Oller *et alt.*, (1995)[25]). For the formulation generatization of several materials, Oller et al. (1995)[26] also proposed the space transformation of the real strains.

---

[39] Oller S., Car E. and Lubliner J. (2001). Definition of a general implicit orthotropic yield criterion. Submitted in *Computer Methods in Applied Mechanics and Engineering.*

The relationship between Green-Lagrange elastic strains in the ¨real anisotropic space¨ E_{IJ}^e and Green-Lagrange elastic strains in the isotropic ¨fictitious space¨ $\overline{E}_{IJ}^e$ are defined (see Figure 9) through the following relationship,

$$\overline{E}_{IJ}^e \overset{def}{=} A_{IJKL}^E E_{KL}^e \tag{2.34}$$

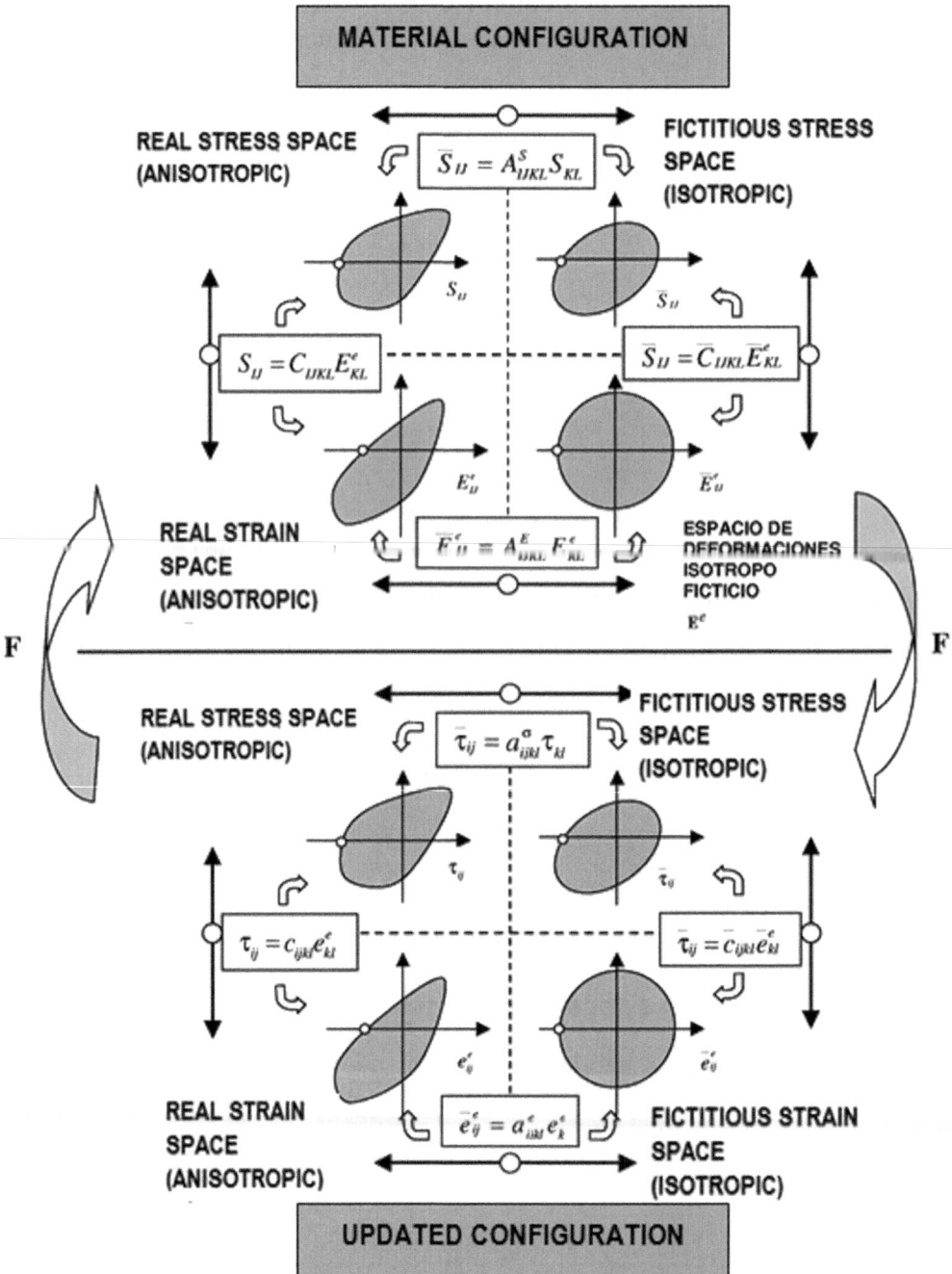

Figure 2.2 – Space transformation. Stresses and strains real and fictitious spaces in large strains.

where $\mathbf{A}^E$ is a fourth-order tensor defined as the linear operator $\mathbf{A}^E : S_2 \to S_2$ which establishes the relationship between Green-Lagrange strain real spaces E_{IJ}^e and fictitious spaces $\overline{E}_{IJ}^e$. This hypothesis implies the non-uniqueness in the elastic strains developed in the two spaces. The *strains transformation tensor* $\mathbf{A}^E$ can be calculated taking into account equations (2.23) and (2.34) as follows,

$$
\begin{aligned}
\overline{\mathbb{C}}^S_{IJMN} \overline{E}^e_{MN} &= A^S_{IJKL} \mathbb{C}^S_{KLRS} E^e_{RS} \\
\overline{\mathbb{C}}^S_{IJMN} A^E_{MNRS} E^e_{RS} &= A^S_{IJKL} \mathbb{C}^S_{KLRS} E^e_{RS} \\
\overline{\mathbb{C}}^S_{IJMN} A^E_{MNRS} &= A^S_{IJKL} \mathbb{C}^S_{KLRS} \\
A^E_{MNRS} &= (\overline{\mathbb{C}}^S_{IJMN})^{-1} A^S_{IJKL} \mathbb{C}^S_{KLRS}
\end{aligned}
\tag{2.35}
$$

where $\overline{\mathbb{C}}^S$ is an arbitrary constitutive tensor defined in the *isotropic fictitious space* and $\mathbb{C}^S$ is the constitutive tensor in the anisotropic real space. The selection of $\overline{\mathbb{C}}^S$ is arbitrary and can be represented by the properties of any known isotropic material, as it is only used for the fictitious space and its influence is later cancelled when it is returned to the real space. Taking into account equation (2.35), the relationship between the constitutive tensors in the anisotropic and the fictitious spaces can be established as,

$$
\mathbb{C}^S_{KLRS} = (A^S_{IJKL})^{-1} \overline{\mathbb{C}}^S_{IJMN} A^E_{MNRS}
\tag{2.36}
$$

The real anisotropic constitutive tensor $\mathbb{C}^S$ and the stress transformation tensor $\mathbf{A}^S$ are expressed in local coordinates. Therefore, it is necessary to express them in the global reference system through a fourth-order tensor rotation $\mathbf{R}$, thus

$$
\begin{aligned}
\mathbb{C}^S_{IJKL} &= R_{IJRS} (\mathbb{C}^S_{RSPQ})_{\text{loc}} R_{PQKL} \\
A^S_{IJKL} &= R_{IJRS} (A^S_{RSPQ})_{\text{loc}} R_{PQKL}
\end{aligned}
\tag{2.37}
$$

where $(\mathbb{C}^S)_{\text{loc}}$ is the fourth-order constitutive tensor in the real anisotropic space expressed in the local reference system and $(\mathbf{A}^S)_{\text{loc}}$ is the space transformation tensor in local orthotropic directions The rotation tensor $\mathbf{R}$ is defined as

$$
R_{IJKL} = r_{IK} r_{JL}
\tag{2.38}
$$

where $r_{IK} = \cos\left[(\vec{e}_I)_{\text{glob}}, (\vec{e}_K)_{\text{loc}}\right]$, being $(\vec{e}_I)_{\text{glob}}$ and $(\vec{e}_K)_{\text{loc}}$ the unit versors corresponding to ith and jth components into the global and local systems, respectively. The angles between the principal directions of the anisotropic material and the global coordinates are considered by the rotation tensor $\mathbf{R}$.

2.7 General definition of the stress space transformation tensor $\mathbf{A}^S$

There are different approaches for the definition of the space transformation tensor $\mathbf{A}^S$ such as the research work carried out by Beten (1981)[22], Oller *et al.* (1995)[26], (1996)[33,34], (1998)[35], Car *et al.* (1999)[36,37] and (2000)[41], etc. Although these definitions provide suitable orthotropic yield criteria, these are hard to fit exactly into the behavior of a desired material.

A new definition of tensor $\mathbf{A}^S$ is formalized to avoid this limitation, through a space mapping theory that can fit exactly any isotropic yield criterion to any other orthotropic one. This is achieved by the following relationship,

$$\left(A^S_{IJKL}\right)^{-1} \equiv B^S_{IJKL} = W_{IJRS}w_{RSKL} \tag{2.39}$$

where $W_{RSKL} = \omega_{RK}\omega_{SL}$ contains the information of the strength thresholds to the elastic limit in each orthotropy, being $\omega_{IJ} = \mathrm{Diag}\left\{\omega_{\{xx\}};\omega_{\{yy\}};\omega_{\{zz\}}\right\} = \mathrm{Diag}\left\{\sqrt{f_x/f};\sqrt{f_y/f};\sqrt{f_z/f}\right\}$, such that taking advantage of the tensor symmetries W_{RSKL}, the following matrix form is obtained,

$$W_{IJKL} = \omega_{IK}\omega_{JL} = \begin{bmatrix} \omega_{\{xx\}}\omega_{\{xx\}} & 0 & 0 & 0 & 0 & 0 \\ 0 & \omega_{\{yy\}}\omega_{\{yy\}} & 0 & 0 & 0 & 0 \\ 0 & 0 & \omega_{\{zz\}}\omega_{\{zz\}} & 0 & 0 & 0 \\ 0 & 0 & 0 & \omega_{\{xx\}}\omega_{\{yy\}} & 0 & 0 \\ 0 & 0 & 0 & 0 & \omega_{\{yy\}}\omega_{\{zz\}} & 0 \\ 0 & 0 & 0 & 0 & 0 & \omega_{\{zz\}}\omega_{\{xx\}} \end{bmatrix} \rightarrow$$

$$\rightarrow W_{IJ} = \omega_{II}\omega_{JJ} = \begin{bmatrix} \omega^2_{xx} & 0 & 0 & 0 & 0 & 0 \\ 0 & \omega^2_{yy} & 0 & 0 & 0 & 0 \\ 0 & 0 & \omega^2_{zz} & 0 & 0 & 0 \\ 0 & 0 & 0 & \omega_{xx}\omega_{yy} & 0 & 0 \\ 0 & 0 & 0 & 0 & \omega_{yy}\omega_{zz} & 0 \\ 0 & 0 & 0 & 0 & 0 & \omega_{zz}\omega_{xx} \end{bmatrix} \tag{2.40}$$

and w_{IJKL} is a tensor aimed at adjusting the function of the proposed orthotropic criterion whose form and symmetries are shown in the following matrix representation,

$$w_{IJKL} = \begin{bmatrix} \alpha_{\{xx\}\{xx\}} & \alpha_{\{xx\}\{yy\}} & \alpha_{\{xx\}\{zz\}} & 0 & 0 & 0 \\ \alpha_{\{yy\}\{xx\}} & \alpha_{\{yy\}\{yy\}} & \alpha_{\{yy\}\{zz\}} & 0 & 0 & 0 \\ \alpha_{\{zz\}\{xx\}} & \alpha_{\{zz\}\{yy\}} & \alpha_{\{zz\}\{zz\}} & 0 & 0 & 0 \\ 0 & 0 & 0 & \beta_{\{xy\}\{xy\}} & 0 & 0 \\ 0 & 0 & 0 & 0 & \beta_{\{yz\}\{yz\}} & 0 \\ 0 & 0 & 0 & 0 & 0 & \beta_{\{zx\}\{zx\}} \end{bmatrix} \rightarrow w_{IJ} =$$

$$= \begin{bmatrix} \alpha_{xx} & \alpha_{xy} & \alpha_{xz} & 0 & 0 & 0 \\ \alpha_{yx} & \alpha_{yy} & \alpha_{yz} & 0 & 0 & 0 \\ \alpha_{zx} & \alpha_{zy} & \alpha_{zz} & 0 & 0 & 0 \\ 0 & 0 & 0 & \beta_{xy} & 0 & 0 \\ 0 & 0 & 0 & 0 & \beta_{yz} & 0 \\ 0 & 0 & 0 & 0 & 0 & \beta_{zx} \end{bmatrix} \; ; \; \mathrm{con} : \alpha_{xy} \neq \alpha_{yx} \; ; \; \alpha_{xz} \neq \alpha_{zx} \; ; \; \alpha_{yz} \neq \alpha_{zy} \tag{2.41}$$

This reduced form of tensors $\mathbf{w}$ and $\mathbf{W}$, due to their symmetry, can simplify how to obtain the space transformation tensor as follows,

$$\left(A^S_{IJ}\right)^{-1} \equiv B^S_{IJ} = W_{IK} w_{KJ} \tag{2.42}$$

The isotropic form is recovered by defining the same equality of the resistance in all directions $f_x = f_y = f_z = f$, such that tensor $\omega_{ij} \equiv \delta_{ij}$ coincides with "Kronecker delta" and the matrix form of the adjusting tensor $w_{IJ} = I_{IJ} = \mathbf{I}$ also coincides with the second-order identity. This statement will be confirmed in the next section.

2.8 Numerical calculation of the adjusting tensor in matrix form w_{IJ}

The adjusting tensor is obtained by forcing the equality compliance expressed in equation (2.25), after adopting the matrix forms of tensors $\mathbf{W}$ and $\mathbf{w}$. Thus,

$$\begin{aligned}
\left[(\mathbf{B}^S)^T \cdot \mathbf{P}^{\text{Ort}} \cdot (\mathbf{B}^S)\right] - \mathbf{P} &= \mathbf{0} \\
\left[(\mathbf{W} \cdot \mathbf{w})^T \cdot \mathbf{P}^{\text{Ort}} \cdot (\mathbf{W} \cdot \mathbf{w})\right] - \mathbf{P} &= \mathbf{0} \\
\left[\mathbf{w}^T \cdot \left(\mathbf{W}^T \cdot \mathbf{P}^{\text{Ort}} \cdot \mathbf{W}\right) \cdot \mathbf{w}\right] - \mathbf{P} &= \mathbf{0}
\end{aligned} \tag{2.43}$$

By solving this quadratic system of equations in $\mathbf{w}$ the symmetric part of this tensor is obtained $\mathbf{w} = w_{IJ}$. Thus, the space transformation tensor $\mathbf{A}^S = (\mathbf{W} \cdot \mathbf{w})^{-1}$ can also be obtained, which leads to an *orthotropic yield criterion* of the $\overline{\mathbf{S}}^T \cdot \mathbf{P} \cdot \overline{\mathbf{S}} - 1 = 0$ type, and which coincides exactly in implicit form to the desired orthotropic criterion $\mathbf{S}^T \cdot \mathbf{P}^{\text{Ort}} \cdot \mathbf{S} - 1 = 0$.

The analytical solution of equation (2.43) can be rather complex depending on the orthotropic yield criterion we wish to approximate. Therefore, a numeric solution should be used for this equation. The solution for this non-linear system of equations in $\mathbf{w}$ is reached by applying the Newton Raphson method (Press *et al.* (1992)[40]) to the following constraint equations,

$$\begin{aligned}
R_{IJ} &= \left[w^T_{IK}\left(W^T_{KR}\, P^{\text{Ort}}_{RS}\, W_{ST}\right)w_{TJ}\right] - P_{IJ} = 0 \\
R_{IJ} &= \left[w_{KI}\left(W_{RK}\, P^{\text{Ort}}_{RS}\, W_{ST}\right)w_{TJ}\right] - P_{IJ} = 0
\end{aligned} \tag{2.44}$$

where R is a symmetric matrix expressing the residue to be eliminated by Newton Raphson method. The residue R elimination can be carried out by linearizing it in the vicinity of the current solution, iteration (i+1). Consequently, such condition is expressed through Taylor series up to the first variation.

$$\mathbf{0} = \mathbf{R}^{i+1} \cong \mathbf{R}^i + \left[\frac{\partial \mathbf{R}}{\partial \mathbf{w}}\right]^i \cdot \Delta \mathbf{w}^{i+1}$$

$$0 = R^{i+1}_{IJ} \cong R^i_{IJ} + \underbrace{\left[\frac{\partial R_{IJ}}{\partial w_{UV}}\right]^i}_{J^i_{IJUV}} \cdot \Delta w^{i+1}_{UV} \tag{2.45}$$

---

[40] Press, W. H., Teulosky, S. A., Vetterling, W. T. and Flannery, B. P. (1992). *Numerical recipes in Fortran 77. The art of scientific computing. Volume I.* Cambridge University Press.

where J_{IJUV} stands for the fourth-order Jacobian operator. By replacing the equation (2.44) into equation (2.45), then,

$$\Delta w_{UV}^{i+1} = -\underbrace{\left\{2\,I_{KIUV}\left(W_{RK}\,\mathrm{P}_{RS}^{\mathrm{Ort}}W_{ST}\right)w_{TJ}\right\}_{i}^{-1}}_{\left\{J_{IJUV}^{i}\right\}^{-1}}\cdot\left\{w_{KI}\left(W_{RK}\,\mathrm{P}_{RS}^{\mathrm{Ort}}\,W_{ST}\right)w_{TJ}\right]-\mathrm{P}_{IJ}\right\}_{i}$$

(2.46)

which by carrying out the following updating the current value of the unknown can be obtained,

$$w_{UV}^{i+1} = w_{UV}^{i} + \Delta w_{UV}^{i+1}$$

(2.47)

This will be the solution sought in the (i+1) iteration, provided that the residue R satisfies the condition imposed to equation (2.44).

Despite the matrix simplifications carried out in equations (2.40), (2.41), (2.42) and (2.43), the equation solution (2.46) is still highly complex as it requires the fourth-order Jacobian tensor inversion $\left\{J_{IJUV}\right\}^{-1}$. A simplified solution for this system of equation is by representing tensors R and **w** as column matrices so that the *Jacobian* operator J_{IJUV} is reduced to a quadratic matrix J_{IJ}. This operational simplification helps to solve the system of non-linear equations. The simplified algorithm solution is detailed as follows,

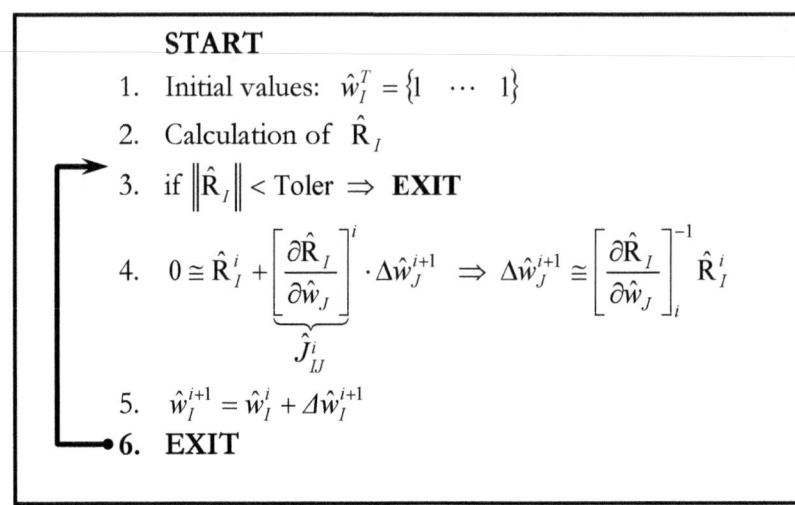

Table 2.1 – The Newton Raphson algorithm for the solution of equation (2.44)

being $\hat{R}_{I}$ a column matrix used to express the residue in that algorithm. Then,

$$\mathrm{R}_{IJ} = \begin{bmatrix} R_{xx} & R_{xy} & R_{xz} & 0 & 0 & 0 \\ R_{yx} & R_{yy} & R_{yz} & 0 & 0 & 0 \\ R_{zx} & R_{zy} & R_{zz} & 0 & 0 & 0 \\ 0 & 0 & 0 & R_{xy} & 0 & 0 \\ 0 & 0 & 0 & 0 & R_{yz} & 0 \\ 0 & 0 & 0 & 0 & 0 & R_{xz} \end{bmatrix} \Rightarrow \hat{\mathrm{R}}_{I} = \begin{Bmatrix} R_{xx} \\ R_{yy} \\ \vdots \\ R_{xy} \\ \vdots \\ R_{xz} \end{Bmatrix}$$

(2.48)

The unknown is also expressed as a column-matrix form as follows,

$$w_{IJ} = \begin{bmatrix} \alpha_{xx} & \alpha_{xy} & \alpha_{xz} & 0 & 0 & 0 \\ \alpha_{yx} & \alpha_{yy} & \alpha_{yz} & 0 & 0 & 0 \\ \alpha_{zx} & \alpha_{zy} & \alpha_{zz} & 0 & 0 & 0 \\ 0 & 0 & 0 & \beta_{xy} & 0 & 0 \\ 0 & 0 & 0 & 0 & \beta_{yz} & 0 \\ 0 & 0 & 0 & 0 & 0 & \beta_{xz} \end{bmatrix} \Rightarrow \hat{w}_I = \begin{Bmatrix} \alpha_{xx} \\ \alpha_{yy} \\ \vdots \\ \beta_{xy} \\ \vdots \\ \beta_{xz} \end{Bmatrix} \tag{2.49}$$

And the Jacobian matrix is built up in a dimensionally compatible form with the two previous definitions as shown below,

$$\hat{J}_{IJ} = \frac{\partial \hat{R}_I}{\partial w_J} = \begin{bmatrix} \dfrac{\partial R_{xx}}{\partial \alpha_{xx}} & \dfrac{\partial R_{xx}}{\partial \alpha_{yy}} & \cdots & \dfrac{\partial R_{xx}}{\partial \beta_{xz}} \\ \dfrac{\partial R_{yy}}{\partial \alpha_{xx}} & \dfrac{\partial R_{yy}}{\partial \alpha_{yy}} & \cdots & \dfrac{\partial R_{yy}}{\partial \beta_{xz}} \\ \vdots & \vdots & \ddots & \vdots \\ \dfrac{\partial R_{xz}}{\partial \alpha_{xx}} & \dfrac{\partial R_{xz}}{\partial \alpha_{yy}} & \cdots & \dfrac{\partial R_{xz}}{\partial \beta_{xz}} \end{bmatrix} \tag{2.50}$$

2.9 Mises-Hill orthotropic criterion verification by the space mapping theory

For illustration purposes a comparison in the principal stress plane $(\sigma_1 - \sigma_2)$ will be shown next, between the Mises-Hill yield criterion obtained through the space mapping theory and its original form. A perfect coincidence can be observed here between the two approaches to obtain the Mises-Hill yield criterion. Therefore, the equation is solved with the Newton-Raphson method (2.42), resulting the $\mathbf{w} = w_{IJ}$ tensor that leading to the definition of a tensor $\mathbf{A}^S$, which adjusts the von Mises isotropic yield criterion to Mises-Hill orthotropic yield criterion by means of the *space mapping* theory.

For example, let us assume a hypothetical orthotropic material in a plane-strain state, which has the following resistances in each of the orthotropy directions: $f_x = 100\,MN/m^2$, $f_y = 200\,MN/m^2$, $f_z = 100\,MN/m^2$, $f_{xy} = 50\,MN/m^2$. The adopted isotropic strength of the comparison is $f = 100\,MN/m^2$.

With this information, tensor W_{IJ} (equation (2.40)) is obtained and then the matrix h defining the orthotropic yield criterion $\mathbf{P}^{Ort} = \mathbf{P}^{Hill}$ (equation (2.18)) is calculated and it also defines the adopted isotropic yield criterion $\mathbf{P} = \mathbf{P}^{Mises}$ (equations (2.10) and (2.12)). After following these previous steps, the matrix form of the residual $\mathbf{R}_{IJ}$ is obtained (equation (2.44), and its order in column matrix form $\mathbf{R}_I$, according to equation (2.48). This way the residual $\mathbf{R}_I(\mathbf{w})$ remains as a function of the adjusting tensor $\mathbf{w}$ and therefore the Jacobian expressed in equation (2.50) can be calculated.

From the solution of Newton-Raphson algorithm (Table 2.1), the following solution for the adjusting tensor is obtained,

$$\hat{w}_I = \left\{ \begin{array}{c} \alpha_{xx} \\ \alpha_{yy} \\ \alpha_{xy} \\ \alpha_{yx} \\ \beta_{xy} \end{array} \right\} = \left\{ \begin{array}{c} -0.75471 \\ 0.53851 \\ 0.98793 \\ 0.49396 \\ -0.61237 \end{array} \right\} \Rightarrow$$

(2.51)

$$\Rightarrow \quad \hat{w}_{IJ} = \begin{bmatrix} \alpha_{xx} & \alpha_{xy} & 0 \\ \alpha_{yx} & \alpha_{yy} & 0 \\ 0 & 0 & \beta_{xy} \end{bmatrix} = \begin{bmatrix} -0.75471 & 0.98793 & 0 \\ 0.49396 & 0.53851 & 0 \\ 0 & 0 & -0.61237 \end{bmatrix}$$

By appropriately substituting this result into equation (2.42), the space-mapping tensor sought is obtained and expressed in quadratic matrix form as,

$$B_{IJ}^S = W_{IK}w_{KJ} = \begin{bmatrix} \omega_{xx}^2 & 0 & 0 \\ 0 & \omega_{yy}^2 & 0 \\ 0 & & \omega_{xx}\omega_{yy} \end{bmatrix} \cdot \begin{bmatrix} \alpha_{xx} & \alpha_{xy} & 0 \\ \alpha_{yx} & \alpha_{yy} & 0 \\ 0 & 0 & \beta_{xy} \end{bmatrix} \Rightarrow$$

(2.52)

$$\Rightarrow \quad A_{IJ}^S = \left(B_{IJ}^S\right)^{-1} = \begin{bmatrix} -0.60207 & 0.55227 & 0 \\ 0.55227 & 0.4219 & 0 \\ 0 & 0 & -1.1547 \end{bmatrix}$$

This tensor, substituted in equation (2.25), leads to an implicit orthotropic yield criterion equivalent to the one expressed in equation (2.25).

In Figure 2.3, on the principal stress space the von Mises isotropic yield criterion and the Mises-Hill orthotropic classic yield criterion for the strength data defined at the beginning of this section can be observed. The same figure shows the adjustment achieved with equation ((2.25).b) through the $A_{IJ}^S = \left(W_{IK}w_{KJ}\right)^{-1}$ tensor, obtaining the latter from equation (2.52). Finally, in the same figure the result obtained with equation ((2.25).b) is represented, but using the space mapping tensor $A_{IJ}^S = \left(W_{IJ}\right)^{-1}$ without adjustment; in other words, obtaining the classic form of the space mapping theory available at the reference in Beten (1981)[22], Oller et al. (1995)[26], (1996)[33,34], (1998)[35], Car et al. (1999)[36,37], (2000)[41], etc.

2.10 Anisotropy in the updated configuration

After thoroughly studying the generalization of the isotropic elastoplastic model in the referential configuration, it is possible to extend the anisotropy treatment to the updated configuration by proceeding as with the referential configuration. The fourth-order tensors are presented below. These tensors define the linear application relating the anisotropic real and isotropic fictitious spaces of the stresses and strains in the updated configuration and will show the validity of this formulation and give a wide generality to this theory.

2.10.1 Transformation of the stress space

The linear relationship between the Kirchhoff isotropic fictitious stress space and the anisotropic real stress space in the updated configuration (see Figure 2.2) is proposed through the following transformation:

$$\overline{\tau}_{ij} \stackrel{def}{=} a^{\tau}_{ijkl}\, \tau_{kl} \tag{2.53}$$

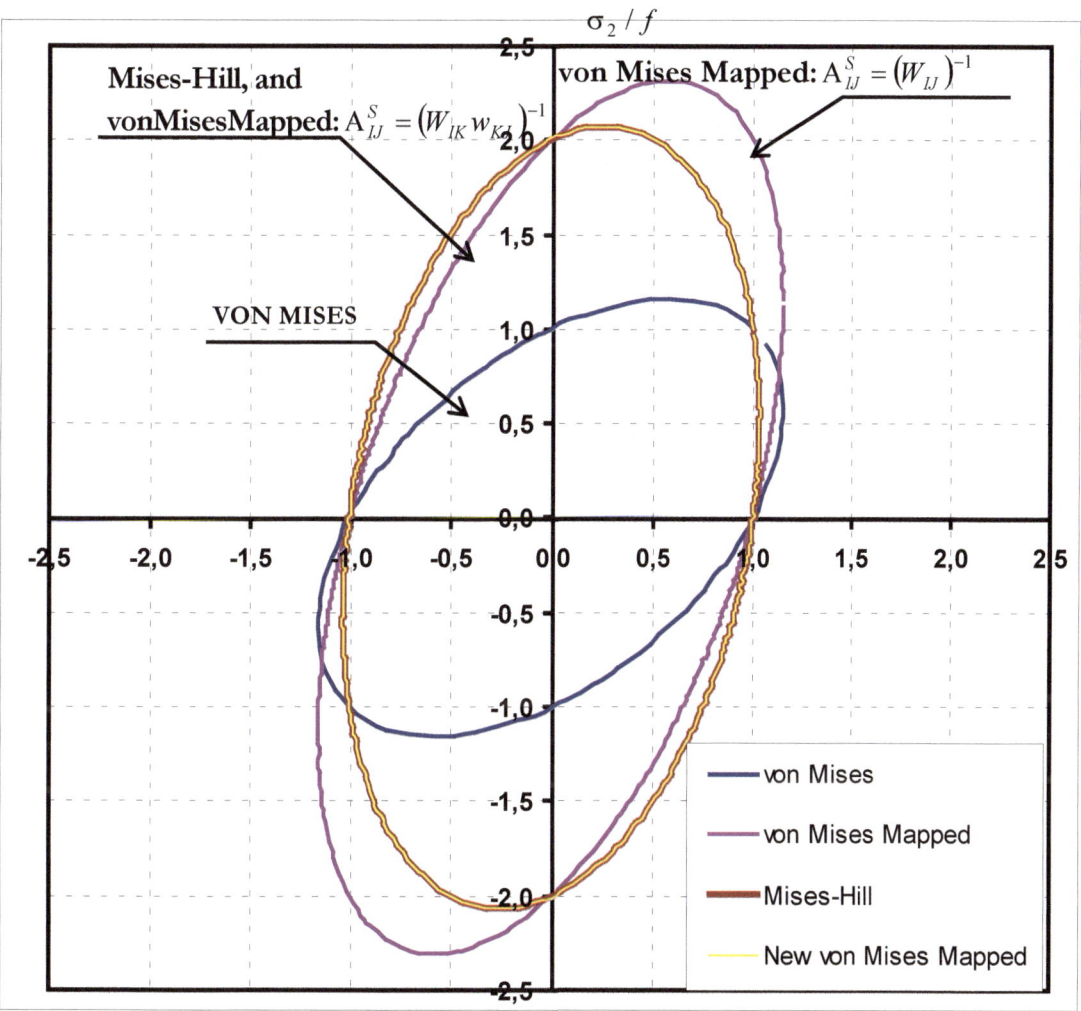

Figure 2.3 – Representation in the plane $(\sigma_1 / f - \sigma_2 / f)$, of von Mises isotropic yield criterion and Mises-Hill orthotropic yield criteron and the approximations to this criterion obtained through the space mapping theory, with $A^{\sigma}_{IJ} = \left(W_{IK}\alpha_{KJ}\right)^{-1}$ and without adjustment $A^{\sigma}_{IJ} = \left(W_{IJ}\right)^{-1}$.

a^{τ} is a fourth-order tensor defined as the linear operator $a^{\tau}:S_2 \to S_2$ which establishes the relationship between the space of the symmetric tensors of the real and fictitious stresses, respectively, in the updated configuration, $\overline{\tau}$ is Kirchhoff stress tensor in the fictitious space and τ is Kirchhoff stress tensor in the anisotropic real space. The fourth-order tensor $a^{\tau} = \underleftarrow{\overline{\phi}}(A^{S})$ comes from the transformation of the stress transformation tensor in the referential configuration. The yield condition of the anisotropic material in the fictitious isotropic space is expressed as:

$$\overline{\mathbb{F}}^{\tau}(\overline{\tau}, g, \alpha, f^{\tau}) = 0 \tag{2.54}$$

where $\overline{\mathbb{F}}^{\tau}$ is the yield function in the *updated configuration* and in *the fictitious isotropic space* and differs from the yield function in the *anisotropic space* $\mathbb{F}^{\tau}$ in the arguments that both functions contain.

In the finite strain context the linear operator $\boldsymbol{a}^{\tau}$ (which relates the real and and the fictitious stress spaces) is not constant. This is due to the configuration change to which the stress transformation tensor is subjected to in the referential configuration $\boldsymbol{a}^{\tau} = \underline{\tilde{\phi}}(\mathbf{A}^{S})$ as a function of the strain gradients $\boldsymbol{F}$, as shown below.

The transformation of the stress tensor when it goes from the referential configuration to the update configuration is expressed as,

$$\overline{\tau} = \vec{\phi}(\overline{\mathbf{S}}) \Rightarrow \overline{\tau}_{ij} = F_{iI}\,\overline{S}_{IJ}\,(F_{jJ})^{T} \tag{2.55}$$

By substituting the Piola-Kirchhoff second tensor in the isotropic fictitious space into the expression given in equation (2.23), then:

$$\overline{\tau}_{ij} = F_{iI}\,(A^{S}_{IJKL}\,S_{KL})(F_{jJ})^{T} \tag{2.56}$$

and the real anisotropic stresses tensor in the referential configuration is obtained by carrying out the *pull-back* of the Kirchhoff anisotropic stress tensor, then:

$$\mathbf{S} = \vec{\phi}(\tau) \Rightarrow S_{KL} = (F_{kK})^{-1}\,\tau_{kl}\,(F_{lL})^{-T} \tag{2.57}$$

By substituting the stress expresion (2.57) into equation (2.56), then:

$$\overline{\tau}_{ij} = \underbrace{\left[F_{iI}\,A^{S}_{IJKL}\,(F_{kK})^{-1}\,(F_{lL})^{-T}\,(F_{jJ})^{T}\right]}_{a^{\tau}_{ijkl}}\tau_{kl} \tag{2.58}$$

From the previous equation it follows that the stress-space transformation tensor in the updated configuration is a function of the strain gradient that expresses the kinematics of the referential system and the transformation tensor of the stress spaces in the referential configuration (see Figure 2.2)

$$a^{\tau}_{ijkl} = \left[F_{iI}\,(F_{kK})^{-1}\,(F_{lL})^{-T}\,(F_{jJ})^{T}\right]A^{S}_{IJKL} \tag{2.59}$$

where $\mathbf{F}$ is the strain gradient tensor.

2.10.2   Transformation of the strain space

As presented in section 2.6.2, the space transformation defined in equation (2.53) can be applied only to proportional materials (equation (2.33)). As in the referential configuration, the elastic strain tensor should be the same in the real and fictitious spaces of the updated configuration. Once again, in order to generalize the formulation to several materials, Car (2000)[41] also proposed a space transformation of real strains.

The relationship between Almansi elastic strains in the "anisotropic real space" e_{ij}^e and Almansi elastic strains in the "isotropic fictitious space" $\bar{e}_{ij}^e$ (see Figure 2.2) is defined through the following relationship,

$$\bar{e}_{ij}^e \overset{def}{=} a_{ijkl}^e \, e_{kl}^e \qquad (2.60)$$

such that as expressed in section 2.6.2 $\boldsymbol{a}^e$ is a fourth-order tensor defined as a linear operator $\boldsymbol{a}^e : S_2 \to S_2$, which establishes the relationship between the Almansi real and fictitious strain spaces, respectively, $\bar{\boldsymbol{e}}$ is the Almansi strain tensor in the fictitious isotropic space and $\boldsymbol{e}$ is the Almansi strain tensor in the anisotropic real space.

The linear operator $\boldsymbol{a}^e$ is defined in the updated configuration and due to the change of configuration it cannot remain constant. In the finite strains context this operator is a function of the stresses transformation tensor in the material configuration $\boldsymbol{a}^e = \underset{\leftarrow}{\phi}(\mathbf{A}^E)$ and of the strain gradients $\mathbf{F}$, as shown below.

The transformation of the Almansi strain tensor when passing from the referential to the updated configuration is expressed as follows,

$$\bar{\boldsymbol{e}} = \vec{\phi}(\boldsymbol{E}) \; \Rightarrow \; \bar{e}_{ij} = (F_{iI})^{-T} \, \bar{E}_{IJ} \, (F_{jJ})^{-1} \qquad (2.61)$$

By substituting the Green-Lagrange strain tensor in the isotropic fictitious space by the analogous expression into the one given in equation (2.34), then:

$$\bar{e}_{ij} = (F_{iI})^{-T} \, (A_{IJKL}^E \, E_{KL})(F_{jJ})^{-1} \qquad (2.62)$$

The Green-Lagrange anisotropic real strain tensor in the referential configuration is obtained by carrying out the Almansi anisotropic strain tensor *pull-back*, then:

$$\boldsymbol{E} = \bar{\phi}(\boldsymbol{e}) \; \Rightarrow \; E_{KL} = (F_{Kk})^T \, e_{kl} \, (F_{lL}) \qquad (2.63)$$

By replacing the strain expression (2.63) into the equation (2.62) then:

$$\bar{e}_{ij} = \underbrace{\left[(F_{iI})^{-T} \, A_{IJKL}^E \, (F_{Kk})^T \, (F_{lL}) \, (F_{jJ})^{-1} \right]}_{a_{ijkl}^e} e_{kl} \qquad (2.64)$$

From the previous equation it follows that the space transformation tensor in the updated configuration is a function of the strain gradient which expresses the kinematic of the referential system and of the strain space transformation tensor in the referential configuration (see Figure 2.2)

$$a_{ijkl}^e = \left[(F_{iI})^{-T} \, (F_{Kk})^T \, (F_{lL}) \, (F_{jJ})^{-1} \right] A_{IJKL}^E \qquad (2.65)$$

where $\mathbf{F}$ is the strain gradient tensor.

---

[41] Car E. (2000). *Modelo constitutivo continuo para el estudio del comportamiento mecánico de los materiales compuestos*. PhD thesis, Universidad Politécnica de Cataluña. Barcelona, España.

2.11 Plastic flow rule. Evolution law of internal variables

Below is the flow and evolution law of the internal variables governing the behavior of a plastic generic anisotropic material. As shown in previous sections, the formulation is established in the referential and updated configurations.

2.11.1 Referential configuration

In the referential configuration the evolution law of the plastic strain in the anisotropic space is given by the following rule of normality

$$\dot{\boldsymbol{E}}^p = \dot{\lambda}\frac{\partial \mathbb{G}^S}{\partial \mathbf{S}} \tag{2.66}$$

Taking into account that all the material anisotropy information is contained in the transformation tensor $\mathbf{A}^S$, the following plastic potential function is proposed for the anisotropic solid,

$$\mathbb{G}^S(\mathbf{S},\boldsymbol{C}) = \overline{\mathbb{G}}^S(\mathbf{S},\mathbf{A}^S,\boldsymbol{C}) = \overline{\mathbb{G}}^S(\overline{\mathbf{S}},\boldsymbol{C}) = \mathcal{K} \tag{2.67}$$

By replacing equation (2.67) into equation (2.66) the following increment of Green-Lagrange plastic strain is obtained,

$$\dot{\boldsymbol{E}}^p = \dot{\lambda}\frac{\partial \mathbb{G}^S}{\partial \mathbf{S}} = \dot{\lambda}\frac{\partial \overline{\mathbb{G}}^S}{\partial \overline{\mathbf{S}}}:\frac{\partial \overline{\mathbf{S}}}{\partial \mathbf{S}} = \dot{\lambda}\frac{\partial \overline{\mathbb{G}}^S}{\partial \overline{\mathbf{S}}}:\mathbf{A}^S = \left(\dot{\overline{\boldsymbol{E}}}^{\,p}\right)^S:\mathbf{A}^S \tag{2.68}$$

where $\left(\dot{\overline{\boldsymbol{E}}}^{\,p}\right)^S = \dot{\lambda}\frac{\partial \overline{\mathbb{G}}^S}{\partial \overline{\mathbf{S}}}$ is the normal plastic flow to the isotropic potential $\overline{\mathbb{G}}^S$. Due to the strain additivity concept in the referential configuration[42] the strain transformation rule can be extended to the strain plastic component, then:

$$\dot{\overline{\boldsymbol{E}}}^{\,p} = \mathbf{A}^E:\dot{\boldsymbol{E}}^p = \dot{\lambda}\,\mathbf{A}^E:\frac{\partial \mathbb{G}^S}{\partial \mathbf{S}} = \dot{\lambda}\,\mathbf{A}^E:\frac{\partial \overline{\mathbb{G}}^S}{\partial \overline{\mathbf{S}}}:\frac{\partial \overline{\mathbf{S}}}{\partial \mathbf{S}} = \dot{\lambda}\,\mathbf{A}^E:\underbrace{\overset{\overline{\mathfrak{R}}^S}{\overbrace{\frac{\partial \overline{\mathbb{G}}^S}{\partial \overline{\mathbf{S}}}}}:\mathbf{A}^S}_{\widetilde{\mathfrak{R}}^S} = \mathbf{A}^E:\left(\dot{\overline{\boldsymbol{E}}}^{\,p}\right)^S:\mathbf{A}^S \tag{2.69}$$

where $\dot{\overline{\boldsymbol{E}}}^{\,p}$ is the temporal change of the fictitious isotropic plastic strain in the referential configuration. Note that $\overline{\mathfrak{R}}^S$ is the normal flow to the potential surface in the isotropic space. This flow affected by A_{IJKL}^S tensor can provide the flow rule associated to the stresses anisotropic space, $\mathfrak{R}^S = \overline{\mathfrak{R}}^S:\mathbf{A}^S$. The $\widetilde{\mathfrak{R}}^S = \mathbf{A}^S:\overline{\mathfrak{R}}^S$ transformation introduces the influence of the elastic anisotropy in the anisotropic flow. The evolution law of the internal variables $\boldsymbol{\alpha} = \{\cdots,\alpha_S,\cdots\}$ is given by the following general rule,

---

[42] NOTE: The additivity of the strains is defined as $\dot{\boldsymbol{E}}^e = \dot{\boldsymbol{E}} - \dot{\boldsymbol{E}}^p$, and being $\mathbf{A}^E$ a linear elastic application from one space to another, the elastic part of the Green-Lagrange strain result: $\mathbf{A}^E:\dot{\boldsymbol{E}}^e = \mathbf{A}^E:\dot{\boldsymbol{E}} - \mathbf{A}^E:\dot{\boldsymbol{E}}^p$, from which $\dot{\overline{\boldsymbol{E}}}^e = \dot{\overline{\boldsymbol{E}}} - \dot{\overline{\boldsymbol{E}}}^p$ is obtained in the fictitious space, where the evolution of plastic law of starin in isotropic space is obtained as, $\dot{\overline{\boldsymbol{E}}}^p = \mathbf{A}^E:\dot{\boldsymbol{E}}^p$.

$$\alpha_S = \dot{\lambda}(\mathbf{H}^m)_S : \frac{\partial \mathbb{G}^S}{\partial \mathbf{S}} = \dot{\lambda}(\mathbf{H}^m)_S : \frac{\partial \overline{\mathbb{G}}^S}{\partial \overline{\mathbf{S}}} : \frac{\partial \overline{\mathbf{S}}}{\partial \mathbf{S}} = \dot{\lambda}(\mathbf{H}^m)_S : \frac{\partial \overline{\mathbb{G}}^S}{\partial \overline{\mathbf{S}}} : \mathbf{A}^S = (\mathbf{H}^m)_S : \left(\dot{\overline{\mathbf{E}}}^{\,p}\right)^S : \mathbf{A}^S \qquad (2.70)$$

where $(\mathbf{H}^m)_S$ is a second-order tensor function of the state of the updated stresses and the updated plastic hardening variable. In the simplest case of the plasticity theory, it takes the form of Piola-Kirchhoff stress second-tensor and in this case $\dot{\alpha}_S$ coincides with the plastic energy density or dissipation. The evolution law of internal variables is then:

$$\overline{\Xi}^p_{\text{mec}} = \dot{\overline{\Psi}}^P = \alpha_S = \dot{\lambda}\, \overline{\mathbf{S}} : \frac{\partial \overline{\mathbb{G}}^S}{\partial \overline{\mathbf{S}}} \qquad (2.71)$$

2.11.2 Updated Configuration

In the updated configuration the evolution law of the plastic strain in the anisotropic space is given by the following rule of normality,

$$L_v(e^p) = \dot{\lambda}\frac{\partial \mathbb{G}^\tau}{\partial \tau} = \dot{\lambda}\frac{\partial g^\tau}{\partial \tau} \qquad (2.72)$$

where $\mathbb{G}^\tau(\tau,g) = g^\tau(\tau) = k$ and $g = g(\tau)$. Taking into account that all the information about the material anisotropy is contained in the transformation tensor a^τ, the following plastic potential function is proposed for the anisotropic solid,

$$\mathbb{G}^\tau(\tau,g) = \overline{\mathbb{G}}^\tau(\tau,a^\tau,g) = \overline{\mathbb{G}}^\tau(\overline{\tau},g) = k \qquad (2.73)$$

By substituting equation (2.73) into equation (2.72) the following increase of Green-Lagrange plastic strain is obtained:

$$d^p = \dot{\lambda}\frac{\partial \mathbb{G}^\tau}{\partial \tau} = \dot{\lambda}\frac{\partial \overline{\mathbb{G}}^\tau}{\partial \overline{\tau}} : \frac{\partial \overline{\tau}}{\partial \tau} = \dot{\lambda}\frac{\partial \overline{\mathbb{G}}^\tau}{\partial \overline{\tau}} : a^\tau = L_v\left(\overline{e}^{\,p}\right)^\tau : a^\tau \qquad (2.74)$$

where $L_v\left(\overline{e}^{\,p}\right)^\tau = \dot{\lambda}\frac{\partial \overline{\mathbb{G}}^\tau}{\partial \overline{\tau}}$ is the normal plastic flow to the isotropic potential $\overline{\mathbb{G}}^\tau$. Due to the strain additivity concept in the updated configuration, as in the referential configuration[42], the transformation rule of the total strains can be extended to the plastic component of the strains, thus

$$\overline{d}^p = a^e : d^p = \dot{\lambda}a^e : \frac{\partial \mathbb{G}^\tau}{\partial \tau} = \dot{\lambda}a^e : \frac{\partial \overline{\mathbb{G}}^\tau}{\partial \overline{\tau}} : \frac{\partial \overline{\tau}}{\partial \tau} = \dot{\lambda}a^e : \underbrace{\overset{\overline{\mathfrak{R}}^\tau}{\overbrace{\frac{\partial \overline{\mathbb{G}}^\tau}{\partial \overline{\tau}}}} : a^\tau}_{\tilde{\mathfrak{R}}^\tau} = a^e : L_v\left(\overline{e}^{\,p}\right)^\tau : a^\tau \qquad (2.75)$$

where $\overline{d}^p$ represents the temporal change of the fictitious isotropic plastic deformation in the updated configuration. Note that $\overline{\mathfrak{R}}^\tau$ is the normal flow to the potential surface in the isotropic space. This flow affected by tensor a^τ_{ijkl} provides the flow rule associated with the stress anisotropic space, $\mathfrak{R}^\tau = \overline{\mathfrak{R}}^\tau : a^\tau$. The transformation $\tilde{\mathfrak{R}}^\tau = a^e : \mathfrak{R}^\tau$ introduces the influence of the elastic anisotropy in the anisotropic flow. The evolution law of the internal variables $\boldsymbol{\alpha} = \{\cdots,\alpha_\tau,\cdots\}$ is given by the following general rule,

$$\alpha_\tau = \dot{\lambda}(\mathbf{h}^m)_\tau : \frac{\partial \mathbb{G}^\tau}{\partial \tau} = \dot{\lambda}(\mathbf{h}^m)_\tau : \frac{\partial \overline{\mathbb{G}}^\tau}{\partial \overline{\tau}} : \frac{\partial \overline{\tau}}{\partial \tau} = \dot{\lambda}(\mathbf{h}^m)_\tau : \frac{\partial \overline{\mathbb{G}}^\tau}{\partial \overline{\tau}} : a^\tau = (\mathbf{h}^m)_\tau : L_\nu\left(\overline{e}^{\,p}\right)^\tau : a^\tau \qquad (2.76)$$

where $(\mathbf{h}^m)_\tau$ is a second-order tensor function of the state of the updated stresses and the updated plastic hardening variable. In the simplest case of the plasticity theory, it takes the form of the Kirchhoff tensor and in this case $\dot{\alpha}_\tau$ coincides with the plastic energy density or dissipation. The evolution law of the internal variable is then:

$$\Xi^P_{\text{mec}} = \dot{\overline{\psi}}^P = \alpha_\tau = \dot{\lambda}\,\overline{\tau} : \frac{\partial \overline{\mathbb{G}}^\tau}{\partial \overline{\tau}} \qquad (2.77)$$

The plastic strain magnitude obtained through the above-mentioned form is equal to the plastic dissipation expressed in equation (2.71). This equality guarantess the formulation objectivity and shows that both the referential and the updated configuration can be used. This statement can be proved by taking into account the "*push-forward*" and "*pull-back*"[43] transport operations; starting from the dissipation in the referential configuration, its conservation is verified in the updated configuration

$$\Xi^P_{\text{mec}} = \dot{\overline{\Psi}}^P = \overline{\mathbf{S}} : \dot{\overline{\mathbf{E}}}^P = (\mathbf{F}^{-1} \cdot \overline{\tau} \cdot \mathbf{F}^{-T}) : (\mathbf{F}^T \cdot \overline{\mathbf{d}}^P \cdot \mathbf{F}) =$$
$$= (\overline{\tau}) : (\underbrace{\mathbf{F}^{-T} \cdot \mathbf{F}^T}_{\mathbf{I}} \cdot \overline{\mathbf{d}}^P \cdot \underbrace{\mathbf{F} \cdot \mathbf{F}^{-1}}_{\mathbf{I}}) = \overline{\tau} : \overline{\mathbf{d}}^P = \dot{\overline{\psi}}^P \qquad (2.78)$$

In the following section it is also showed how the dissipation is also the same in an isotropic fictitious space and in a real anisotropic space.

2.12 Definition of the dissipation in the isotropic fictitious space. Unicity of the dissipation

This section will show how the plastic dissipation of the anisotropic constitutive model developed in the previous sections is the same, whether it is considered in the anisotropic real space or in the isotropic fictitious space. As mentioned before, the advantage of using the constitutive model in the isotropic fictitious space is that all the algorithms developed for isotropic materials can be used. Consequently, this formulation brings advantages in the implementation of the finite element code.

2.12.1   Referential configuration

The magnitude of Helmholtz free energy for the isothermic process is the same for the anisotropic space and for the isotropic fictitious one,

---

[43] NOTE: Consider the validity of the following tensorial product operation of configuration transformation:
$(\mathbf{A} \cdot \mathbf{B}) : \mathbf{C} = \mathbf{B} : (\mathbf{A}^T \cdot \mathbf{C}) = \mathbf{A} : (\mathbf{C} \cdot \mathbf{B}^T)$.

$$\Psi(\boldsymbol{E}^e,\theta,\alpha^m) = \Psi^e(\boldsymbol{E}^e,\theta) + \Psi^p(\alpha^m) = \frac{1}{2\,m^0}\left[\boldsymbol{E}^e : \mathbb{C}^S : \boldsymbol{E}^e\right] + \Psi^p(\alpha^m)$$

$$\Rightarrow \mathbf{S} = m^0\,\frac{\partial\Psi(\boldsymbol{E}^e,\theta,\alpha^m)}{\partial\boldsymbol{E}^e} = \mathbb{C}^S : \boldsymbol{E}^e$$

$$\overline{\Psi}(\overline{\boldsymbol{E}}^e,\theta,\alpha^m) = \overline{\Psi}^e(\overline{\boldsymbol{E}}^e,\theta) + \overline{\Psi}^p(\alpha^m) = \frac{1}{2\,m^0}\left[\overline{\boldsymbol{E}}^e : \overline{\mathbb{C}}^S : \overline{\boldsymbol{E}}^e\right] + \overline{\Psi}^p(\alpha^m) \qquad (2.79)$$

$$\Rightarrow \overline{\mathbf{S}} = m^0\,\frac{\partial\overline{\Psi}(\overline{\boldsymbol{E}}^e,\theta,\alpha^m)}{\partial\overline{\boldsymbol{E}}^e} = \overline{\mathbb{C}}^S : \overline{\boldsymbol{E}}^e =$$

$$= \left[\mathbf{A}^S : \mathbb{C}^S : (\mathbf{A}^E)^{-1}\right] : (\mathbf{A}^E : \boldsymbol{E}^e) = \mathbf{A}^S : \mathbf{S}$$

where $\Psi(\boldsymbol{E}^e,\theta,\alpha^m) = \overline{\Psi}(\overline{\boldsymbol{E}}^e,\theta,\alpha^m)$ is the total free energy in the *referential configuration in anisotropic and isotropic spaces, respectively*, and $\Psi^p(\alpha^m) = \overline{\Psi}^p(\alpha^m)$ is the plastic component of such free energy. The constitutive equation in the *referential configuration* in the *isotropic space* is obtained by considering The Clasius-Duhem expression and the transformation equations of stresses (2.23) and strain spaces (2.34). The dissipation's general form is given by:

$$\Xi^p_{\mathrm{mec}} = \mathbf{S} : \dot{\boldsymbol{E}}^p - m^0\,\frac{\partial\Psi}{\partial\alpha^m}\dot{\alpha}^m - \frac{1}{\theta}\dot{\mathbf{q}}\cdot\nabla\theta \geq 0 \qquad (2.80)$$

The first term of the dissipation expression in the referential configuration can also be written as a function of the transformation tensors of stresses $\mathbf{A}^S$ and strain spaces $\mathbf{A}^E$ as well as of the stress and strain fictitious spaces as follows,

$$\mathbf{S} : \dot{\boldsymbol{E}}^p = \left[(\mathbf{A}^S)^{-1} : \overline{\mathbf{S}}\right] : \left[(\mathbf{A}^E)^{-1} : \dot{\overline{\boldsymbol{E}}}^p\right] \qquad (2.81)$$

By substituting the plastic fictitious strain $\dot{\overline{\boldsymbol{E}}}^p$ in this latter by its expresion (equation (2.69)), then,

$$\mathbf{S} : \dot{\boldsymbol{E}}^p = \dot{\lambda}(\mathbf{A}^S)^{-1} : \overline{\mathbf{S}} : (\mathbf{A}^E)^{-1} : \mathbf{A}^E : \frac{\partial\overline{\mathbb{G}}^S}{\partial\overline{\mathbf{S}}} : \mathbf{A}^S$$

$$\mathbf{S} : \dot{\boldsymbol{E}}^p = \dot{\lambda}\,\overline{\mathbf{S}} : \frac{\partial\overline{\mathbb{G}}^S}{\partial\overline{\mathbf{S}}} = \overline{\mathbf{S}} : \dot{\overline{\boldsymbol{E}}}^p \qquad (2.82)$$

Taking into consideration the above-mentioned equation and the conservation of energy, the dissipation expression for isothermic processes can be obtained,

$$\Xi^p_{\mathrm{mec}} = \mathbf{S} : \dot{\boldsymbol{E}}^p - m^0\,\frac{\partial\Psi}{\partial\alpha^m}\dot{\alpha}^m = \overline{\mathbf{S}} : \dot{\overline{\boldsymbol{E}}}^p - m^0\,\frac{\partial\overline{\Psi}}{\partial\alpha^m}\dot{\alpha}^m \equiv \overline{\Xi}^p_{\mathrm{mec}} \geq 0 \qquad (2.83)$$

The expression above shows that the dissipation is unique regardless of whether the model is formulated in the anisotropic space or in its equivalent isotropic space. The computational implementation of this model for anisotropy is very advantageous. While dealing with the material in the isotropic fictitious space, it is only necessary to transform the stress and strain spaces into an isotropic fictitious space and then use the algorithms studied in classic literature for isotropic materials.

2.12.2   Updated configuration

The Helmholtz free energy expression in the anisotropic space and in the isotropic fictitius space in the updated configuration for an isothermic process is obtained as:

$$\psi(e^e,\theta,\alpha^m)=\psi^e(e^e,\theta)+\psi^p(\alpha^m)=\frac{1}{2m}\Big[e^e:\mathbf{c}^\tau:e^e\Big]+\psi^p(\alpha^m)$$

$$\Rightarrow \boldsymbol{\tau}=m\frac{\partial\psi(e^e,\theta,\alpha^m)}{\partial e^e}=\mathbf{c}^\tau:e^e$$

$$\overline{\psi}(\overline{e}^e,\theta,\alpha^m)=\psi^e(\overline{e}^e,\theta)+\overline{\psi}^p(\alpha^m)=\frac{1}{2m}\Big[\overline{e}^e:\overline{\mathbf{c}}^\tau:\overline{e}^e\Big]+\overline{\psi}^p(\alpha^m) \qquad (2.84)$$

$$\Rightarrow \overline{\boldsymbol{\tau}}=m\frac{\partial\overline{\psi}(\overline{e}^e,\theta,\alpha^m)}{\partial\overline{e}^e}=\overline{\mathbf{c}}^\tau:\overline{e}^e=$$

$$=\Big[a^\tau:\mathbf{c}^\tau:(a^e)^{-1}\Big]:(a^e:e^e)=a^\tau:\boldsymbol{\tau}$$

where $\psi(e^e,\theta,\alpha^m)=\overline{\psi}(\overline{e}^e,\theta,\alpha^m)$ is the total free energy in the *updated configuration in the anisotropic and isotropic space, respectively,* and $\psi^p(\alpha^m)=\overline{\psi}^p(\alpha^m)$ is the plastic amount of such free energy. The constitutive equation in *the updated configuration in the isotropic space* is obtained by considering the Clasius-Duhem expression and the transformation equations of stress (2.53) and strains spaces (2.60). The general form for the dissipation is given by:

$$\Xi^p_{mec}=\boldsymbol{\tau}:d^p-m\frac{\partial\psi}{\partial\alpha^m}\dot{\alpha}^m-\frac{1}{\theta}\dot{\mathbf{q}}\cdot\nabla\theta\geq0 \qquad (2.85)$$

The first term of the dissipation expression in the updated configuration can also be written as a function of the transformation tensors of the stress a^τ and strain spaces a^e and stress and strain fictitious spaces as,

$$\boldsymbol{\tau}:d^p=\Big[(a^\tau)^{-1}:\overline{\boldsymbol{\tau}}\Big]:\Big[(a^e)^{-1}:\overline{d}^p\Big] \qquad (2.86)$$

By substituting the plastic fictitious strain $\overline{d}^p$ in this latter equation by its expression (equation (2.69)), then,

$$\boldsymbol{\tau}:d^p=\dot{\lambda}(a^\tau)^{-1}:\overline{\boldsymbol{\tau}}:(a^e)^{-1}:a^e:\frac{\partial\overline{\mathbb{G}}^\tau}{\partial\overline{\boldsymbol{\tau}}}:a^\tau$$

$$\boldsymbol{\tau}:d^p=\dot{\lambda}\overline{\boldsymbol{\tau}}:\frac{\partial\overline{\mathbb{G}}^\tau}{\partial\overline{\boldsymbol{\tau}}}=\overline{\boldsymbol{\tau}}:\overline{d}^p \qquad (2.87)$$

Considering the latter equation and the conservation of energy, the dissipation expression for isothermic processes is,

$$\Xi^p_{mec}=\boldsymbol{\tau}:d^p-m\frac{\partial\psi}{\partial\alpha^m}\dot{\alpha}^m=\overline{\boldsymbol{\tau}}:\overline{d}^p-m\frac{\partial\overline{\psi}}{\partial\alpha^m}\dot{\alpha}^m\equiv\overline{\Xi}^p_{mec}\geq0 \qquad (2.88)$$

This expression shows that the dissipation is unique regardless of whether the model is formulated in the anisotropic space or its equivalent isotropic space. As in the referential configuration, the computational implementation of this model for anisotropy is advantageous. When dealing with the material in an isotropic fictitious space it is only necessary to carry out the stress and strain space transformation into an isotropic fictitious space and then use the algorithms developed in the classic literature for isotropic materials.

2.13 Tangent constitutive equation

The Newton-Raphson method, which uses tangent stiffness matrices, is one of the most widely used techniques for the numeric approximation of the equations describing the elastoplastic solids behavior. Therefore, an increment relationship between stresses and strains is required to obtain the tangent linear operator. Below are presented the continuous tangent elastoplastic linear operators establishing the relationship between the total stresses and strains increment in the referential and updated configurations.

Materials with inelastic behavior require the time step first-order numerical integration of coupled systems of differential equations (Simo and Taylor (1985)[44]). The result of the integration algorithm is a function of a non-linear response defining the stress tensor as a function of the strain history until the current time step. This integration algorithm can deal with the equivalent elastoplastic problem as an equivalent linearized problem in time.

The tangent operator involved in the linearized problem is obtained through a linearization of the consistent response function with the integration algorithm of the constitutive equation. The use of these tangent operators keeps the quadratic convergence of iterative solutions based on Newton methods (Simo and Taylor (1985)[44]), (Crisfield (1991)[45]). The convergence speed is directly influenced by the accuracy obtained in the tangent-stiffness matrix system and it is extremely important for the general analysis accuracy (Ortiz and Popov (1985)[46]).

The elastoplastic operator presented in this section is independent from the constitutive equation integration process and therefore it does not keep the characteristic quadratic convergence of the solution schemes based on Newton-Raphson method. The expression of the consistent tangent elastoplastic linear operator and the integration algorithm of the constitutive equation can be found In Car (2000)[41]. The quadratic convergence of the Newton-Raphson non-linear characteristic problem can be obtained by using this operator.

2.13.1   Referential configuration

The tangent constitutive equation in the referential configuration is obtained by the temporal variation of Piola-Kirchhoff stress, this is:

$$\dot{S}_{IJ} = \frac{\partial S_{IJ}}{\partial E_{KL}^e} \dot{E}_{KL}^e \tag{2.89}$$

Taking into account that Piola Kirchhoff stress in the anisotropic space can be written as a function of the stress in the isotropic fictitious space, then:

[44] Simo J. and Taylor R. (1985). Consistent tangent operators for rate-dependent elastoplasticity. *Computer Methods in Applied Mechanics and Engineering.* 48, 101-118.

[45] Crisfield M. (1991). *Non linear finite element analysis of solids and structures.* John Wiley & Sons Ltd.

[46] Ortiz M. and Popov E. (1985). Accuracy and stability of integration algorithms for elastoplastic constitutive relations. *International Journal of Numerical Methods in Engineering.* Vol. 21, pp. 1561-1576.

$$\dot{S}_{IJ} = \underbrace{\frac{\partial S_{IJ}}{\partial \overline{S}_{RS}}}_{(A^S_{IJRS})^{-1}} \frac{\partial \overline{S}_{RS}}{\partial \overline{E}^e_{MN}} \underbrace{\frac{\partial \overline{E}^e_{MN}}{\partial E^e_{KL}}}_{A^E_{MNKL}} \dot{E}^e_{KL}$$

$$\dot{S}_{IJ} = (A^S_{IJRS})^{-1} \overline{\mathbb{C}}_{RSMN} A^E_{MNKL} \dot{E}^e_{KL}$$

$$\dot{S}_{IJ} = (A^S_{IJRS})^{-1} \overline{\mathbb{C}}_{RSMN} (\dot{\overline{E}}_{MN} - \dot{\overline{E}}^p_{MN}) \tag{2.90}$$

By considering the plastic consistency condition, the constitutive equation in the isotropic fictitious space can be obtained as follows:

$$\dot{\overline{S}}_{IJ} = \overline{\mathbb{C}}^{ep}_{IJKL} \dot{\overline{E}}_{KL} \quad \text{or alternatively,} \quad \dot{\overline{\mathbf{S}}} = \overline{\mathbb{C}}^{ep} : \dot{\overline{\mathbf{E}}} \tag{2.91}$$

where $\overline{\mathbb{C}}^{ep}$ is the tangent elastoplastic constitutive tensor in the referential configuration in the isotropic fictitious space and its expression is given by the following equation,

$$\overline{\mathbb{C}}^{ep}_{IJKL} = \overline{\mathbb{C}}_{IJKL} - \frac{\left(\overline{\mathbb{C}}_{IJRS} \widetilde{\mathfrak{R}}^S_{RS} \right) \otimes \left(\dfrac{\partial \overline{\mathbb{F}}^S}{\partial \overline{S}_{RS}} \overline{\mathbb{C}}_{RSKL} \right)}{\underbrace{\dfrac{\partial \overline{\mathbb{F}}^S}{\partial \overline{S}_{PQ}} \overline{\mathbb{C}}_{PQLN} \widetilde{\mathfrak{R}}^S_{LN} - \sum_m \dfrac{\partial \overline{\mathbb{F}}^S}{\partial \alpha^m_S}(h^m_{TU})_S \widetilde{\mathfrak{R}}^S_{TU}}_{A}} \tag{2.92}$$

Taking into account equation (2.90), which relates the Piola-Kirchhoff stress tensor in the real and fictitious spaces, the anisotropic stress variation can be obtained as follows:

$$\dot{S}_{IJ} = (A^S_{IJKL})^{-1} \overline{\mathbb{C}}^{ep}_{KLRS} \dot{\overline{E}}_{RS} \tag{2.93}$$

By combining this equation with equation (2.34), the elastoplastic anisotropic constitutive tensor expression is obtained as:

$$\dot{S}_{IJ} = \underbrace{(A^S_{IJKL})^{-1} \overline{\mathbb{C}}^{ep}_{KLRS} A^E_{RSPQ}}_{\overline{\mathbb{C}}^{ep}_{IJPQ}} \dot{E}_{PQ} \quad \text{or alternatively,} \quad \dot{\mathbf{S}} = \overline{\mathbb{C}}^{ep} : \dot{\mathbf{E}} \tag{2.94}$$

where $\mathbb{C}^{ep}_{IJPQ} = (A^S_{IJKL})^{-1} \overline{\mathbb{C}}^{ep}_{KLRS} A^E_{RSPQ}$ is the anisotropic elastoplastic tensor depending on the tangent-isotropic fictitious elastoplastic constitutive tensor.

2.13.2   Spatial configuration

The tangent constitutive equation in the spatial configuration is obtained by considering the temporal variation of Kirchhoff stress, then:

$$L_v(\tau_{ij}) = \frac{\partial \tau_{ij}}{\partial e^e_{kl}} L_v(e^e_{kl}) \tag{2.95}$$

Considering that the Kirchhoff stress in the anisotropic space can be written using the stress in the isotropic fictitious space, then,

$$L_v(\tau_{ij}) = \underbrace{\frac{\partial \tau_{ij}}{\partial \overline{\tau}_{rs}}}_{(a_{ijrs}^{\tau})^{-1}} \frac{\partial \overline{\tau}_{rs}}{\partial \overline{e}_{mn}^e} \underbrace{\frac{\partial \overline{e}_{mn}^e}{\partial e_{kl}^e}}_{a_{mnkl}^e} L_v(e_{kl}^e)$$

$$L_v(\tau_{ij}) = (a_{ijrs}^{\tau})^{-1} \overline{c}_{rsmn} a_{mnkl}^e L_v(e_{kl}^e)$$

$$L_v(\tau_{ij}) = (a_{ijrs}^{\tau})^{-1} \overline{c}_{rsmn} \left[L_v(\overline{e}_{mn}) - L_v(\overline{e}_{mn}^p) \right] \quad ; \quad \overline{d}_{mn}^p \equiv L_v(\overline{e}_{mn}^p)$$

(2.96)

By considering the plastic consistency condition, the constitutive equation in the isotropic fictitious space is obtained,

$$L_v(\tau_{ij}) = \overline{c}_{ijkl}^{ep} L_v(\overline{e}_{kl}) \qquad \text{or alternatively,} \qquad L_v(\tau) = \overline{\mathbf{c}}^{ep} : L_v(\overline{e})$$

(2.97)

where $\overline{\mathbf{c}}^{ep}$ is the tangent-elastoplastic constitutive tensor in the updated configuration (the isotropic fictitious space) and its expression is given by the following equation,

$$\overline{\mathbf{c}}_{ijkl}^{ep} = \overline{\mathbf{c}}_{ijkl} - \frac{\left(\overline{\mathbf{c}}_{ijrs} \widetilde{\Re}_{rs}^\tau \right) \otimes \left(\dfrac{\partial \overline{\mathbb{F}}^\tau}{\partial \tau_{rs}} \overline{\mathbf{c}}_{rskl} \right)}{\underbrace{\dfrac{\partial \overline{\mathbb{F}}^\tau}{\partial \tau_{pq}} \overline{\mathbf{c}}_{pqln} \widetilde{\Re}_{ln}^\tau - \sum_m \dfrac{\partial \overline{\mathbb{F}}^\tau}{\partial \alpha_\tau^m} (h_{tu}^m)_\tau \widetilde{\Re}_{tu}^\tau}_{A}}$$

(2.98)

Taking into account equation (2.96), which relates the Kirchhoff stress tensor in the real and fictitious spaces, it is possible to obtain the real anisotropic stress variation as:

$$L_v(\tau_{ij}) = (a_{ijkl}^{\tau})^{-1} \overline{c}_{klrs}^{ep} L_v(\overline{e}_{rs})$$

(2.99)

By combining this equation with equation (2.60), the anisotropic elastoplastic constitutive tensor equation is obtained as:

$$L_v(\tau_{ij}) = \underbrace{(a_{ijkl}^{\tau})^{-1} \overline{c}_{klrs}^{ep} a_{rspq}^e}_{c_{ijpq}^{ep}} L_v(\overline{e}_{pq}) \qquad \text{or alternatively,} \qquad L_v(\tau) = \mathbf{c}^{ep} : L_v(e)$$

(2.100)

where $c_{ijpq}^{ep} = (a_{ijkl}^{\tau})^{-1} \overline{c}_{klrs}^{ep} a_{rspq}^e$ is the anisotropic elastoplastic tensor which depends on the isotropic fictitious tangent elastoplastic constitutive tensor in the updated configuration.

The tangent elastoplastic constitutive tensors previously obtained cannot be used to obtain the quadratic convergences in the solution of increment problems based on Newton methods. The reason is that the integration procedure of the constitutive equation was not considered during its formulation, so it is not consistent with the integration procedure.

The constitutive elastoplastic tangent fictitious tensors given by equations (2.92) and (2.98) turned out to be symmetric for the associated plasticity case, whereas the anisotropic tangent elastoplastic tensor in the referential or updated configuration given by equations (2.94) and (2.100) is not generally symmetric (see Car (2000)[41]).

3 MIXING THEORY

3.1 Introduction

Composite materials are made up of different types of organic and inorganic substances. Their atomic equilibrium depends on various kinds of interatomic bonds creating crystalline and amorphous materials.

The mechanical characteristics of composite materials depend on their own intrinsic properties: macroscopic structure, bond type, crystalline structure, etc. The materials' behavior is also influenced by their extrinsic properties such as the characteristics of the manufacturing process, the size, defects and distribution of micropores, microcracks, initial stress states, etc. Regarding the constitutive behavior simulation, only a contribution to studies to improve the composite extrinsic properties can be given.

The whole composite behavior is conditioned by the constitutive law of each of the compounding substances as a function of the proportion of their total volume and morphological distribution.

There are different theories for the constitutive behavior simulation of composite materials (see a synthesis in Car (2000)[1], Zalamea (2000)[2]. The "Mixing Theory" (Trusdell and Toupin (1960)[3]) is one of them and it is considered suitable for the composite materials' linear behavior simulation and, with some modifications, it can also represent the materials' non-linear behavior. Furthermore, the classic form of this theory says that the component materials coexisting in a point of a solid must have the same strain (parallel participation of the components). This hypothesis involves a strong limitation for the composite materials' behavior prediction. To solve this problem "a general formulation of the mixing theory" is presented in this chapter based on a compatibility equation adapted to the composite behavior (serial-parallel participation of the components).

In this chapter a formulation adapted to fiber-reinforced matrix materials is developed. The compatibility equation among the compounding materials automatically provides their degree of kinematic participation. This capacity along with the anisotropy treatment of composite materials (Oller et al. (1995)[4] Car et al. (2001)[5]) significantly broadens the chances of working on different fiber-reinforced matrix composite materials.

---

[1] Car E. (2000). *Modelo constitutivo continuo para el estudio del comportamiento mecánico de los materiales compuestos*. PhD thesis, Universidad Politécnica de Cataluña. Barcelona, España.
[2] Zalamea F. (2000). *Tratamiento numérico de materiales compuestos mediante la teoría de homogeneización*. PhD thesis, Universidad Politécnica de Cataluña. Barcelona, España.
[3] Trusdell, C. and Toupin, R. (1960). *The classical Field Theories*. Handbuch der Physik III/I. Springer Verlag, Berlin.
[4] Oller S., Botello S., Miquel J., and Oñate E. (1995). An anisotropic elastoplastic model based on an isotropic formulation. *Engineering Computation*, 12 (3), pp. 245-262.
[5] Car E., Oller S., Oñate E. (2001). A large strain plasticity model for anisotropic materials — Composite material application. *International Journal of Plasticity*, 17, pp. 1437-1463.

In addition to this general formulation of the mixing theory, another formulation is introduced to extend the validity of the theory to composite materials where relative movements occur among their components. Two techniques are presented in this chapter to solve this problem:

1. The first solution is based on a new modification of the composite compatibility equation to introduce inelastic incompatible strains among the components.

2. The second technique is based on the modification of the composite materials' properties limiting their participation to the stress transmission capacity through the contact interphases between the components.

The classic mixing theory was originally studied by Trusdell and Toupin (1960)[3] in 1960. They set the foundations for further studies (Green and Naghdi (1965)[6], Ortiz and Popov (1982)[7,8], Oller et al. (1996)[9], Oller and Oñate (1996)[10]). The theory here presented is more general than the classic one. The constitutive behavior of a highly anisotropic "*n-phases*" composite material is represented without the classic compatibility equation limitation of the original theory. A serial or parallel behavior relationship between the compounding substances is allowed.

This generalized mixing theory could be seen as the "constitutive model manager" of each composite component for the interaction evaluation of the different behavior laws of the various composite phases. With this behavior, or substances in this particular case, each substance can keep its isotropic or anisotropic, linear or non-linear, original constitutive law and condition the composite global behavior at the same time.

As mentioned before, the classic mixing theory is only suitable for the mechanical behavior simulation of certain composite materials that have a parallel response (the same strain and without relative movements among them) of their components. Under this category fall materials with long-fiber-reinforced matrices aligned with load action. For a different load orientation or any other type of composite materials such as tissues with various fiber orientations among them, it is necessary to carry out modifications in the classic theory. One of the objectives of this chapter is to present these modifications, which started in previous research studies in 1995 (see references[11,12,13,14]). These works proposed a "Serial-Parallel Model" requiring experimental correlation to adjust the composite kinematic coupling at each point. Originally, this theory took into consideration the

[6] Green A. and Naghdi P. (1965). A dynamical theory of interacting continua. *Journal Engineering Science*, 3 3-231.

[7] Ortiz M. and Popov E. (1982). A physical model for the inelasticity of concrete. *Proc. Roy. Soc. London*, A383, 101-125.

[8] Ortiz M. and Popov E. (1982). Plain concrete as a composite material. *Mechanics of Materials*, 1, 139-150.

[9] Oller S., Oñate E., Miquel J., and Botello S. (1996). A plastic damage constitutive model for composite materials. *Int. J. Slolids and Structures*, 33 (17), 2501-2518.

[10] Oller S., Oñate E. (1996). A Hygro-Thermo-Mechanical constitutive model for multiphase composite materials. *Int. J. Solids and Structures*. Vol.33, (20-22), 3179-3186.

[11] Oller S., Neamtu L., Oñate E. (1995). Una Generalización de la teoría de mezclas clásica para el tratamiento de compuestos en Serie-Paralelo. *I Congreso Nacional de materiales compuestos MATCOMP 95*. Ed. F. París. J. Cañas. Materiales Compuestos 95 - AEMAC, pp. 433-438, Sevilla, España.

[12] Neamtu L., Oller S., Oñate E. (1997). A generalized mixing theory Elasto-Damage-Plastic model for finite element analysis of composites. Fifth International Conference on Computational Plasticity – COMPLAS V. Ed. D. R. Owen, E. Oñate, E. Hinton. CIMNE, pp. 1214-1212. Barcelona.

[13] Oñate E., Neamtu L. Oller S. (1997). Generalization of the classical mixing theory for analysis of composite materials. *International Conference on Advances in Computational Engineering Science (ICES'97)*. Ed S.N. Atluri and G. Yagawa. Tech. Science Press, pp. 625-630 - Georgia U.S.A.

[14] Oñate E., Neamtu L., Oller S. (1997). Un modelo constitutivo para el análisis por el M.E.F. para materiales compuestos. *II Congreso Nacional de materiales compuestos. "MATCOMP 97"*. Ed. J. Güemes y C. Navarro. Materiales Compuestos 97 - AEMAC, pp 206-211 - Madrid. España.

components' orientation and a kinematic behavior ranging from an extreme serial behavior to a extreme parallel one was assigned. Currently, this formulation has been transformed into a wider and more versatile version which adjusts automatically the kinematic behavior of each component of the composite material through an interpretation of the fiber orientation (see diagram in Figure 3.1). Below is a brief description of the formulation's general guidelines.

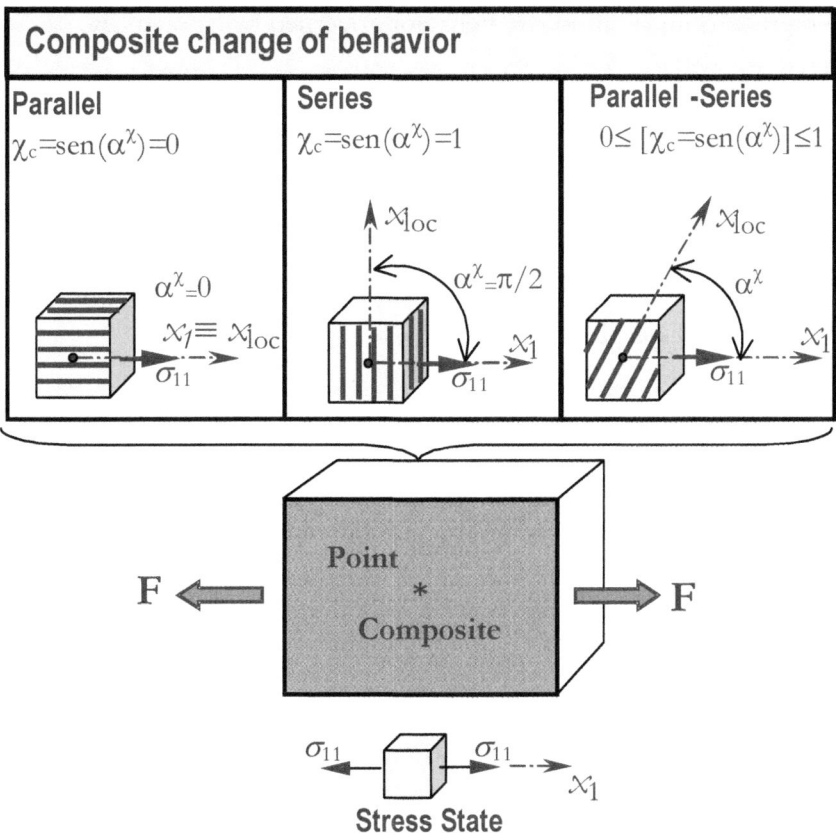

Figure 3.1 – Simplified representation of the "series-parallel behavior" of a point in a composite solid.

3.2 Classic mixing theory

The classic mixing theory of basic substances is based on the mechanics of the local continuum solid and is suitable for the behavior explanation of a point of a composite solid. It is based on the interaction principle of compounding substances of the composite material. The following basic assumptions are considered:

   i. A set of compounding substances participate in each infinitesimal volume of the composite.

   ii. Each component contributes in the composite behavior proportionally to their volumetric participation.

   iii. All the components have the same strain compatibility or closure equation.

    iv. Each component volume is much smaller than the total composite volume.

The second hypothesis involves a homogenous distribution of all the component substances at each point of the composite. The different component substance interaction and their corresponding constitutive law determine the composite material's behavior and it depends basically on the volume percentage of each component and its distribution in the composite. Materials with different behaviors can be combined (elastic, elastoplastic, etc.); each one represents an evolutionary behavior governed by its own law.

The third hypothesis says that in the absence of atomic diffusion[*] the following compatibility condition is satisfied under the assumption of small strains for each of the composite material phases:

$$\varepsilon_{ij} = \left(\varepsilon_{ij}\right)_1 = \left(\varepsilon_{ij}\right)_2 = \cdots = \left(\varepsilon_{ij}\right)_n \tag{3.1}$$

where ε_{ij} and $\left(\varepsilon_{ij}\right)_n$ represent the composite material strain and the *ith* component strain of such composite material, respectively.

3.2.1 Free energy expression

The composite material's free energy is given by the additive composition of the free energy of each of the component materials considered as a function of its volumetric participation, thus

$$m\,\Psi(\boldsymbol{\varepsilon}^e, \theta, \boldsymbol{\alpha}^m) = \sum_{c=1}^{n} k_c\, m_c\, \Psi_c\left(\boldsymbol{\varepsilon}^e, (\boldsymbol{\varepsilon}^p)_c, \theta, (\boldsymbol{\alpha}^m)_c\right) \quad \Rightarrow \quad \Psi = \sum_{c=1}^{n} k_c\, \frac{m_c}{m}\, \Psi_c \tag{3.2}$$

where $\Psi_c = \frac{1}{2m_c}(\varepsilon_{ij}^e\, C_{ijkl}\, \varepsilon_{kl}^e) + \Psi_c^p$ is the free energy per unit mass and volume corresponding to each of the substances of the mixture, k_c is the volumetric participation coefficient, $(\varepsilon^p)_c$ is the plastic strain of each phase and $(\alpha^m)_c$ are the internal variables defining the physical behavior of any generic component.

The weighting factor or volumetric participation coefficient k_c gives the contribution of each phase and is obtained by the volumetric participation of each of the component materials with respect to the total volume.

$$k_c = \frac{dV_c}{dV_0} \tag{3.3}$$

where V_c represents the material *cth* component volume and V_0 is the total volume of the composite material. The volumetric participation coefficients of the different components of the composite material must satisfy the following condition:

$$\sum_{c=1}^{n} k_c = 1 \tag{3.4}$$

in which the free energy can be recovered for single phase materials and the mass conservation can be guaranteed. Following a similar procedure using simple materials based

---

[*] **Note:** The atomic diffusion phenomena are produced at temperatures close to the melting point. In the analysis a temperature below the melting point is considered.

on the Clausius-Duhem inequality and applying Coleman's method[10,15,16,17,18], in which a positive dissipation is guaranteed, the following entropy expression is obtained,

$$\eta = -m\frac{\partial\Psi(\varepsilon_{ij}^e,\theta,\alpha_i)}{\partial\theta} = -\sum_{c=1}^{n} k_c m_c \frac{\partial\Psi_c(\varepsilon_{ij}^e,\theta,\alpha_i)_c}{\partial\theta} = \sum_{c=1}^{n} k_c \eta_c \qquad (3.5)$$

where η_c is each one of the material phases entropy. The constitutive equation can also be obtained by applying Coleman's method and the Clasius-Duhem inequality.

$$\sigma_{ij} = m\frac{\partial\Psi(\varepsilon_{ij}^e,\theta,\alpha_i)}{\partial\varepsilon_{ij}} = \sum_{c=1}^{n} k_c m_c \frac{\partial\Psi_c(\varepsilon_{ij}^e,\theta,\alpha_i)_c}{\partial\varepsilon_{ij}} = \sum_{c=1}^{n} k_c (\sigma_{ij})_c \qquad (3.6)$$

The secant constitutive equation for composite materials (equation (3.6)) is:

$$\sigma_{ij} = \mathbb{C}_{ijkl}^S \varepsilon_{kl}^e = \sum_{c=1}^{n} k_c (\sigma_{ij})_c = \sum_{c=1}^{n} k_c (\mathbb{C}_{ijkl}^S)_c (\varepsilon_{kl}^e)_c \qquad (3.7)$$

Taking into account the compatibility condition expressed by equation (3.1), the strain for each component is given by

$$(\varepsilon_{kl})_c = \varepsilon_{kl} = (\varepsilon_{kl}^e)_c + (\varepsilon_{kl}^p)_c + (\varepsilon_{kl}^\theta)_c \Rightarrow (\varepsilon_{kl}^e)_c = \varepsilon_{kl} - (\varepsilon_{kl}^p)_c - (\varepsilon_{kl}^\theta)_c \qquad (3.8)$$

where $(\varepsilon_{kl}^e)_c$, $(\varepsilon_{kl}^p)_c$ and $(\varepsilon_{kl}^\theta)_c$ represent the elastic, plastic and thermal strain participation. The composite material plastic strain is obtained by developing both members of the equality expression in (3.7) taking into account the equation (3.8) and the elastic strain in the composite material are $(\varepsilon_{kl}^e) = \varepsilon_{kl} - (\varepsilon_{kl}^p) - (\varepsilon_{kl}^\theta)$, then[*],

$$\mathbb{C}_{ijkl}^\sigma (\varepsilon_{kl} - \varepsilon_{kl}^p - \varepsilon_{kl}^\theta) = \sum_{c=1}^{n} k_c (\mathbb{C}_{ijkl}^\sigma)_c \left(\varepsilon_{kl} - (\varepsilon_{kl}^p)_c - (\varepsilon_{kl}^\theta)_c\right) \Rightarrow \qquad (3.9)$$

$$\varepsilon_{kl}^p = (\mathbb{C}_{ijkl}^\sigma)^{-1}\left[\sum_{c=1}^{n} k_c (\mathbb{C}_{ijkl}^\sigma)_c \left((\varepsilon_{kl}^p)_c + (\varepsilon_{kl}^\theta)_c\right)\right] - \varepsilon_{kl}^\theta \qquad (3.10)$$

The tangent constitutive tensor is obtained by the stress variation with respect to the strains and is given by:

$$\mathbb{C}_{ijkl}^\sigma = \frac{\partial\sigma_{ij}}{\partial\varepsilon_{kl}} = m\frac{\partial^2\Psi(\varepsilon_{ij}^e.\theta,\alpha_i)}{\partial\varepsilon_{ij}\partial\varepsilon_{kl}} = \sum_{c=1}^{n} k_c (\mathbb{C}_{ijkl}^\sigma)_c \qquad (3.11)$$

The conjugated tensor of the thermic dilatancy coefficient is obtained by the variation of the Helmholtz free energy with respect to strains and temperatures,

[15] Lubliner, J., Oliver, J., Oller, S., & Oñate, E. (1989). A plastic-damage model for concrete. *International Journal of Solids and Structures*, 25(3), 299 - 326.
[16] Lubliner, J. (1985). *Thermomechanics of Deformable Bodies*. Department of Civil Engineering, University of California, Berkeley, U.S.A.
[17] Lubliner, J. (1990). *Plasticity theory*. MacMillan, New York.
[18] Maugin, G. A. (1992). *The thermomechanics of plasticity and fracture*. Cambridge University Press.

[*] Note: For the derivative expression it is also necessary to take into account the strain compatibility condition given by equation (3.1) and that $\mathbb{C}_{ijkl}^\sigma \varepsilon_{ij}^e = \sum_{c=1}^{n} k_c (\mathbb{C}_{ijkl}^\sigma)_c (\varepsilon_{ij}^e)_c$

$$\beta_{ij} = -m\frac{\partial^2\Psi(\varepsilon_{ij}^e.\theta,\alpha_i)}{\partial\varepsilon_{ij}\partial\theta} = \sum_{c=1}^{n} k_c(\beta_{ij})_c \tag{3.12}$$

The classical mixing theory, based on the hypothesis that the strain is the same for all the composite components, is strictly valid only if it is applied to material components working in parallel. These materials are characterized by the fact that their stress state is the result of the sum of the stresses of each component, which are weighted proportionally to their volume in each phase with respect to the total, for example long fiber matrices, reinforced concrete, etc. As for short reinforced matrices, the equality hypothesis among all the component strains is no longer valid. In order to solve this problem there are two alternatives: to define another closure equation (equation (3.1)), suitable for the material phenomena simulation, or to carry out a modification of each one of the component's properties and keep the strain equality hypothesis in each one of the composite components.

3.3 Classical theory modification. Serial-parallel model

Due to the classic mixing theory limitation, various modifications have been made. An alternative was presented by Oller *et al.* (1995)[19] and Oñate *et al.* (1997)[20], who proposed the classical mixing theory generalization in small strains to represent the composite component's behavior participating in a combination of serial-parallel behaviors. This involves an automatic adjustment of the composite properties taking into consideration each component's topology and distribution. Thus, each point of a solid can have a different strain.

Figure 3.2 shows a simplified flow description of the meaning of the parallel and serial behaviors of a generic composite material.

Figure 3.2 – Simplified description of the parallel and serial behaviors of a laminated composite material.

The main hypothesis of this mixing theory generalization is based on the strain definition of the whole set as a weighted addition of the parallel and series contributions, therefore:

$$\varepsilon_{ij} = (1-\aleph)\varepsilon_{ij}^{par} + \aleph\,\varepsilon_{ij}^{ser} \tag{3.13}$$

[19] Oller, S., Neamtu, L., & Oñate, E. (1995). Una generalización de la teoría de mezclas clásica para el tratamiento de compuestos en serie/paralelo. Congreso Nacional de Materiales Compuestos, pp. 433 - 438.
[20] Oñate, E., Neamtu, L., & Oller, S. (1997). Un modelo constitutivo para el análisis por el "MEF" para materiales compuestos. In J. Güemes & C. Navarro (Eds.), Materiales Compuestos '97, pp. 206 - 211.

where ε_{ij}^{par} and ε_{ij}^{ser} represent the parallel and serial strains, respectively, and $0 \leq \aleph \leq 1$ is the serial-parallel coupling parameter weighting the relation between the two strain types-behaviors. This parameter has the information of the preferred direction of the material behavior in one point and relates it with the direction of the larger main stress in the point. The basic behavior coupling is described in implicit form under the hypothesis of a closed energetic model [20].

The parallel and serial strain components are approximated by the following expressions:

$$\varepsilon_{ij}^{par} \cong \frac{1}{n} \sum_{c=1}^{n} (\varepsilon_{ij})_c \qquad y \qquad \varepsilon_{ij}^{ser} = \sum_{c=1}^{n} k_c (\varepsilon_{ij})_c \tag{3.14}$$

As observed, the parallel strain sets a weak form of the compatibility equation (3.1), and matches fully with components distributed strictly in parallel. Taking into account equations (3.13) and (3.14), the strain of a point is obtained and its behavior is the result of a composition of series and parallel behaviors at the same time.

$$\varepsilon_{ij} = (1-\aleph)\left[\frac{1}{n} \sum_{c=1}^{n} (\varepsilon_{ij})_c\right] + \aleph\left[\sum_{c=1}^{n} k_c (\varepsilon_{ij})_c\right]$$

$$\frac{\partial \varepsilon_{ij}}{\partial (\varepsilon_{kl})_c} = \left[(1-\aleph)+\aleph \cdot k_c n\right] \cdot I_{ijkl} = \hat{\aleph}_c \cdot I_{ijkl} \Rightarrow \begin{cases} \text{Parallel} : \aleph = 0 \Rightarrow \hat{\aleph}_c = 1 \\ \text{Serial} \quad : \aleph = 1 \Rightarrow \hat{\aleph}_c = k_c n \end{cases} \tag{3.15}$$

where $(\varepsilon_{ij})_c$ is the strain of each composite component, n is the number of component materials participating at each point of a composite solid, and ε_{ij} is the total strain of a point of the composite. Equation (3.15) can be interpreted as a relaxation of the serial-parallel basic behaviors and reciprocal influences due to the $\aleph$. According to this hypothesis, the coupling assumes a weighted internal redistribution of the free energy of the system between the two basic behaviors.

The free energy is described in the following addition which is the classic for composite materials (Oller $et.$ $alt.$ (1995)[19]),

$$m \Psi(\varepsilon_{ij}^e, \theta, \alpha_i) = \sum_{c=1}^{n} k_c m_c \Psi_c\left((\varepsilon_{ij}^e)_c, \theta, (\alpha_i)_c\right)$$

$$\Psi_c = \frac{1}{2m_c}\left((\varepsilon_{ij})_c - (\varepsilon_{ij}^p)_c\right)(\mathbb{C}_{ijkl}^\sigma)_c\left((\varepsilon_{kl})_c - (\varepsilon_{kl}^p)_c\right) + \Psi_c^p \Rightarrow \frac{\partial \Psi_c}{\partial (\varepsilon_{ij}^p)_c} = \frac{1}{m_c}(\mathbb{C}_{ijkl}^\sigma)_c\left((\varepsilon_{kl})_c - (\varepsilon_{kl}^p)_c\right) \tag{3.16}$$

where Ψ is the free energy of the whole composite material, Ψ_c is the free energy of one of the component materials, α_i represents the set of internal variables of the composite material, θ is the temperature and k_c is the cth component coefficient within the composite.

The system entropy is obtained by the Clausius-Duhem inequality and is given by Oller et $al.$ (1995)[19]

$$\eta = -m\frac{\partial \Psi(\varepsilon_{ij}^e, \theta, \alpha_i)}{\partial \theta} = \sum_{c=1}^{n} k_c m_c \frac{\partial \Psi_c\left((\varepsilon_{ij}^e)_c, \theta, (\alpha_i)_c\right)}{\partial \theta} = \sum_{c=1}^{n} k_c \eta_c \tag{3.17}$$

The constitutive law is obtained by the Clausius-Duhem inequality and the Coleman method (Lubliner (1990)[17]), leading to the following hyperstatic form:

$$
\begin{aligned}
\sigma_{ij} &= m\frac{\partial \Psi(\varepsilon_{ij}^e,\theta,\alpha_i)}{\partial \varepsilon_{ij}} = \sum_{c=1}^{n} m_c\, k_c\, \frac{\partial \Psi_c}{\partial(\varepsilon_{kl})_c}\frac{\partial(\varepsilon_{kl})_c}{\partial \varepsilon_{ij}} = \sum_{c=1}^{n} k_c(\sigma_{kl})_c\left(\hat{\aleph}_c \cdot I_{klij}\right)^{-1} \\
&= \sum_{c=1}^{n} k_c(\mathbb{C}_{klrs}^{\sigma})_c\left((\varepsilon_{rs})_c - (\varepsilon_{rs}^p)_c\right)\cdot\left(\hat{\aleph}_c \cdot I_{klij}\right)^{-1} = \sum_{c=1}^{n} k_c\hat{\aleph}_c^{-1}\cdot(\mathbb{C}_{ijrs}^{\sigma})_c\cdot\left((\varepsilon_{rs})_c - (\varepsilon_{rs}^p)_c\right)
\end{aligned}
\tag{3.18}
$$

The stress obtained as shown above is exact for the material behavior in parallel whereas it will only be an approximation for materials behavior in series, as in this latter the composite stress expression tends to have the average stress of all its component stresses. The tangent constitutive tensor is obtained by considering the stress change with respect to the strain change and it is given by:

$$
\begin{aligned}
\mathbb{C}_{ijkl}^{\sigma} &= \frac{\partial \sigma_{ij}}{\partial \varepsilon_{kl}} = m\frac{\partial \Psi^2}{\partial \varepsilon_{ij}\partial \varepsilon_{kl}} = \sum_{c=1}^{n} k_c\frac{\partial(\sigma_{ij})_c}{\partial(\varepsilon_{rs})_c}\frac{\partial(\varepsilon_{rs})_c}{\partial \varepsilon_{kl}} = \\
&= \sum_{c=1}^{n} k_c\left[(\mathbb{C}_{klrs}^{\sigma})_c\cdot\left(\hat{\aleph}_c\cdot I_{klij}\right)^{-1}\right]\cdot\left(\hat{\aleph}_c\cdot I_{rskl}\right)^{-1} = \sum_{c=1}^{n} k_c\,\hat{\aleph}_c^{-2}(\mathbb{C}_{klrs}^{\sigma})_c
\end{aligned}
\tag{3.19}
$$

As shown above, a simplified formulation is established as a function of the principal stress direction with respect to the orientation of the behavior of the dominant material. Due to this limitation, the formulation has been extended to the tensor level where all the stress tensor directions acting at that point can be related to the material's behavior directions at the same point. This new formulation leads to a coupling fourth-order tensor $N_{ijkl} \equiv \mathbf{N}$ which can be written by its symmetries as a square matrix; and for orthotropic materials it is reduced to a diagonal square matrix (Oñate et al. (1997)[13]. Taking into consideration this new behavior hypothesis, the following composite strain definition is established

$$
\boldsymbol{\varepsilon} = (\mathbf{I}-\mathbf{N}):\boldsymbol{\varepsilon}^{par} + \mathbf{N}:\boldsymbol{\varepsilon}^{ser}
\tag{3.20}
$$

where $\mathbf{I}$ is the identity matrix and $\mathbf{N}$ is the serial-parallel coupling diagonal matrix where all its elements satisfy the $0 \le N_I \le 1$ condition, and it can also be obtained by experimental tests. For this new hypothesis, the free energy does not change with respect to the one defined in equation (3.16), therefore

$$
m\Psi(\varepsilon_{ij}^e,\theta,\alpha_i) = \sum_{c=1}^{n} k_c\, m_c\, \Psi_c\left((\varepsilon_{ij}^e)_c,\theta,(\alpha_i)_c\right)
\tag{3.21}
$$

Such that for each component it can also be written in the classic form,

$$
\Psi_c = \frac{1}{2m_c}\left((\varepsilon_{ij})_c - (\varepsilon_{ij}^p)_c\right)\mathbb{C}_{ijkl}^{\sigma}\left((\varepsilon_{kl})_c - (\varepsilon_{kl}^p)_c\right)+\Psi_c^p
\tag{3.22}
$$

Under these new hypotheses, the constitutive equation and the tangent constitutive tensor are:

$$\boldsymbol{\sigma} = \sum_{c=1}^{n} k_c \left[(\mathbf{I} - \mathbf{N}) + n \cdot k_c \cdot \mathbf{N} \right]^{-1} : (\boldsymbol{\sigma})_c$$

$$\mathbb{C}^{\sigma} = \sum_{c=1}^{n} k_c \left[(\mathbf{I} - \mathbf{N}) + n \cdot k_c \cdot \mathbf{N} \right]^{-1} : (\mathbb{C}^{\sigma})_c : \left[(\mathbf{I} - \mathbf{N}) + n \cdot k_c \cdot \mathbf{N} \right]^{-1} \qquad (3.23)$$

This model has the advantage of being a general formulation and is suitable for analyzing an equivalent material using $\aleph$ or $\mathbf{N}$ regardless of the behavior of each of the component materials. However, although it might look like an advantage, there is a problem with obtaining strain $(\varepsilon_{ij})_c$ for each component. This can be solved by establishing a new formulation coupling taking into account the component "c" participation in the others,

$$(\varepsilon_{ij})_c = (\varepsilon_{ij}^e)_c + (\varepsilon_{ij}^p)_c = \varepsilon_{ij} - \left\{ (1 - \aleph) \left[\frac{1}{n} \sum_{\forall m \neq c} (\varepsilon_{ij})_m \right] + \aleph \left[\sum_{\forall m \neq c} k_m (\varepsilon_{ij})_m \right] \right\}$$

$$\text{with:} \qquad (\varepsilon_{ij}^p)_c = \int_t \dot{\lambda}_c \frac{\partial (\mathbb{G}^{\sigma})_c}{\partial (\sigma_{ij})_c} dt \qquad (3.24)$$

As observed in the above expression, this problem can only be solved iteratively, by making an initial guess of the strains in each component, and then correcting them iteratively until a stationary state of their magnitudes is achieved.

3.4 The generalized mixing theory

In previous sections the classic mixing theory was defined as well as a modification to be applied on composite materials that invalidated the strain compatibility hypothesis due to their serial-parallel behavior. In this section a new formulation, which is currently on a developmental stage and is based on the classic mixing theory generalization, is established for the solution of any reinforced matrix composite material. It does not have the classic theory limitation of tackling the strain compatibility. On the other hand, this new approach of the problem leads to an automatic adjustment of the composite material closure or compatibility equation. This is the most important improvement achieved with respect to the formulation presented in section 3.3.

As shown before, the mixing theory of basic substances is based on the mechanics of the local continuum solid and is suitable for explaining the behavior of a point of a composite solid with components which have strain compatibility. It is based on the interaction principle of the substances making up the material and the following hypotheses are assumed:

i.   In each composite infinitesimal volume there is participation of a finite number of component substances.

ii. Each component material contribution in the composite behavior is proportional to their volumetric participation.

iii. All the components follow a general compatibility equation adapted to the topology of the serial-parallel composite. This is the fundamental hypothesis differentiating itself from the classic mixing theory.

iv. The volume occupied by each component is much lower than the total composite volume.

The **parallel behavior** basic hypothesis involves the following concepts (the classic mixing theory) summarized as follows:

$$\begin{cases} \boldsymbol{\sigma} = k_1\,\boldsymbol{\sigma}_1 + \cdots + k_c\,\boldsymbol{\sigma}_c = \sum_{c=1}^{n} k_c\,\boldsymbol{\sigma}_c \\[2mm] \boldsymbol{\varepsilon} = \boldsymbol{\varepsilon}_1 = \cdots = \boldsymbol{\varepsilon}_c \\[2mm] \Psi_c^e = \dfrac{1}{2m_c}\left[\boldsymbol{\varepsilon}_c - \boldsymbol{\varepsilon}_c^p\right] : (\mathbb{C}^S)_c : \left[\boldsymbol{\varepsilon} - \boldsymbol{\varepsilon}_c^p\right] \\[2mm] \boldsymbol{\sigma}_c = (\mathbb{C}^\sigma)_c : \left[\boldsymbol{\varepsilon}_c - \boldsymbol{\varepsilon}_c^p\right] \equiv (\mathbb{C}^\sigma)_c : \left[\boldsymbol{\varepsilon} - \boldsymbol{\varepsilon}_c^p\right] \\[2mm] \mathbb{C}^{\mathrm{par}} = \sum_{c=1}^{n} k_c\,(\mathbb{C}^\sigma)_c \end{cases} \tag{3.25}$$

The complementary form of the parallel formulation leads to **the serial behavior hypothesis,** which involves the summarized concepts as follows:

$$\begin{cases} \boldsymbol{\sigma} = \boldsymbol{\sigma}_1 = \cdots = \boldsymbol{\sigma}_c \\[2mm] \boldsymbol{\varepsilon} = k_1\,\boldsymbol{\varepsilon}_1 + \cdots + k_c\,\boldsymbol{\varepsilon}_c = \sum_{c=1}^{n} k_c\,\boldsymbol{\varepsilon}_c = \\[2mm] \quad = k_1\,(\boldsymbol{\varepsilon}_1^e + \boldsymbol{\varepsilon}_1^p) + \cdots + k_c\,(\boldsymbol{\varepsilon}_c^e + \boldsymbol{\varepsilon}_c^p) = \sum_{c=1}^{n} k_c\,(\boldsymbol{\varepsilon}_c^e + \boldsymbol{\varepsilon}_c^p) = \\[2mm] \quad = \sum_{c=1}^{n} k_c\,(\boldsymbol{\phi}_c : \boldsymbol{\varepsilon}^e + \boldsymbol{\varepsilon}_c^p) \begin{cases} \boldsymbol{\varepsilon}^e = \sum_{c=1}^{n} k_c\,(\boldsymbol{\phi}_c : \boldsymbol{\varepsilon}^e) \; \Rightarrow \; \sum_{c=1}^{n} k_c\,\boldsymbol{\phi}_c = \mathbf{I}\;;\; \mathrm{con}: \boldsymbol{\varepsilon}_c^e \overset{\mathrm{def}}{=} \boldsymbol{\phi}_c : \boldsymbol{\varepsilon}^e \\[2mm] \boldsymbol{\varepsilon}^p = \sum_{c=1}^{n} k_c\,\boldsymbol{\varepsilon}_c^p \end{cases} \\[6mm] \overline{\Psi}_c^e = \dfrac{1}{2m_c}\,\boldsymbol{\sigma}_c : (\mathbb{C}^\sigma)_c^{-1} : \boldsymbol{\sigma}_c \\[2mm] \boldsymbol{\sigma}_c = (\mathbb{C}^\sigma)_c : \left[\boldsymbol{\varepsilon}_c - \boldsymbol{\varepsilon}_c^p\right] \equiv (\mathbb{C}^\sigma)_c : \left[\boldsymbol{\phi}_c : \boldsymbol{\varepsilon}^e\right] \equiv (\mathbb{C}^\sigma)_c : \left[\boldsymbol{\phi}_c : (\boldsymbol{\varepsilon} - \boldsymbol{\varepsilon}^p)\right] \\[2mm] \boldsymbol{\varepsilon}^p = \sum_{c=1}^{n} k_c\,\boldsymbol{\varepsilon}_c^p \; \Rightarrow (\mathbb{C}^{\mathrm{ser}})^{-1} : \boldsymbol{\sigma} = \sum_{c=1}^{n} k_c\,(\mathbb{C}^\sigma)_c^{-1} : \boldsymbol{\sigma} \; \Rightarrow \\[2mm] \Rightarrow \quad \mathbb{C}^{\mathrm{ser}} = \left[\sum_{c=1}^{n} k_c\,(\mathbb{C}^\sigma)_c^{-1}\right]^{-1} \end{cases} \tag{3.26}$$

By the **third hypothesis** of this generalized mixing theory (see hypothesis iii) the relationship between the composite strain and each component strain can be established. This compatibility equation links the parallel (3.25) and serial behavior hypotheses (3.26). Consequently, this theory will be called the serial-parallel compatibility hypothesis.

$$\left(\varepsilon_{ij}\right)_c = \underbrace{(1-\chi_c)\cdot I_{ijkl}\,\varepsilon_{kl}}_{\left(\varepsilon_{ij}^{\mathrm{par}}\right)_c} + \underbrace{\chi_c\cdot\left[(\phi_{ijkl})_c\cdot(\varepsilon_{kl}-\varepsilon_{kl}^p)+(\varepsilon_{kl}^p)_c\right]}_{\left(\varepsilon_{ij}^{\mathrm{ser}}\right)_c}$$

$$\left(\varepsilon_{ij}\right)_c = \left[(1-\chi_c)\cdot I_{ijkl}\,\varepsilon_{kl} + \chi_c\cdot(\phi_{ijkl})_c\cdot\varepsilon_{kl}\right] - \chi_c\cdot\underbrace{\left[(\phi_{ijkl})_c\,\varepsilon_{kl}^p - (\varepsilon_{kl}^p)_c\right]}_{(\hat{\varepsilon}_{kl}^p)_c} \tag{3.27}$$

$$\frac{\partial(\varepsilon_{ij})_c}{\partial\varepsilon_{kl}} = (1-\chi_c)\cdot I_{ijkl} + \chi_c\cdot(\phi_{ijkl})_c$$

They can also be expressed in the two following forms:

$$\left(\varepsilon_{ij}\right)_c = \left[\left(1-\chi_c\right)\cdot I_{ijkl} + \chi_c\cdot\left(\varphi_{ijkl}\right)_c\right]:\varepsilon_{kl} - \chi_c(\hat{\varepsilon}^p_{kl})_c \quad \text{or also}: \quad \left(\boldsymbol{\varepsilon}\right)_c = \left[\left(1-\chi_c\right)\cdot \mathbf{I}_4 + \chi_c\cdot\left(\boldsymbol{\varphi}\right)_c\right]:\boldsymbol{\varepsilon} - \chi_c\left(\hat{\boldsymbol{\varepsilon}}^p\right)_c$$

Being $(\hat{\varepsilon}^p_{kl})_c$ a plastic strain defined for operational purposes and without any physical meaning, obtained as a result of the composite plastic strain average and distributed among its components according to their respective stiffness ratio $\left(\varphi_{ijkl}\right)_c\varepsilon^p_{kl}$, and the real plastic strain component $(\varepsilon^P_{kl})_c$. There is also in the same expression $\left(\varepsilon_{ij}\right)_c$, which is the strain in the *cth component* that can be broken down into one parallel participation component $\left(\varepsilon^{\mathrm{par}}_{ij}\right)_c$ another one serial $\left(\varepsilon^{\mathrm{ser}}_{ij}\right)_c$, and ε_{kl} the total strain in the composite. The serial-parallel coupling parameter $0\le\left[\chi_c = \sin\alpha_\chi\right]\le 1$, depends on the angle $0\le\left[\alpha_\chi = (\mathrm{x}^{\mathrm{f}}_{\mathrm{Loc}}, \mathrm{x}^\sigma_1)\right]\le\pi/2$, between the principal stress orientation (x^σ_1) and the fiber orientation $(x^{\mathrm{f}}_{\mathrm{Loc}})$. The value of this parameter is 0 for a pure parallel behavior and 1 for a pure serial behavior (see Figure 3.1). Physically, this parameter is used to locate the reinforcement position with respect to the action.

Another tensor participating in the compatibility equation (3.27) is responsible for the *cth* component strain when all the composite components have serial distribution and it is obtained from the elastic part of equations (3.26) as follows,

$$\begin{aligned}
\boldsymbol{\varepsilon}^e &= k_1\,\boldsymbol{\varepsilon}^e_1 + \cdots + k_c\,\boldsymbol{\varepsilon}^e_c = \sum_{c=1}^n k_c\,\boldsymbol{\varepsilon}^e_c = \\[4pt]
&= k_1\,((\mathbb{C}^\sigma)^{-1}_1:\boldsymbol{\sigma}_1) + \cdots + k_c\,((\mathbb{C}^\sigma)^{-1}_c:\boldsymbol{\sigma}_c) = (\mathbb{C}^{\mathrm{ser}})^{-1}:\boldsymbol{\sigma} \\[4pt]
&= k_1\,\underbrace{((\mathbb{C}^\sigma)^{-1}_1:(\mathbb{C}^{\mathrm{ser}}):\boldsymbol{\varepsilon}^e)}_{\boldsymbol{\varphi}_1} + \cdots + k_c\,\underbrace{((\mathbb{C}^\sigma)^{-1}_c:(\mathbb{C}^{\mathrm{ser}}):\boldsymbol{\varepsilon}^e)}_{\boldsymbol{\varphi}_c} = ((\mathbb{C}^{\mathrm{ser}})^{-1}:(\mathbb{C}^{\mathrm{ser}}):\boldsymbol{\varepsilon}^e) \\[4pt]
&= k_1\,(\boldsymbol{\varphi}_1) + \cdots + k_c\,(\boldsymbol{\varphi}_c) = \underbrace{\left[\sum_{c=1}^n k_c\,\boldsymbol{\varphi}_c\right]}_{\mathbf{I}} \quad \text{with:} \quad \left(\varphi_{ijkl}\right)_c = \left(\mathbb{C}^\sigma_{ijrs}\right)^{-1}_c \mathbb{C}^{\mathrm{ser}}_{rskl}
\end{aligned} \tag{3.28}$$

In this equation $\left(\mathbb{C}^\sigma_{ijrs}\right)_c$ is the constitutive tensor of the *cth* component and $\mathbb{C}^{\mathrm{ser}}_{rskl} = \left[\sum_{c=1}^n k_c\,(\mathbb{C}^\sigma_{rskl})^{-1}_c\right]^{-1}$ is the composite constitutive tensor when its n components work in pure serial.

Note that these extreme conditions are met in equation (3.27). In other words, for a reinforcement fiber **parallel** alignment $(\chi_c = 0)$ (classic mixing theory), the strain for the *cth* component is obtained $\left(\varepsilon_{ij}\right)_c = \left(\varepsilon^{\mathrm{par}}_{ij}\right)_c = \left(1-\chi_c\right)\cdot I_{ijkl}\,\varepsilon_{kl} = \varepsilon_{ij}$ which is equal for all the other components, whereas for serial configuration materials $(\chi_c = 1)$, the strain obtained $\left(\varepsilon_{ij}\right)_c = \left(\varepsilon^{\mathrm{ser}}_{ij}\right)_c = \chi_c\cdot\left[\left(\varphi_{ijkl}\right)_c\cdot\left(\varepsilon_{kl} - \varepsilon^P_{kl}\right) + \left(\varepsilon^P_{kl}\right)_c\right] \ne \varepsilon_{ij}$ for the *cth* component is different from the other components.

The free energy of parallel-serial composite materials is obtained as a result of the weighted sum of the free energies of each component, which are defined by the principle of uncoupled elasticity where the elastic part is expressed as follows:

$$\Psi^e = \sum_{c=1}^n k_c\,\frac{m_c}{m}\,\Psi^e_c \quad ; \quad \Psi^e_c = \frac{1}{2\,m_c}\left[\left(\varepsilon_{ij}\right)_c - \left(\varepsilon^p_{ij}\right)_c\right]\left(\mathbb{C}^\sigma_{ijkl}\right)_c\left[\left(\varepsilon_{kl}\right)_c - \left(\varepsilon^p_{kl}\right)_c\right] \tag{3.29}$$

Or can be written in compact form as,

$$\Psi^e = \sum_{c=1}^{n} k_c \frac{m_c}{m} \Psi_c^e = \sum_{c=1}^{n} k_c \frac{1}{2m} \left[(\varepsilon)_c - (\varepsilon^p)_c \right] : (\mathbf{C}^\sigma)_c : \left[(\varepsilon)_c - (\varepsilon^p)_c \right] \tag{3.30}$$

where the coefficient of the volumetric participation of each component is defined as $k_c = v_c / v$, v_c being the volume of the cth component and v the total volume. Moreover, in equation (3.29), the tensor $\left(\varepsilon_{ij}^p \right)_c$ represents the unrecoverable part of the cth strain. The constitutive equation for a simple "serial-parallel" component is obtained as,

$$(\sigma_{ij})_c = m_c \frac{\partial \Psi_c}{\partial (\varepsilon_{ij})_c} = m_c \frac{\partial}{\partial (\varepsilon_{ij})_c} \left[\frac{1}{2m_c} \left((\varepsilon_{rs})_c - (\varepsilon_{rs}^p)_c \right) (\mathbb{C}_{rstu}^\sigma)_c \left((\varepsilon_{tu})_c - (\varepsilon_{tu}^p)_c \right) + \Psi_c^p \right] =$$
$$= (\mathbb{C}_{ijtu}^\sigma)_c \left((\varepsilon_{tu})_c - (\varepsilon_{tu}^p)_c \right)$$

The constitutive equation definition for composite materials is obtained by the potential defined in equation (3.29), where the dissipation condition of the Thermodynamics second principle is guaranteed[21]. Thus, taking into consideration the previous equation for a single component and the derivative of the strain of a component with respect to the total (3.27), the constitutive equation of the composite material is obtained,

$$\sigma_{ij} = m \frac{\partial \Psi}{\partial \varepsilon_{ij}} = m \sum_{c=1}^{n} k_c \frac{m_c}{m} \frac{\partial \Psi_c}{\partial (\varepsilon_{kl})_c} \frac{\partial (\varepsilon_{kl})_c}{\partial \varepsilon_{ij}} =$$
$$= \sum_{c=1}^{n} k_c \left\{ (\mathbb{C}_{klrs}^\sigma)_c \left[(1 - \chi_c) \cdot I_{klij} + \chi_c \cdot (\phi_{klij})_c \right] \left[(\varepsilon_{rs})_c - (\varepsilon_{rs}^p)_c \right] \right\} \tag{3.31}$$

Or, expressed in compact form,

$$\boldsymbol{\sigma} = m \frac{\partial \Psi}{\partial \boldsymbol{\varepsilon}} = m \sum_{c=1}^{n} k_c \frac{m_c}{m} \frac{\partial \Psi_c}{\partial (\boldsymbol{\varepsilon})_c} : \frac{\partial (\boldsymbol{\varepsilon})_c}{\partial \boldsymbol{\varepsilon}} =$$
$$= \sum_{c=1}^{n} k_c \left[(1 - \chi_c) \cdot \mathbf{I}_4 + \chi_c \cdot (\boldsymbol{\phi})_c \right]^T : \left\{ (\mathbf{C}^\sigma)_c : \left[(\boldsymbol{\varepsilon})_c - (\boldsymbol{\varepsilon}^p)_c \right] \right\} \tag{3.32}$$

where $\mathbf{I}_4$ is the fourth-order identity tensor. The constitutive tensor of the composite material is then obtained,

$$\mathbb{C}_{tuij}^\sigma = m \frac{\partial \Psi}{\partial \varepsilon_{tu} \partial \varepsilon_{ij}} = m \frac{\partial}{\partial \varepsilon_{tu}} \left(\frac{\partial \Psi}{\partial \varepsilon_{ij}} \right) = \frac{\partial}{\partial \varepsilon_{tu}} (\sigma_{ij}) =$$
$$= \sum_{c=1}^{n} k_c \left\{ (\mathbb{C}_{klrs}^\sigma)_c \left[(1 - \chi_c) \cdot I_{klij} + \chi_c \cdot (\phi_{klij})_c \right] \right\} \cdot \left[(1 - \chi_c) \cdot I_{rstu} + \chi_c \cdot (\phi_{rstu})_c \right] \tag{3.33}$$

Or, in compact form,

$$\mathbf{C}^\sigma = m \frac{\partial \psi}{\partial \boldsymbol{\varepsilon} \otimes \partial \boldsymbol{\varepsilon}} = m \frac{\partial}{\partial \boldsymbol{\varepsilon}} \left(\frac{\partial \Psi}{\partial \boldsymbol{\varepsilon}} \right) = \frac{\partial}{\partial \boldsymbol{\varepsilon}} (\boldsymbol{\sigma}) =$$
$$= \sum_{c=1}^{n} k_c \left[(1 - \chi_c) \cdot \mathbf{I}_4 + \chi_c \cdot (\boldsymbol{\phi})_c \right]^T : \left\{ (\mathbf{C}^\sigma)_c : \left[(1 - \chi_c) \cdot \mathbf{I}_4 + \chi_c \cdot (\boldsymbol{\phi})_c \right] \right\} \tag{3.34}$$

---

[21] Malvern L. (1969). *Introduction to the mechanics of a continuous medium.* Prentice-Hall.

The stress and constitutive tensor previously defined show that the kinematic definition carried out through equation (3.27) leads to an implicit change of the material properties. In other words, the kinematic definition can be presented as a modification of the constitutive tensor (see equation (3.34)).

3.5 Large strains classic mixing theory

The classical mixing theory is extended to the field of finite strains in this section. This extension must be carried out as composite materials are generally subjected to stresses that cause large stains. The importance of this statement is evident when observing the reinforcement phase alignment with respect to the stress direction in the material under large displacements and strains. At the end of this chapter an example of this phenomenon will be presented as well as the importance of completing this formulation of the mixing theory to deal with large strains, otherwise serious mistakes can be made.

3.5.1 Closure or compatibility equation

As observed in previous sections, the third hypothesis based on the classic mixing theory assumes that in the absence of atomic diffusion, the strain is identical for all the composite components. This hypothesis must be verified both *in the referential configuration* and the *spatial configuration* for each phase. In the referential configuration the strain compatibility condition is given by Trusdell & Toupin (1960)[3] and Oñate *et al.* (1991)[22],

$$E_{IJ} \equiv (E_{IJ})_1 = (E_{IJ})_2 = \cdots = (E_{IJ})_n \tag{3.35}$$

where $E_{ij} = \frac{1}{2}(C_{IJ} - G_{IJ})$ is the Green-Lagrange strain tensor, $\mathbf{G}$ is the material metric tensor that is expressed for an orthogonal coordinate system as $G_{IJ} = I_{IJ} = \delta_{IJ}$ and $\mathbf{C} = \mathbf{F}^T\mathbf{F}$ is the Cauchy-Green right strain tensor.

In the updated configuration, the strain compatibility condition is proposed as:

$$e_{ij} \equiv (e_{ij})_1 = (e_{ij})_2 = \cdots = (e_{ij})_n \tag{3.36}$$

where $e_{ij} = \frac{1}{2}(g^{ij} - b_{ij}^{-1})$ is the Almansi strain tensor, $\mathbf{b} = \mathbf{F}\mathbf{F}^T$ is the Cauchy-Green left strain tensor, g^{ij} is the spatial metric tensor given by $g^{ij} = I_{ij} = \delta_{ij}$ and $F_{iJ} = \dfrac{\partial x_i}{\partial X_J}$ is the strain gradient.

Taking into account the definition of the Cauchy-Green right tensor and the compatibility equation (3.35), the closure equation is obtained as a function of the strain gradients:

$$F_{iJ} \equiv (F_{iJ})_1 = (F_{iJ})_2 = \cdots = (F_{iJ})_n \tag{3.37}$$

This hypothesis is valid only if the mixing theory is applied for materials with parallel behavior. In other words, the composite tensile state is the result of the addition of the stresses of each component proportional to the volume they occupy with respect to the total, for example large-fiber matrix, reinforced concrete, etc. For short-fiber reinforced

[22] Oñate, E., Oller, S., Botello, S., & Canet, J. (1991). Methods for analysis of composite material structures (in Spanish). Technical Report 11, CIMNE, Barcelona, Spain.

materials, the assumption that the strains of all the components are equal is not valid. To solve this problem, there are two alternatives: to define another closure equation as shown in the previous section or to carry out a correction of each of the component properties and keeping the closing equations of the classic mixing theory (3.35) and (3.36).

3.5.2 Free energy function

The composite materials that satisfy equations (3.35) and (3.36) also satisfy the basic additive condition of the free energy of its components both in the referential and updated configurations (Trusdell & Tupin (1960)[3]):

$$m^0 \, \Psi(\mathbf{E}^e, \theta, \alpha^m) = \sum_{c=1}^{n} k_c \, m_c^0 \, \Psi_c(\mathbf{E}, (\mathbf{E}^p)_c, \theta, (\alpha^m)_c)$$

$$m \, \psi(\mathbf{e}^e, \theta, \alpha^m) = \sum_{c=1}^{n} k_c \, m_c \, \psi_c(\mathbf{e}, (\mathbf{e}^p)_c, \theta, (\alpha^m)_c)$$

(3.38)

where Ψ_c and ψ_c are the free energies corresponding to each of the n component substances of the mixture defined in the referential and the updated configurations, respectively, k_c is the coefficient of the volumetric participation, $(\mathbf{E}^p)_c$ and $(\mathbf{e}^p)_c$ are the plastic strains in each phase in the referential and updated configurations, respectively, and $(\alpha^m)_c$ are the internal variables of the cth component defining the irreversible physical behavior of each component substance.

The mixing theory of basic substances in large strains says that all the substances participating in the composite behave proportionally to the volume they occupy with respect to the total volume. The coefficient of volumetric participation is defined in the referential configuration as:

$$k_c = \frac{dV_c}{dV_0}$$

(3.39)

in which V_c is the component volume and V_0 is the composite total volume in the referential equation. On the other hand, the continuity equation [23] says that

$$J = \det \mathbf{F} = \frac{dv_0}{dV_0}$$

(3.40)

where v_0 is the composite volume in the updated configuration. Taking into account closure equation (3.37), it follows that:

$$\frac{dv_c}{dV_c} = \det \mathbf{F}_c = J \implies dV_c = \frac{1}{J} dv_c$$

(3.41)

This equation establishes that the relationship between the volume of one component in the spatial configuration and the volume of a component in the referential equation is given by the strain gradient determinant. Taking into account equations (3.41) and (3.39) the relation of the volumetric participation in the updated configuration can be written as:

---

[23] Malvern, L. (1969). *Introduction to the Mechanics of a Continuous Medium.* Prentice-Hall.

$$k_c = \frac{1}{J}\frac{dv_c}{dV_0} = \frac{dv_c}{dv_0} \tag{3.42}$$

This equation shows that the coefficient of volumetric participation is constant in both configurations. The coefficient of volumetric participation of each component must satisfy the following condition with respect to the continuity equation (3.40) or mass conservation:

$$\sum_{c=1}^{n} k_c = 1 \tag{3.43}$$

3.5.3 The constitutive equation

To simulate the composite behavior, once the free energy function is defined, it is necessary to define the composite constitutive equation. Bellow a brief presentation of the constitutive equation is given, for both the referential and the updated configurations. A formulation of small elastic strains and large plastic strains will be considered in which the free energy will be uncoupled in its elastic and inelastic parts. The elastic part is a quadratic expression in the strains.

3.5.3.1 The referential configuration

The composite stress definition $\mathbf{S}$ is obtained by the hyperelastic model formulation to guarantee the dissipation condition of the second Principle of Thermodynamics (Malvern $(1969)^{23}$), then:

$$\mathbf{S} = m^0 \frac{\partial \Psi(\mathbf{E}^e, \theta, \alpha^m)}{\partial \mathbf{E}} = \sum_{c=1}^{n} m_c^0 k_c \frac{\partial \Psi_c(\mathbf{E}, (\mathbf{E}^P)_c, \theta, (\alpha^m)_c)}{\partial \mathbf{E}} =$$

$$= \sum_{c=1}^{n} k_c (\mathbf{S})_c = \sum_{c=1}^{n} k_c ((\mathbb{C}^S)_c (\mathbf{E}^e)_c) \tag{3.44}$$

And the constitutive tensor of the composite is given by:

$$\mathbb{C}^S = \frac{\partial \mathbf{S}}{\partial \mathbf{E}} = m^0 \frac{\partial^2 \Psi(\mathbf{E}^e, \theta, \alpha^m)}{\partial \mathbf{E} \otimes \partial \mathbf{E}} = \sum_{c=1}^{n} k_c (\mathbb{C}^S)_c \tag{3.45}$$

where $(\mathbb{C}^S)_c$ represents the tangent constitutive anisotropic real tensor for the cth component.

3.5.3.2 Updated configuration

A similar procedure carried out in the referential configuration is presented below. The composite stress definition $\boldsymbol{\tau}$ is obtained by a formulation of the hyperelastic model to guarantee the dissipation condition established by the "Second Principle of the Thermodynamics" (Malvern $(1969)^{23}$). Then,

$$\boldsymbol{\tau} = m \frac{\partial \psi(\mathbf{e}^e, \theta, \alpha^m)}{\partial \mathbf{e}} = \sum_{c=1}^{n} m_c k_c \frac{\partial \psi_c(\mathbf{e}, (\mathbf{e}^P)_c, \theta, (\alpha^m)_c)}{\partial \mathbf{e}} =$$

$$= \sum_{c=1}^{n} k_c (\boldsymbol{\tau})_c = \sum_{c=1}^{n} k_c ((\mathbf{c}^\tau)_c (\mathbf{e}^e)_c) = J \boldsymbol{\sigma} \tag{3.46}$$

Being $\boldsymbol{\tau}$ and $\boldsymbol{\sigma}$ Kirchhoff and Cauchy stresses, respectively, and J the Jacobian, (see equation (3.41)) and the constitutive tensor of the composite in the updated, then,

$$\mathbf{c}^{\tau} = \frac{\partial \boldsymbol{\tau}}{\partial \mathbf{e}} = m \frac{\partial^2 \psi(\mathbf{e}^e, \theta, \alpha^m)}{\partial \mathbf{e} \otimes \partial \mathbf{e}} = \sum_{c=1}^{n} k_c (\mathbf{c}^{\tau})_c \qquad (3.47)$$

where $(\mathbf{c}^{\tau})_c$ represents the anisotropic tangent constitutive real tensor for the cth component.

The flow chart in Figure 3.3 describes the algorithm for the solution by using the finite element method in non-linear problems with finite strains for multiphase composite materials. Each phase has its own constitutive model and is independent from the other composite phases.

The algorithm starts in the referential configuration and, by using the tensor operations from the referential configuration to the updated one ("push-forward") (Car, (2000)[1]), the constitutive equation solution is tackled for each of the phases of the composite material. Each of these phases can have different types of constitutive behavior such as plasticity, damage, etc., which in turn can be isotropic or anisotropic. Once the stress of each of the components is determined, it is necessary to obtain the composite material stress by equation (3.7). Then the internal forces at each point of the material can be obtained. Once the internal forces are determined, it is necessary to verify their equilibrium with respect to the external applied forces.

3.6 Generalized mixing theory formulated in large strains

An extension of the classical mixing theory definition (section 3.5) for large strains is given below. Therefore, a formulation is proposed based on the non-compliance hypothesis of the compatibility equation (3.36) and its consequences. On the other hand, this change in the fundamental hypothesis of the mixing theory involves the following definition of the free energy functions in the material and updated configuration.

$$m^0 \, \Psi(\mathbf{E}^e, \theta, \alpha^m) = m^0 \sum_{c=1}^{n} k_c \, \Psi_c((\mathbf{E})_c, (\mathbf{E}^p)_c, \theta, (\alpha^m)_c)$$

$$m \, \psi(\mathbf{e}^e, \theta, \alpha^m) = m \sum_{c=1}^{n} k_c \, \psi_c((\mathbf{e})_c, (\mathbf{e}^p)_c, \theta, (\alpha^m)_c) \qquad (3.48)$$

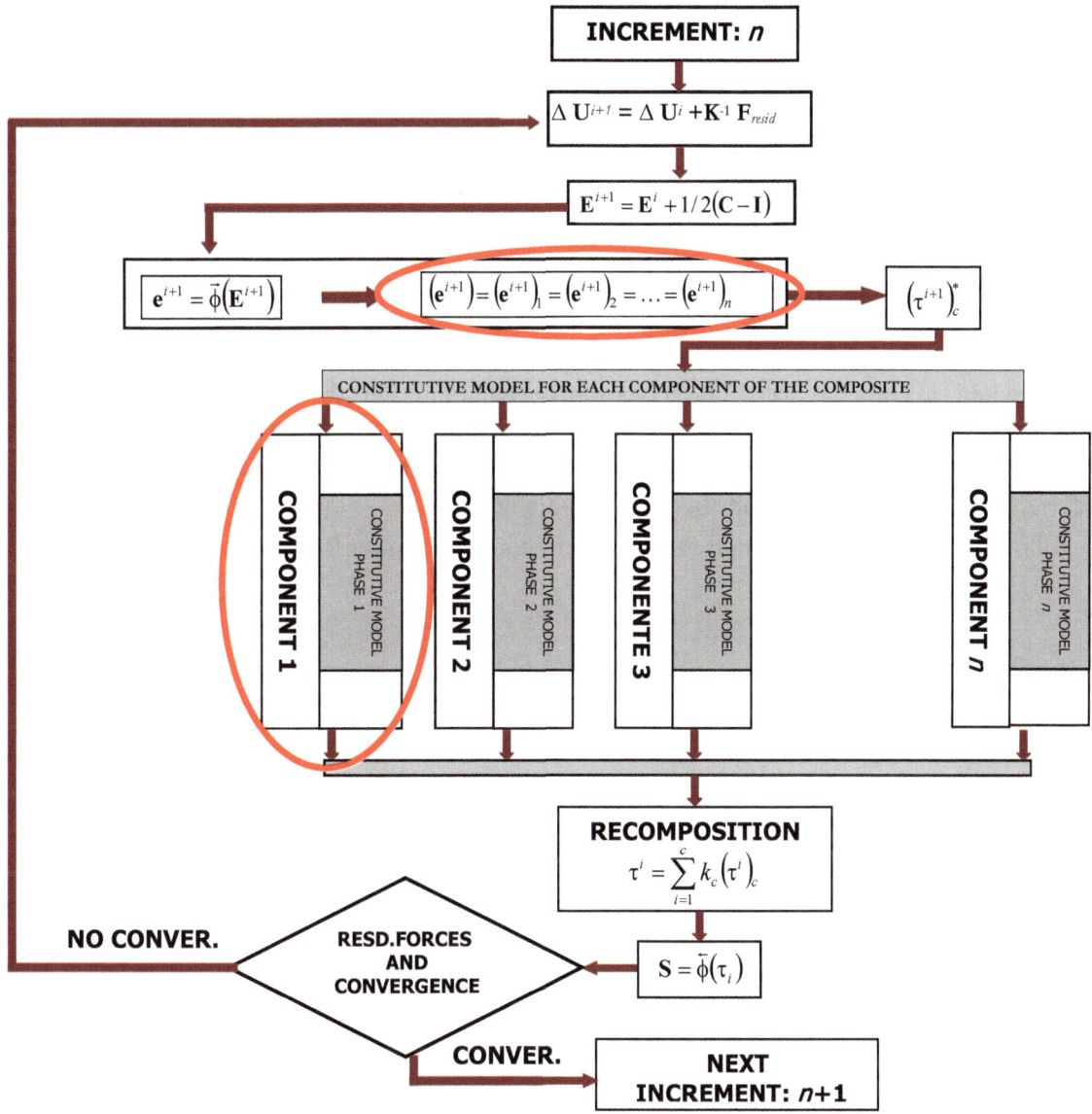

Figure 3.3 – Solution scheme of a non-linear multiphase problem. Mixed algorithm for the classical mixing theory in the referential and updated configurations.

where Ψ_c and ψ_c are the free energies for each of the n component substances defined in the referential and updated configuration, respectively, $k_c = \dfrac{1}{J}\dfrac{dv_c}{dV_0} = \dfrac{dv_c}{dv_0}$ is the coefficient of volumetric participation as a function of J, $(\mathbf{E})_c = \left[(1-\chi_c)\cdot\mathbf{I}_4 + \chi_c\cdot(\mathbf{\Phi})_c\right]:\mathbf{E} - \chi_c\cdot(\hat{\mathbf{E}}^P)_c$ and $(\mathbf{e})_c = \left[(1-\chi_c)\cdot\mathbf{I}_4 + \chi_c\cdot(\mathbf{\phi})_c\right]:\mathbf{e} - \chi_c\cdot(\hat{\mathbf{e}}^P)_c$ are the total strains in the referential and updated configurations, respectively, $(\mathbf{E}^P)_c$ and $(\mathbf{e}^P)_c$ are the plastic strains of each phase in the two kinematic configurations previously formulated and $(\alpha^m)_c$ are the internal variables of the *cth* component defining the irreversible physical behavior of each component substances[15,24], $(\hat{\mathbf{E}}^p)_c$ and $(\hat{\mathbf{e}}^p)_c$ are the plastic strains defined for operational purposes having no physical meaning (see equation (3.27)). The definition of each component strain participate the variables already introduced in section 3.4. The only difference is how the serial behavior factor is obtained for each configuration, then

$$
\begin{cases}
\left(\Phi_{ijkl}\right)_c = \left(\mathbb{C}^S_{ijrs}\right)_c^{-1} \mathbb{C}^{ser}_{rskl} & \text{Referential Configuration} \\
\left(\varphi_{ijkl}\right)_c = \left(c^{\tau}_{ijrs}\right)_c^{-1} c^{ser}_{rskl} & \text{Current Configuration}
\end{cases}
\tag{3.49}
$$

3.6.1 Constitutive equation

A similar formulation to the one developed for the large-strains classic mixing theory is presented in this section. The constitutive equations in the referential and updated configurations for the generalized mixing theory are given below. Note that the main difference is the strain definition in each configuration $(\mathbf{E})_c$, as in this new formulation the strain is not the same for all the components $\mathbf{E} \neq (\mathbf{E})_1 \neq \cdots \neq (\mathbf{E})_c$ and, consequently, the following expression for the stress in the referential configuration is obtained as,

$$
\begin{aligned}
\mathbf{S} = m^0 \frac{\partial \Psi(\mathbf{E}^e, \theta, \alpha^m)}{\partial \mathbf{E}} &= \sum_{c=1}^{n} m^0_c k_c \frac{\partial \Psi_c}{\partial (\mathbf{E})_c} : \frac{\partial (\mathbf{E})_c}{\partial \mathbf{E}} = \\
&= \sum_{c=1}^{n} k_c \, (\mathbf{S})_c : \frac{\partial (\mathbf{E})_c}{\partial \mathbf{E}} = \\
&= \sum_{c=1}^{n} k_c \left[(1-\chi_c) \cdot \mathbf{I}_4 + \chi_c \cdot (\Phi)_c \right]^T : \left\{ (\mathbb{C}^S)_c : \left[(1-\chi_c) \cdot \mathbf{I}_4 + \chi_c \cdot (\Phi)_c \right] : (\mathbf{E}^e)_c \right\}
\end{aligned}
\tag{3.50}
$$

And in the updated configuration it gets the following form,

$$
\begin{aligned}
\tau = m \frac{\partial \psi(\mathbf{e}^e, \theta, \alpha^m)}{\partial \mathbf{e}} &= \sum_{c=1}^{n} m_c \, k_c \frac{\partial \psi_c}{\partial (\mathbf{e})_c} : \frac{\partial (\mathbf{e})_c}{\partial \mathbf{e}} = \\
&= \sum_{c=1}^{n} k_c \, (\tau)_c : \frac{\partial (\mathbf{e})_c}{\partial \mathbf{e}} = \\
&= \sum_{c=1}^{n} k_c \left[(1-\chi_c) \cdot \mathbf{I}_4 + \chi_c \cdot (\phi)_c \right]^T : \left\{ (\mathbf{c}^{\tau})_c : \left[(1-\chi_c) \cdot \mathbf{I}_4 + \chi_c \cdot (\phi)_c \right] : (\mathbf{e}^e)_c \right\} = J \, \sigma
\end{aligned}
\tag{3.51}
$$

[24] Oller S. (2001). *Fractura mecánica – Un enfoque global*. CIMNE – Ediciones UPC. Barcelona

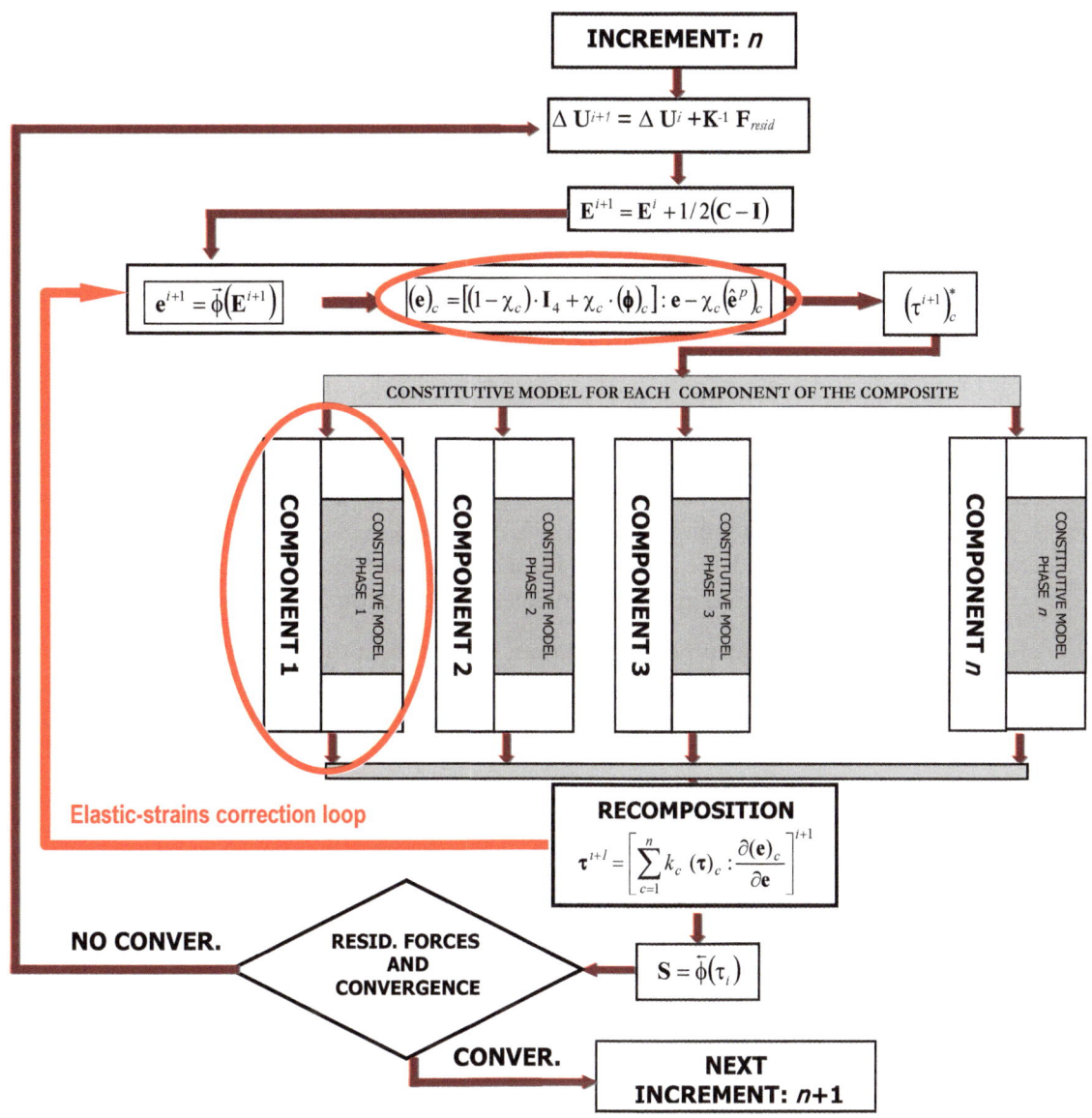

Figure 3.4 – Solution scheme of a non-linear multiphase problem. Mixed algorithm for the generalized mixing theory in the referential and updated configurations.

Following the analysis, the constitutive tensor of the referential configuration is obtained,

$$\mathbb{C}^S = \frac{\partial \mathbf{S}}{\partial \mathbf{E}} = m^0 \frac{\partial^2 \Psi(\mathbf{E}^e, \theta, \alpha^m)}{\partial \mathbf{E} \otimes \partial \mathbf{E}} = \sum_{c=1}^{n} k_c \left[(1 - \chi_c) \cdot \mathbf{I}_4 + \chi_c \cdot (\mathbf{\Phi})_c \right]^T : \left\{ (\mathbb{C}^S)_c : \left[(1 - \chi_c) \cdot \mathbf{I}_4 + \chi_c \cdot (\mathbf{\Phi})_c \right] \right\}$$

(3.52)

This tensor is expressed in the updated configuration as,

$$\mathbf{c}^S = \frac{\partial \mathbf{\tau}}{\partial \mathbf{e}} = m \frac{\partial^2 \psi(\mathbf{e}^e, \theta, \alpha^m)}{\partial \mathbf{e} \otimes \partial \mathbf{e}} = \sum_{c=1}^{n} k_c \left[(1 - \chi_c) \cdot \mathbf{I}_4 + \chi_c \cdot (\mathbf{\phi})_c \right]^T : \left\{ (\mathbf{c}^\tau)_c : \left[(1 - \chi_c) \cdot \mathbf{I}_4 + \chi_c \cdot (\mathbf{\phi})_c \right] \right\}$$

(3.53)

The algorithmic solution of this problem starts in the referential configuration and through the "push-forward" operations the constitutive equation solution is tackled for each of the composite phases from the referential configuration to the updated configuration. Each of these phases can have different types of behaviors. Like in the classic mixing theory, once the stress of each component is determined, it is necessary to obtain the composite material stress by equation (3.50) or (3.51), as well as the internal forces at each point of the material (see Figure 3.4). After determining the internal forces it is necessary to verify their equilibrium with the applied external forces.

3.7 Mixing theory modification for short length reinforcement

The classic mixing theory formulation is oriented towards large-fiber reinforced matrix materials in which the kinematic condition of the classic or generalized formulation is satisfied. However, as the fiber aspect relation[*] decreases, the fiber-matrix compatibility condition is no longer satisfied. This is due to the increasing sliding effect and the limited stresses transmission between the fiber and the matrix at the fiber ends. This situation creates a concentration of stresses and strains on the fiber and on the surrounding matrix as a result of the discontinuity. The fiber's "effectiveness" in the composite material decreases as their length decreases too. Figure 3.5 shows a matrix strain surrounding a discontinuum fiber embedded in it and subjected to a traction load parallel to the fiber.

In long-fiber reinforced composite materials, the strain in the matrix and the fibers is the same. Moreover, the stress along the reinforcement does not change except at its ends where the strain is lower than in the matrix. As for short-fiber reinforced matrix, this phenomenon is essential for the mechanical property determination of the composite. This problem has been studied by Cox (1952)[25], using the so called "shear delay analysis" based on the assumption that both the fiber and the matrix remain elastic.

This phenomenon is explained in Figure 3.5. In section AA, the whole composite strain is only due to the matrix strain. In section BB, precisely at the fiber end, experimental tests show that the stress transmission from the matrix to the fiber is carried out gradually with null stress at the tip and the stress is increased gradually along the fiber until both the matrix and the fiber strains are the same. According to this, the maximum value of the axial stress is found on the fiber's central zone. The *transfer length* l_c is defined as the reinforcement length required guaranteeing the fiber-matrix compatibility and the stress transfer from the matrix to the fiber. Any reinforcement length lower than this magnitude will not participate completely in the stress transfer mechanisms (Jayatilaka (1969)[26]).

---

[*] NOTE: It is s defined as the aspect relation with respect to coefficient $l/2r$ where l and r are the length and the radius of a short fiber, respectively.

[25] Cox, H. L. (1952). The elasticity and the strength of paper and other fibrous materials. Br. J. Appl. Phys., 3, 72 - 79.

[26] Jayatilaka, A. (1979). *Fracture of engineering brittle materials.* Applied Science Publishers.

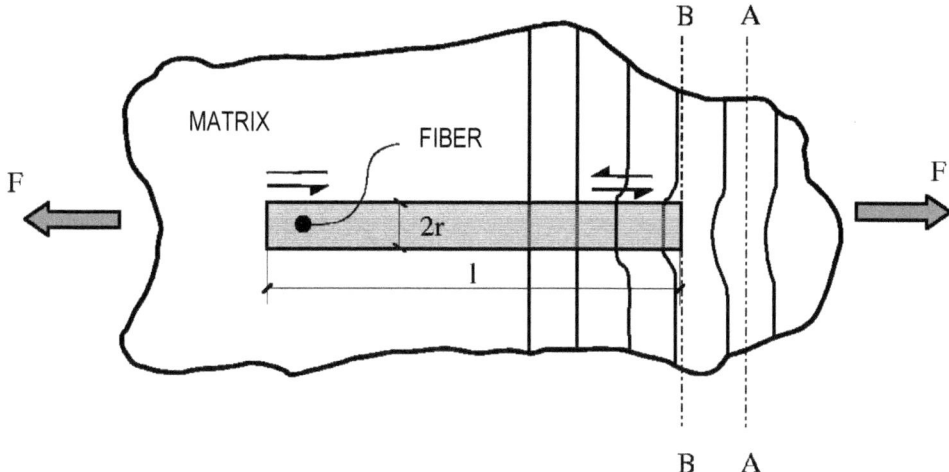

Figure 3.5 – Strain around a discontinuum fiber embedded in a matrix subjected to traction.

In Figure 3.6 the stress distribution in a reinforced fiber is shown. The maximum tangent stress is found at the fiber ends and it is almost zero on the central zone. As observed in the figure, the axial stress at the fiber's ends drops to zero, and results in an average stress on the fiber with length l lower than in the continuum fiber subjected to the same external loads. The reinforcement "effectiveness" decreases proportionally to the fiber length, as the whole fiber cannot work at maximum stress. Therefore, in short-fiber reinforced composite materials, the fiber's length l must be greater than the critical length transfer l_c in order to take advantage of their maximum capacity.

Due to these local phenomena, short-fiber reinforced composite materials do not satisfy completely the compatibility condition expressed in equation (3.36) such as the different strains between the matrix and the fibers. Consequently, and in order to represent the constitutive behavior of these materials, it is necessary to formulate another strain closure equation (Oller *et. alt.* (1995)[11]), or to maintain the classic mixing theory based on the strain equality hypothesis in all the components and to carry out a correction of each of the components' properties (Car *et al.* (1997)[27]).

---

[27] Car, E., Oller, S., & Oñate, E. (1997). An elastoplastic constitutive coupled model with mechanical and hygrometric damage. Applied to flexible pavements. U. de Brasilia (Ed.), *XVIII CILAMCE Congresso Ibero Latino-Americano de Métodos Computacionais Em Engenharia* (pp. 2100 - 2108). Brasilia.

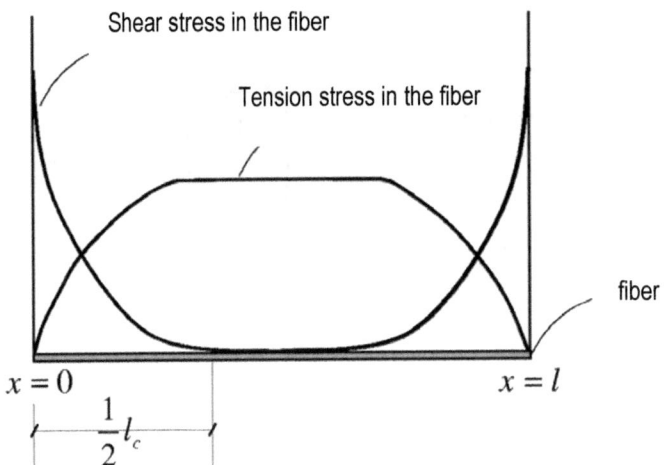

Figure 3.6 – Axial stress distribution in the fiber and shear fiber-matrix interphases.

3.7.1 Fiber axial stress distribution

As mentioned in previous sections, stress transfer is carried out from the matrix to the fibers in the interphase zone where the tangent stresses occur. For the determination of an analytical expression of the stress distribution on a fiber, it is necessary to consider the equilibrium in the stress transfer zone (see Figure 3.7).

The fiber equilibrium in the longitudinal direction x is given by:

$$\sigma_f \pi r^2 + 2\tau\pi r\, dx = (\sigma_f + d\sigma_f)\pi r^2 \quad\Rightarrow\quad \frac{\partial \sigma_f}{\partial x} = \frac{2\tau}{r} \tag{3.54}$$

Or, in terms of the fiber load,

$$\frac{\partial P_f}{\partial x} = 2\tau\pi r \tag{3.55}$$

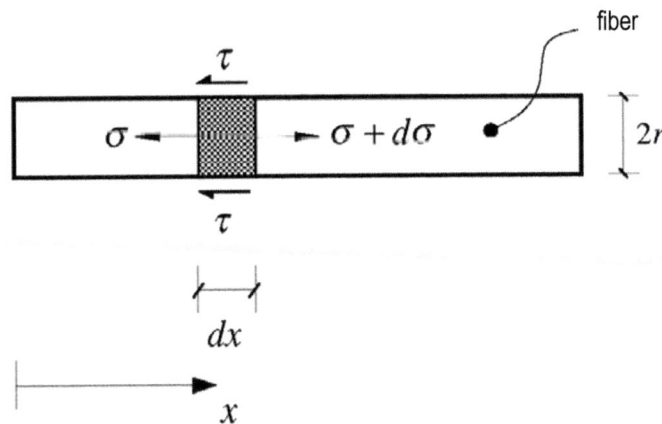

Figure 3.7 – Stresses at the fiber ends.

where σ_f is the fiber stress in the x direction, $d\sigma_f$ is the fiber stress increment in $x + dx$ and τ is the tangent stress in the fiber-matrix interphase. The tangent stress τ is produced due to the different strains between the fiber and the matrix and therefore it depends on the difference between the fiber and matrix displacement fields. The equilibrium between the matrix and the short fiber can be described through the following differential equation in the longitudinal fiber (Jayatilaka, (1979)[26]).

$$\frac{\partial^2 P_f}{\partial x^2} = H\left[\frac{P_f}{C_f^\sigma A_f} - E_m\right] \tag{3.56}$$

where P_f is the maximum interaction force between the reinforcement and the matrix, H is a constant that depends on the fiber's topology distribution, C_f^σ is the Young reinforcement modulus, A_f is the reinforcement's average transversal section and E_m is the longitudinal strain in the matrix. By solving the differential equation (3.56) the force in the fiber can be obtained,

$$P_f = C_1\, senh(\beta x) + C_2 \cosh(\beta x) + C_f^\sigma A_f E_m \tag{3.57}$$

C_1 and C_2 being the constants resulting from the boundary conditions $P_f = 0$ in $x = 0$ and $x = l$; β is a coefficient given by the following expression[26]

$$\beta = \sqrt{\frac{H}{C_f A_f}} = \sqrt{\frac{G_c}{C_f^\sigma}\frac{2\pi}{A_f \ln\left(\frac{r'}{r}\right)}} \tag{3.58}$$

in which G_c is the composite transversal elastic module and r' is the average distance between the reinforcement fibers (see Figure 3.8).

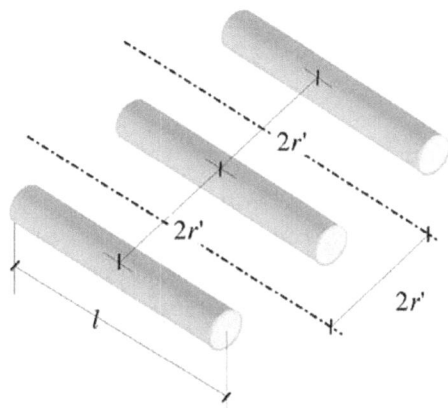

Figure 3.8 – Reinforcement aspect relation.

Once the integration constants are obtained, the resulting stress equation is,

$$\sigma_f(x) = C_f^\sigma E_m\left[1 - \frac{\cosh\left(\beta\left(\frac{l}{2} - x\right)\right)}{\cosh\left(\beta\frac{l}{2}\right)}\right] \qquad \forall\ 0 \le x \le \frac{l}{2} \tag{3.59}$$

This equation sets the axial stress distribution along the fiber. This distribution is shown schematically in Figure 3.6. There is no stress constant value in the reinforcement central zone but if the reinforcement is long enough, then the assumption that $\sigma_f \cong C_f E_m$ can be admitted. The maximum stress value is produced in $x = l/2$ and is given by

$$(\sigma_f)_{max} = \sigma_f(x = \frac{l}{2}) = C_f^\sigma E_m \left[1 - \frac{1}{\cosh\left(\beta\frac{l}{2}\right)} \right] \tag{3.60}$$

3.7.2 Tangent stress distribution in the interface

The tangent stress distribution in the interface zone is obtained by formulating the equilibrium (inside the fiber) between the axial stresses and the adherence to the matrix. Thus, taking into account equations (3.54) and (3.59), then:

$$\tau_f(x) = \frac{C_f \, Er\beta}{2} \, \frac{\text{senh}\left(\beta\left(\frac{l}{2} - x\right)\right)}{\cosh\left(\beta\frac{l}{2}\right)} \tag{3.61}$$

In the equation above the tangent stress distribution function is defined in the fiber-matrix interface. A schematic form of this distribution is shown in Figure 3.6. The shear stress value is zero in the fiber's central zone and coincides with the maximum value of the axial stress. There are no differentiated strains between the fiber and the matrix, which explains the zero value of the tangent stresses. The maximum tangent stress is obtained at the fiber end and is given by:

$$(\tau_f)_{max} = \tau_f(x = \frac{l}{2}) = \frac{C_f^\sigma \, Er\beta}{2} \, \tanh\left(\frac{\beta l}{2}\right) \tag{3.62}$$

One way to incorporate the short-fiber reinforced contribution to the mixing theory is through the average stress along the fiber, therefore:

$$\overline{\sigma}_f = \frac{1}{l}\int_0^l \sigma_f(x)\,dx = C_f^\sigma \left[1 - \frac{\tanh\left(\beta\frac{l}{2}\right)}{\left(\beta\frac{l}{2}\right)} \right] E_m = \widetilde{C}_f^\sigma E_m \tag{3.63}$$

$\widetilde{C}_f^\sigma$ being the average Young's modulus of the reinforcement or the homogenized modulus. Equation (3.63) shows that Young's modulus of fibers' reinforcement is a function of their length and other geometric parameters. In the large-fiber case, the average elastic module tends to Young's module reinforcement value, whereas in this case it is strongly affected by the capacity of the matrix-reinforcement interface to transfer stresses.

The definition of an average Young's modulus of the reinforcement with a lower magnitude than the real one gives some mechanical characteristics to the composite that depend not only on its own intrinsic properties but on the whole matrix-reinforcement properties. The interface properties among the components are key factors in their participation. In other words, the mechanical properties of a point in a solid do not depend only on themselves but on the whole set.

3.7.3 Short-fiber constitutive model

Short-fiber reinforced matrices of composite materials are generally subjected to higher stresses than long-fiber reinforced matrices. Generally, the mechanical properties of the short-fiber reinforced composite materials are lower than the continuum reinforcement of composite materials[*].

The previous definition for Young's modulus leads to the redefinition of the elastic part of the free energy function in the referential or updated configuration for the reinforcement phase as follows:

$$\Psi(\mathbf{E}^e,\theta,\alpha^m)=\Psi^e(\mathbf{E}^e,\theta)+\Psi^p(\alpha^m)=\frac{1}{2m^0}\left[\mathbf{E}^e:\widetilde{\mathbb{C}}_f^S:\mathbf{E}^e\right]+\Psi^p(\alpha^m)$$

$$\psi(\mathbf{e}^e,\theta,\alpha^m)=\psi^e(\mathbf{e}^e,\theta)+\psi^p(\alpha^m)=\frac{1}{2m}\left[\mathbf{e}^e:\widetilde{\mathbf{c}}_f^\tau:\mathbf{e}^e\right]+\psi^p(\alpha^m)$$

(3.64)

This same concept, based on the stress homogenization along the fiber (equation (3.63)), can be extended to "3-D" through a simplification, obtaining the following approximated constitutive tensor for the short fiber,

$$\widetilde{\mathbb{C}}_f^S=\mathbb{C}_f^S\left[1-\frac{\tanh\left(\beta\frac{l}{2}\right)}{\left(\beta\frac{l}{2}\right)}\right]$$

(3.65)

where the constitutive tensor of the reinforcement in the referential configuration $\mathbb{C}_f^S$ is orthotropic. This constitutive tensor transported to the updated configuration is affected by the fiber's length change, transversal section change and separation change of the reinforcement fibers and can be obtained by a "push forward" of the constitutive tensor in the referential configuration (equation (3.65)), thus:

$$\widetilde{\mathbf{c}}_f^\tau=\vec{\phi}(\widetilde{\mathbb{C}}_f^S)$$

(3.66)

Consequently, the formulation here presented can deal with the reinforcement effectivity loss in the response due to the short length that prevents the total transfer of the stresses from the matrix.

From the second principle of the thermodynamics, the stress in the referential or updated configurations is obtained as,

$$\left.\begin{array}{l}\mathbf{S}=m^0\dfrac{\partial\Psi}{\partial\mathbf{E}^e}=\underbrace{\left[1-\dfrac{\tanh\left(\beta\frac{l}{2}\right)}{\left(\beta\frac{l}{2}\right)}\right]}_{\varsigma}\cdot\mathbb{C}_f^S:\mathbf{E}^e=\widetilde{\mathbb{C}}_f^S:\mathbf{E}^e\\[2em]\boldsymbol{\tau}=m\dfrac{\partial\psi}{\partial\mathbf{e}^e}=\widetilde{\mathbf{c}}_f^\tau:\mathbf{e}^e\end{array}\right\}\quad\boldsymbol{\tau}=\vec{\phi}(\mathbf{S})$$

(3.67)

In these equations for the stress in the two configurations, factor ς represents the correction due to short length reinforcement in the composite material.

3.8 Composite constitutive equation

The analysis of short-fiber reinforced composite materials by the classic or generalized mixing theory is insufficient because it cannot take into account the phenomena occurring

---

[*] NOTE: Continuum reinforcement is considered that presenting a greater length than necessary for the stresses transfer from the matrix to the reinforcement.

in the interface zone among both materials. As mentioned before, any kinematic definition for the components' behavior is only valid for materials with continuum reinforcement such as long-fiber reinforced matrices, neglecting the phenomena occurring at the fibers' ends. For short-fiber reinforced composite materials the compatibility equation is not satisfied. Therefore, the classic mixing theory must be modified because it does not satisfy the compatibility equation, by defining a different closure equation (Oller *et al* (1995)[11]) or by a correction of the properties of each component keeping the closing equation of the classic mixing theory (Car *et al.* (1998)[28]). This method above leads to more straightforward formulation.

3.8.1 Free energy for short reinforced composite materials

The free energy of a fiber-reinforced composite material in the referential configuration is given by the sum of free energies of each of the phases that make up the matrix weighted as a function of their volumetric participation, therefore:

$$m^0 \, \Psi(\mathbf{E}^e, \theta, \alpha^m) = \underbrace{\sum_{c_m=1}^{n_m} k_{c_m} m_{c_m}^0 \, \Psi_{c_m} ((\mathbf{E})_{c_m}, (\mathbf{E}^p)_{c_m}, \theta, (\alpha^m)_{c_m}) +}_{\text{Matrix components}}$$

$$+ \underbrace{\sum_{c_r=1}^{n_r} k_{c_r} m_{c_r}^0 \, \varsigma_{c_r} \, \Psi_{c_r} ((\mathbf{E})_{r_m}, (\mathbf{E}^p)_{c_r}, \theta, (\alpha^m)_{c_r})}_{\text{Fiber components}}$$

(3.68)

the composite free energy in the updated configuration is obtained,

$$m \, \psi(\mathbf{e}^e, \theta, \alpha^m) = \underbrace{\sum_{c_m=1}^{n_m} k_{c_m} m_{c_m} \, \psi_{c_m} ((\mathbf{e})_{c_m}, (\mathbf{e}^p)_{c_m}, \theta, (\alpha^m)_{c_m}) +}_{\text{Matrix components}}$$

$$+ \underbrace{\sum_{c_r=1}^{n_r} k_{c_r} m_{c_r} \, \varsigma_{c_r} \, \psi_{c_r} ((\mathbf{e}^p)_{c_r}, (\mathbf{e}^p)_{c_r}, \theta, (\alpha^m)_{c_r})}_{\text{Fiber components}}$$

(3.69)

where Ψ_{c_m} and Ψ_{c_r} are the free energies of the components of the matrix and reinforcement in the referential configuration, ψ_{c_m} and ψ_{c_r} are the free energies of the components of the matrix and the reinforcement in the updated configuration, k_{c_m} and k_{c_r} are the coefficients of volumetric participation of the components of the matrix and the reinforcement, $(\mathbf{E})_{c_{m,r}} = \left[(1 - \chi_{c_{m,r}}) \cdot \mathbf{I}_4 + \chi_{c_{m,r}} \cdot (\mathbf{\Phi})_{c_{m,r}} \right] : \mathbf{E} - \chi_{c_{m,r}} \cdot (\hat{\mathbf{E}}^p)_{c_{m,r}}$ is the strain in the referential configuration and $(\mathbf{e})_{c_{m,r}} = \left[(1 - \chi_{c_{m,r}}) \cdot \mathbf{I}_4 + \chi_{c_{m,r}} \cdot (\phi)_{c_{m,r}} \right] : \mathbf{e} - \chi_{c_{m,r}} \cdot (\hat{\mathbf{e}}^p)_{c_{m,r}}$ the strain in the updated configuration, the first one for the matrix and the second one for the reinforcement phase. $(\mathbf{E}^p)_{c_m}$ and $(\mathbf{E}^p)_{c_r}$ are the plastic strains in the referential configuration of the components of the fiber, and matrix $(\mathbf{e}^p)_{c_m}$ and $(\mathbf{e}^p)_{c_r}$ are the plastic strains in the updated configuration of the components of the fiber and the matrix,

---

[28] Car, E., Oller, S., & Oñate, E. (1998). Un modelo constitutivo elasto plástico acoplado con daño mecánico e higrométrico. Aplicación a pavimentos flexibles. Rev. Int. de Ingeniería de Estructuras, 3(1), 19 - 37

$(\hat{\mathbf{E}}^P)_{c_{m,r}}$ and $(\hat{\mathbf{e}}^P)_{c_{m,r}}$ are the plastic strains defined for operative purposes (without any physical meaning), ς_{c_r} represents the correction as a function of the reinforcement phase length in the composite material, , $\alpha_{c_m}^m$ and $\alpha_{c_r}^m$ are the internal plastic variables of the components of the matrix and the reinforcement defining the physical behavior of each substance.

Short-fiber reinforcement is less efficient than long-fiber reinforcement, therefore it follows that the mechanical properties of short-fiber reinforced composite materials are lower. The expression of the constitutive tensor of the material shown in equation (3.11) in small strains is generalized for short reinforcements and large strains as follows:

$$\mathbb{C}_{ijkl}^S = \frac{\partial S_{ij}}{\partial E_{kl}} = m^0 \frac{\partial^2 \Psi(E_{ij}^e . \theta, \alpha_i)}{\partial E_{ij} \partial E_{kl}} =$$

$$= \underbrace{\sum_{c_m=1}^{n_m} k_{c_m} (\hat{\mathbb{C}}_{ijkl}^S)_{c_m}}_{\text{Matrix components}} + \underbrace{\sum_{c_r=1}^{n_r} k_{c_r} \varsigma_{c_r} (\hat{\mathbb{C}}_{ijkl}^S)_{c_r}}_{\text{Fiber components}}$$

(3.70)

And in the updated configuration the constitutive tensor is defined as,

$$c_{ijkl}^S = \frac{\partial \tau_{ij}}{\partial e_{kl}} = m \frac{\partial^2 \psi(e_{ij}^e . \theta, \alpha_i)}{\partial e_{ij} \partial e_{kl}} =$$

$$= \underbrace{\sum_{c_m=1}^{n_m} k_{c_m} (\hat{c}_{ijkl}^\tau)_{c_m}}_{\text{Matrix Components}} + \underbrace{\sum_{c_r=1}^{n_r} k_{c_r} \varsigma_{c_r} (\hat{c}_{ijkl}^\tau)_{c_r}}_{\text{Fiber components}}$$

(3.71)

When working with the "generalized mixing theory" the constitutive tensors for the matrix and the reinforcement are defined as:

$$(\hat{\mathbb{C}}^S)_{c_{m,r}} = \left\{ \left[(1 - \chi_{c_{m,r}}) \cdot \mathbf{I}_4 + \chi_{c_{m,r}} \cdot (\boldsymbol{\Phi})_{c_{m,r}} \right]^T : \langle \mathbb{C}^S \rangle_{c_{m,r}} : \left[(1 - \chi_{c_{m,r}}) \cdot \mathbf{I}_4 + \chi_{c_{m,r}} \cdot (\boldsymbol{\Phi})_{c_{m,r}} \right] \right\}$$

$$(\hat{\mathbf{c}}^\tau)_{c_{m,r}} = \left\{ \left[(1 - \chi_{c_{m,r}}) \cdot \mathbf{I}_4 + \chi_{c_{m,r}} \cdot (\boldsymbol{\phi})_{c_{m,r}} \right]^T : \langle \mathbf{c}^\tau \rangle_{c_{m,r}} : \left[(1 - \chi_{c_{m,r}}) \cdot \mathbf{I}_4 + \chi_{c_{m,r}} \cdot (\boldsymbol{\phi})_{c_{m,r}} \right] \right\}$$

And when using the "classic mixing theory" the constitutive tensors are defined simply as $(\hat{\mathbb{C}}^S)_{c_{m,r}} = (\mathbb{C}^S)_{c_{m,r}}$ and $(\hat{\mathbf{c}}^\tau)_{c_{m,r}} = (\mathbf{c}^\tau)_{c_{m,r}}$. In previous definitions the composite material constitutive tensors are represented in the referential and updated configurations, n_m is the number of component materials making up the composite matrix and n_r is the number of component materials that make up the reinforcement phase. For continuum reinforcement factor ς_{c_r} tends to the unit and makes the expression for short fibers coincide with the long-fiber one.

The stress equation in the composite material in the referential and updated configurations is defined as:

$$S_{ij} = m^0 \frac{\partial \Psi(E_{ij}^e, \theta, \alpha_i)}{\partial E_{ij}} = \sum_{c_m=1}^{n_m} \underbrace{k_{c_m} (S_{kl})_{c_m} \frac{\partial (E_{kl})_{c_m}}{\partial E_{ij}}}_{\text{Matrix components}} + \sum_{c_r=1}^{n_r} \underbrace{k_{c_r} (S_{kl})_{c_r} \frac{\partial (E_{kl})_{c_r}}{\partial E_{ij}}}_{\text{Fiber components}} = $$

$$= \sum_{c_m=1}^{n_m} \underbrace{k_{c_m} (\hat{\mathbb{C}}_{ijkl}^S)_{c_m} (E_{kl}^e)_{c_m}}_{\text{Matrix components}} + \sum_{c_r=1}^{n_r} \underbrace{k_{c_r} \varsigma_{c_r} (\hat{\mathbb{C}}_{ijkl}^S)_{c_r} (E_{kl}^e)_{c_r}}_{\text{Fiber components}}$$

(3.72)

$$\tau_{ij} = m \frac{\partial \psi(e_{ij}^e, \theta, \alpha_i)}{\partial e_{ij}} = \sum_{c_m=1}^{n_m} \underbrace{k_{c_m} (\tau_{kl})_{c_m} \frac{\partial (e_{kl})_{c_m}}{\partial e_{ij}}}_{\text{Matrix components}} + \sum_{c_r=1}^{n_r} \underbrace{k_{c_r} (\tau_{kl})_{c_r} \frac{\partial (e_{kl})_{c_r}}{\partial e_{ij}}}_{\text{Fiber components}} = $$

$$= \sum_{c_m=1}^{n_m} \underbrace{k_{c_m} (\hat{c}_{ijkl}^\tau)_{c_m} (e_{kl}^e)_{c_m}}_{\text{Matrix components}} + \sum_{c_r=1}^{n_r} \underbrace{k_{c_r} \varsigma_{c_r} (\hat{c}_{ijkl}^\tau)_{c_r} (e_{kl}^e)_{c_r}}_{\text{Fiber components}}$$

(3.73)

If the "classic mixing theory" is used to solve short-fiber reinforced composites, the material stiffness and strength modification can also be interpreted as a modification of the closure equation or compatibility equation of strains in the referential configuration, therefore:

$$E_{ij} = \underbrace{(E_{ij}')_{1_m} = (E_{ij}')_{2_m} = \cdots = (E_{ij}')_{n_m}}_{\text{Strain in the components of the matrix.}} \equiv \underbrace{(E_{ij}')_{1_r} = (E_{ij}')_{2_r} = \cdots = (E_{ij}')_{n_r}}_{\text{Strain in the components of the fiber.}}$$

$$E_{ij} = \underbrace{(E_{ij}^e + E_{ij}^p)_{1_m}}_{(E_{ij}')_{1_m}} = \cdots = \underbrace{(E_{ij}^e + E_{ij}^p)_{n_m}}_{(E_{ij}')_{n_m}} \equiv \underbrace{(\vartheta_{1_r} E_{ij}^e + E_{ij}^p)_{1_r}}_{(E_{ij}')_{1_r}} = \cdots = \underbrace{(\vartheta_{n_r} E_{ij}^e + E_{ij}^p)_{n_r}}_{(E_{ij}')_{n_r}}$$

(3.74)

From the equation above, the elastic strain of the reinforcement components is given by the following expression:

$$(E_{ij}^e)_{n_r} = \frac{1}{\vartheta_{n_r}} \left(E_{ij} - (E_{ij}^p)_{n_r} \right) \quad \Rightarrow \quad (E_{ij}^e)_{n_r} = \varsigma_{n_r} \left(E_{ij} - (E_{ij}^p)_{n_r} \right)$$

(3.75)

According to this equation the strains in each component -short fibers or particles- are affected by a $\vartheta_{n_r} = 1/\varsigma_{n_r}$ factor, as a function of the reinforcement length and the fibers' topology distribution within the composite matrix. Taking into account equation (3.44), the stress expression in each of the reinforcement components in the referential configuration is given by:

$$
\begin{aligned}
\mathbf{S} &= m^0 \frac{\partial \Psi(\mathbf{E}^e, \theta, \alpha^m)}{\partial \mathbf{E}} = \sum_{c=1}^{n} m_c^0 \, k_c \frac{\partial \Psi_c((\mathbf{E})_c, (\mathbf{E}^P)_c, \theta, (\alpha^m)_c)}{\partial (\mathbf{E})_c} : \frac{\partial (\mathbf{E})_c}{\partial \mathbf{E}} = \\
&= \sum_{c_m=1}^{n_m} k_{c_m} \, (\mathbf{S})_{c_m} : \frac{\partial (\mathbf{E})_c}{\partial \mathbf{E}} + \sum_{c_r=1}^{n_r} k_{c_r} \, (\mathbf{S})_{c_r} : \frac{\partial (\mathbf{E})_c}{\partial \mathbf{E}} = \\
&= \sum_{c_m=1}^{n_m} k_{c_m} \, (\hat{\mathbb{C}}^S)_{c_m} \, (\mathbf{E}^e)_{c_m} + \sum_{c_r=1}^{n_r} k_{c_r} \, \varsigma_{c_r} \, (\hat{\mathbb{C}}^S)_{c_r} \, (\mathbf{E}^e)_{c_r} \\
&= \sum_{c_m=1}^{n_m} k_{c_m} \, (\hat{\mathbb{C}}^S)_{c_m} \left(\mathbf{E} - (\mathbf{E}^P)_{c_m} \right) + \sum_{c_r=1}^{n_r} k_{c_r} \, \varsigma_{c_r} \, (\hat{\mathbb{C}}^S)_{c_r} \left(\mathbf{E} - (\mathbf{E}^P)_{c_r} \right)
\end{aligned}
\tag{3.76}
$$

The change of properties of each component in the updated configuration can also be interpreted as a modification in the closure equation:

$$
\begin{aligned}
e_{ij} &= \underbrace{\left(e'_{ij}\right)_{1_m} = \left(e'_{ij}\right)_{2_m} = \cdots = \left(e'_{ij}\right)_{n_m}}_{\text{Strain in the components of the matrix}} \equiv \underbrace{\left(e'_{ij}\right)_{1_r} = \left(e'_{ij}\right)_{2_r} = \cdots = \left(e'_{ij}\right)_{n_r}}_{\text{Strain in the components of the fiber}} \\[2mm]
e_{ij} &= \underbrace{\left(e_{ij}^e + e_{ij}^p\right)_{1_m}}_{\left(e'_{ij}\right)_{1_m}} = \cdots = \underbrace{\left(e_{ij}^e + e_{ij}^p\right)_{n_m}}_{\left(e'_{ij}\right)_{n_m}} \equiv \underbrace{\left(\vartheta_{1_r} e_{ij}^e + e_{ij}^p\right)_{1_r}}_{\left(e'_{ij}\right)_{1_r}} = \cdots = \underbrace{\left(\vartheta_{n_r} e_{ij}^e + e_{ij}^p\right)_{n_r}}_{\left(e'_{ij}\right)_{n_r}}
\end{aligned}
\tag{3.77}
$$

From this latter it follows that the elastic strain of the reinforcement components is given by the following expression:

$$
\left(e_{ij}^e\right)_{n_r} = \frac{1}{\vartheta_{n_r}} \left(e_{ij} - (e_{ij}^p)_{n_r}\right) \quad \Rightarrow \quad \left(e_{ij}^e\right)_{n_r} = \varsigma_{n_r} \left(e_{ij} - (e_{ij}^p)_{n_r}\right)
\tag{3.78}
$$

According to this equation, like its counterpart equation in the referential configuration, the strains in each reinforcement component -short fibers or particles- are affected by a $\vartheta_{n_r} = 1/\varsigma_{n_r}$ factor as a function of the reinforcement length and fiber topology distribution within the composite matrix. Taking into account equation (3.46), it can be noted that the stress expression in each of the reinforcement components in the referential configuration is given by the following expression:

$$
\begin{aligned}
\boldsymbol{\tau} &= m \frac{\partial \psi(\mathbf{e}^e, \theta, \alpha^m)}{\partial \mathbf{e}} = \sum_{c=1}^{n} m_c k_c \frac{\partial \psi_c(\mathbf{e}, (\mathbf{e}^P)_c, \theta, (\alpha^m)_c)}{\partial (\mathbf{e})_c} : \frac{\partial (\mathbf{e})_c}{\partial \mathbf{e}} = \\
&= \sum_{c_m=1}^{n_m} k_{c_m} \, (\boldsymbol{\tau})_{c_m} : \frac{\partial (\mathbf{e})_c}{\partial \mathbf{e}} + \sum_{c_r=1}^{n_r} k_{c_r} \, (\boldsymbol{\tau})_{c_r} : \frac{\partial (\mathbf{e})_c}{\partial \mathbf{e}} = \\
&= \sum_{c_m=1}^{n_m} k_{c_m} \, (\hat{\mathbf{c}}^\tau)_{c_m} \, (\mathbf{e}^e)_{c_m} + \sum_{c_r=1}^{n_r} k_{c_r} \, \varsigma_{c_r} \, (\hat{\mathbf{c}}^\tau)_{c_r} \, (\mathbf{e}^e)_{c_r} \\
&= \sum_{c_m=1}^{n_m} k_{c_m} \, (\hat{\mathbf{c}}^\tau)_{c_m} \left(\mathbf{e} - (\mathbf{e}^P)_{c_m} \right) + \sum_{c_r=1}^{n_r} k_{c_r} \, \varsigma_{c_r} \, (\hat{\mathbf{c}}^\tau)_{c_r} \left(\mathbf{e} - (\mathbf{e}^P)_{c_r} \right)
\end{aligned}
\tag{3.79}
$$

Taking into consideration the compatibility condition given in equations (3.74) and (3.77), and the stress expression in the referential and updated configurations given by equations (3.76) and (3.79), it can be observed that there is a proportionality factor between the strain of the reinforcement phase components and the strain of the whole set. The product of this proportionality factor by the component constitutive tensor leads to a modification of the constitutive tensor of the component $\varsigma_{c_r} \hat{\mathbf{c}}_{c_r}$. A new constitutive tensor is then obtained for the reinforcement component and the compatibility condition of the classic mixing theory for the component stress calculation can be used.

3.9 Fiber mechanical properties in the mixing theory – linear behavior in small strains

The purpose of this section is to set the mechanical properties of the reinforcement phase in transversal direction to the stress' main direction. The mixing theory of basic substances in its decoupled form for small strains ($\sigma = \tau = S$ and $\varepsilon = e = E$) establishes that the transversal elastic module of a composite material is given by the weighted sum of the transversal elastic modules of the components,

$$\mathbb{C}_2 = k_m (\mathbb{C}_2)_m + k_f (\mathbb{C}_2)_f \tag{3.80}$$

where $\mathbb{C}_2$, $(\mathbb{C}_2)_m$ and $(\mathbb{C}_2)_f$ are transversal elastic modules of the composite, the matrix and the fiber, respectively. The assumption of a parallel material must be completed by a transversal elastic module of a serial material where all the stresses are the same $\left[(\sigma_2)_f = (\mathbb{C}_2)_f : (\varepsilon_2)_f\right] \equiv \left[(\sigma_2)_m = (\mathbb{C}_2)_m : (\varepsilon_2)_m\right] \equiv \left[\sigma_2 = \mathbb{C}_2 : \varepsilon_2\right]$, from where the following is obtained:

$$\varepsilon_2 = k_m (\varepsilon_2)_m + k_f (\varepsilon_2)_f \quad \Rightarrow \quad \mathbb{C}_2 = \frac{(\mathbb{C}_2)_m \cdot (\mathbb{C}_2)_f}{k_m (\mathbb{C}_2)_f + k_f (\mathbb{C}_2)_m} \tag{3.81}$$

Experimental evidence shows that for the behavior representation of fiber reinforced matrix composites, the fibers cannot be considered as material which only has longitudinal stiffness. Thus, the reinforcement in transversal direction with respect to a given stress enhances the stiffness of the whole set. The transversal elastic module of the reinforcement is obtained by considering the equations equality (3.80) and (3.81) for the one-dimensional case

$$(\mathbb{C}_2)_f \cong \frac{\dfrac{(\mathbb{C}_2)_m \cdot (\mathbb{C}_{2\approx1})_f}{k_m (\mathbb{C}_{2\approx1})_f + k_f (\mathbb{C}_2)_m} - k_m (\mathbb{C}_2)_m}{k_f} \tag{3.82}$$

where $(\mathbb{C}_{2\approx1})_f$ represents a first approximation of the transversal elasticity module through its length size $(\mathbb{C}_2)_f \approx (\mathbb{C}_1)_f$. Another approximation of the transversal elastic module can be obtained by taking into account a modification of the previous equation which considers Poisson's effect produced by the fiber's lateral contraction.

$$(\mathbb{C}_2)_f \cong \frac{\dfrac{(\mathbb{C}'_2)_m \cdot (\mathbb{C}_{2\approx1})_f}{k_m (\mathbb{C}_{2\approx1})_f + k_f (\mathbb{C}'_2)_m} - k_m (\mathbb{C}'_2)_m}{k_f} \tag{3.83}$$

where $(\mathbb{C}_2')_m = (\mathbb{C}_2)_m / 1 - v^2$. Analogously, taking into consideration the equation proposed by Halpin and Tsai[29] the fiber transversal elastic module is given by

$$(\mathbb{C}_2)_f \cong \frac{(\mathbb{C}_2)_m \dfrac{1 + \xi \eta k_f}{1 - \eta k_f} - k_m (\mathbb{C}_2)_m}{k_f} \tag{3.84}$$

where η is a coefficient that results from a function of the matrix's elastic module, the reinforcement longitudinal module and an experimental parameter ξ (Hull, (1987)[30]), (Barbero, (1998)[31]):

$$\eta \cong \frac{\dfrac{(\mathbb{C}_1)_m}{(\mathbb{C}_2)_m} - 1}{\dfrac{(\mathbb{C}_1)_m}{(\mathbb{C}_2)_m} + \xi} \tag{3.85}$$

3.10 Comparative example. "Micromodel" vs. "mixing theory" with anisotropy in large strains

This example has been chosen among others because it shows the capacities of the "classic mixing theory with anisotropy and large strains", presented in previous sections, as compared to the results obtained by a micromodel where each of the component materials is individualized.

The example consists in subjecting a unit size structure of a composite material to traction in which the reinforcement and matrix phases are discretized. Then the results obtained by this micromodel are compared to the results obtained by the macro module proposed work. In Figure 3.9 the unit size piece is shown, where both phases of the composite material have been discretized. The boundary conditions imposed can also be observed. The finite element mesh is set up by 5701 triangular finite elements of 3 nodes and 2940 nodes.

As an alternative to the mesh described before for the numerical simulation through the model previously described, the same unit size piece modeled by only one single finite element of 4 nodes and 2 x 2 points of integration is analyzed. The sliding phenomenon between the fiber and matrix -fiber-matrix displacement (FMD)- or debounding, will not be considered in this example as it will be described further in this book.

The mechanical properties of the materials making up the composite are shown in Table 3.1 and Table 3.2.

---

[29] Halpin J. and Tsai S. W. (1969). *Effects of environmental factors on composite materials.* Air Force Materials Lab, No. 67-423.

[30] Hull, D. (1987). *Materiales compuestos.* Editorial Reverté, España.

[31] Barbero, E. J. (1998). *Introduction to Composite Materials Design.* Taylor and Francis.

Young module	13,00 Mpa
Poisson coefficient	0,325
Yield stress	43,323 Mpa
Post-yield behavior law	Exponential with softening
Fracture energy	10 N/m
V_m	76%

Table 3.1 – Epoxy resin properties, macromodel and micromodel.

Young module	239,551 Mpa
Poisson coefficient	0,0
Yield stress	3000 Mpa
Post-Yield behavior law	Linear with hardening
V_f	24%

Table 3.2 – Fiber carbon properties, macromodel and micromodel.

The micromodel is made up of two materials: fiber and matrix. They are considered isotropic and homogenous and have mechanical properties that coincide with the mechanical properties of the macromodel components. In Figure 3.10 the micromodel material distribution is shown schematically.

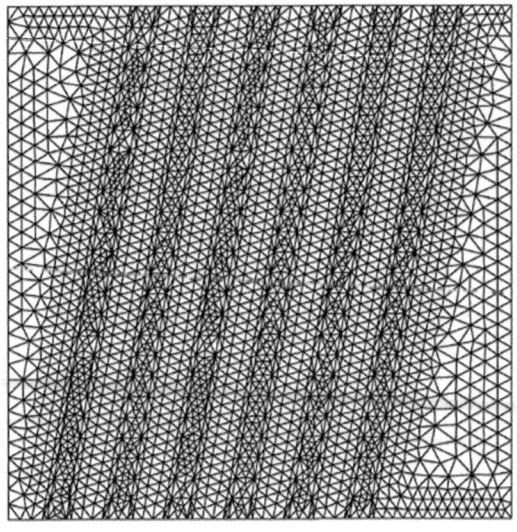

Figure 3.9 – Micromodel Finite element mesh.

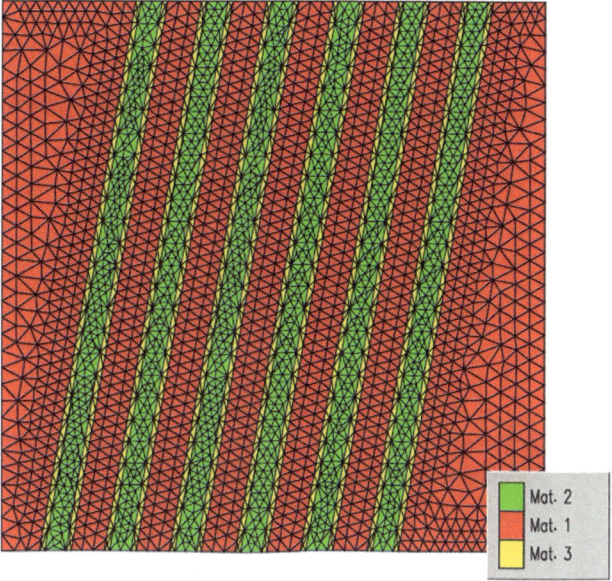

Figure 3.10 – Micromodel materials.

The numerical test consists in imposing displacements on the upper part of the unit size structure producing traction. This traction on the specimen leads to the reinforcement alignment with the load direction. In Figure 3.11 the deformed shape in its final state is shown. It can also be observed that the fibers have aligned themselves with the direction of the applied stress. This alignment of the reinforcement phase with the stress direction makes it necessary to introduce the theory of large strains in the constitutive proposed model (see section 3.5).

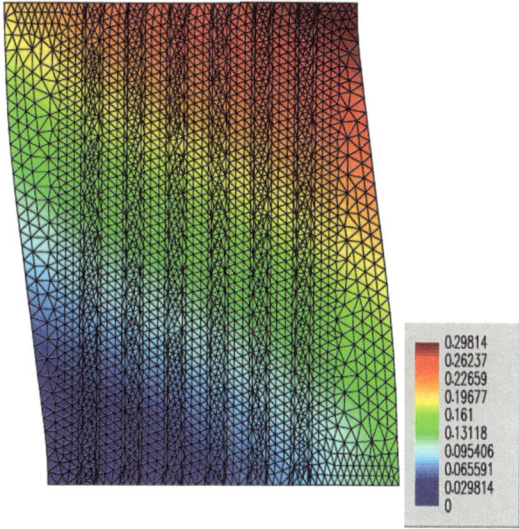

Figure 3.11 – Micromodel displacements contours and deformed shape.

The advantage of using a micromodel is that a detailed analysis of the mechanical processes can be done during the load application. Figure 3.11 shows the shear stress on the material for different loading cases. Figure 3.12-1 shows the stress states in a loading phase in which the stresses above the elastic limits of the component materials are not verified (see the plasticity internal variable in Figure 3.14). It can also be observed in the same figure that the matrix zone among the fibers is the one presenting a higher tensional state. As the displacements increase (Figure 3.12-2, Figure 3.12-3 and Figure 3.12-4) a homogenization of the matrix stress state is observed.

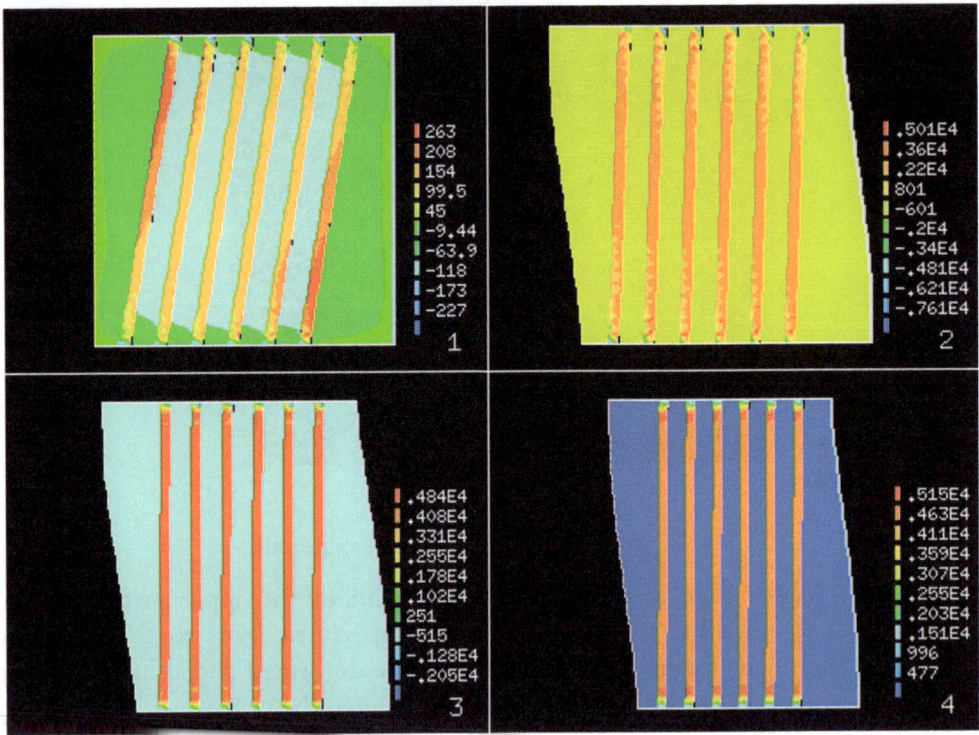

Figure 3.12 – Stress contours σ_{xy} for different loading stages.

Figure 3.13 shows the stresses in the micromodel in the direction of the imposed displacements. Figure 3.13-1 corresponds to a stress state in a loading step in which the composite materials stresses above the elastic limit are not verified (see Figure 3.14). Figure 3.13-2, Figure 3.13-3 and Figure 3.13-4 show the stress state in the direction of the imposed displacement as displacements increase. It can also be observed that in the first loading steps the matrix has a homogenous stress state in the direction of the applied stresses. Figure 3.13-2 shows that the reinforcement increases considerably as it aligns itself with the direction of the stress applied.

Figure 3.14 shows the plasticity contours in each composite component. It also shows that as the displacement increases, the irreversible strains in the matrix are verified in the areas between reinforcements (see Figure 3.14-2 and Figure 3.14-3). In Figure 3.14-4 it can be observed that the elastic limit has been exceeded, consequently leading to irreversible strains.

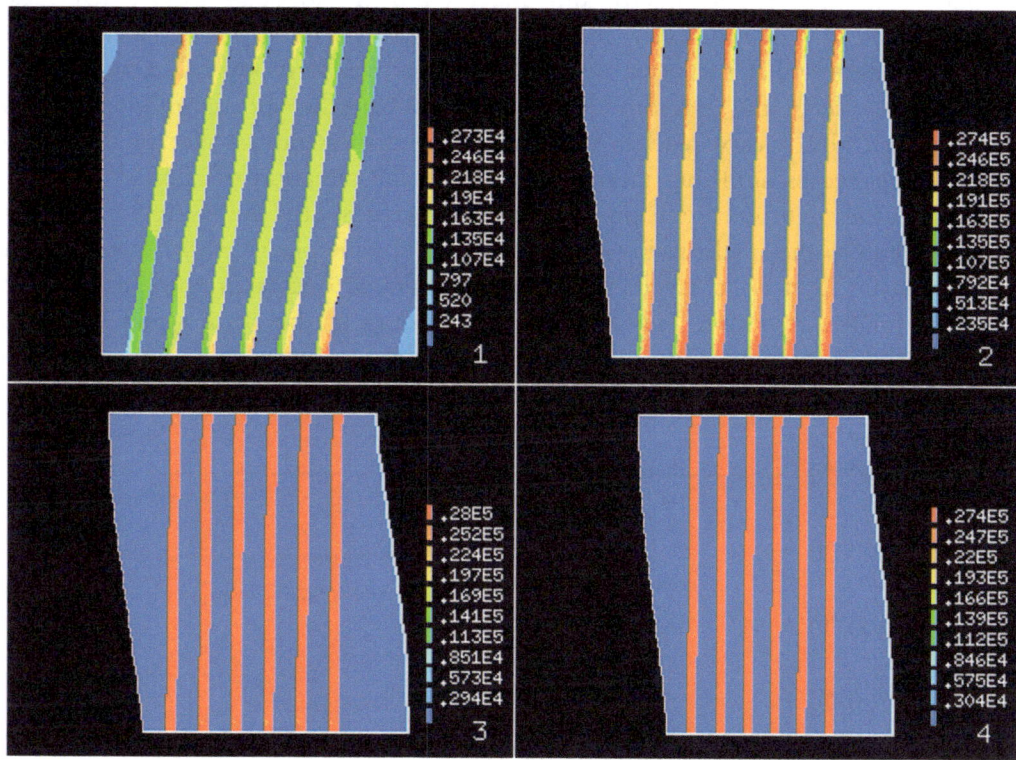

Figure 3.13 – Stress contours σ_{yy} for different loading stages.

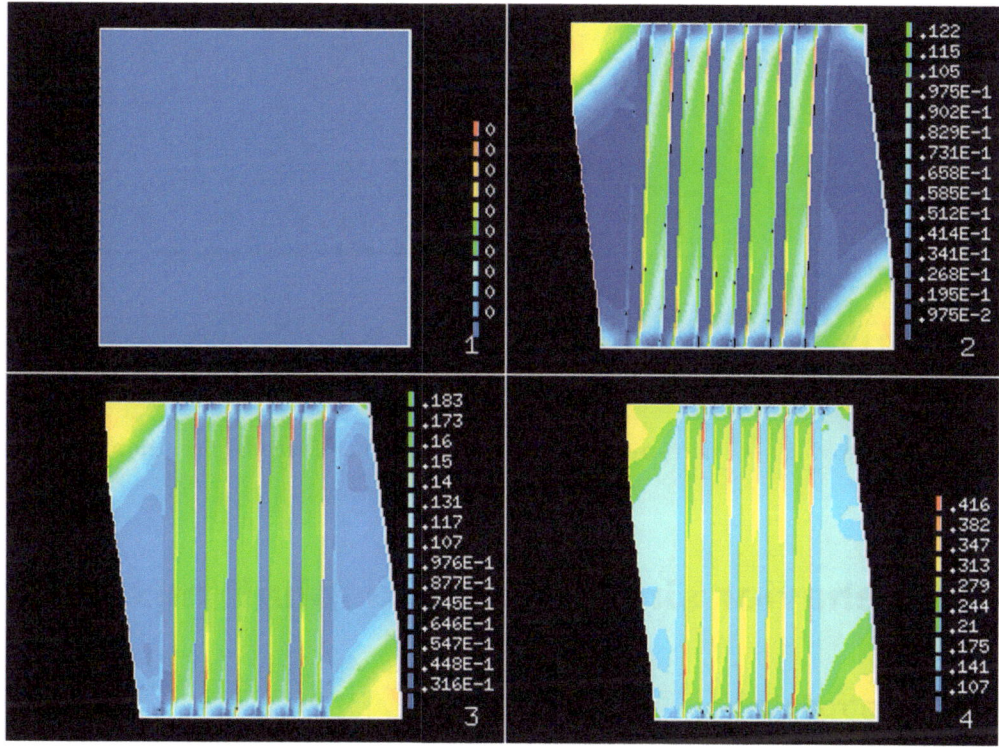

Figure 3.14 – Internal plasticity variable contours for different loading stages.

Figure 3.15 shows the micro and macro models loading-displacement response. Different values of the transversal module of the reinforcement phase are considered. The same figure shows that the value of the transversal elastic module of this phase plays a fundamental role in the macromodel response. When the shear modulus is zero, it is observed that the matrix reaches its limit of proportionality while the tension in the composite decreases until the fibers coincide with the direction of applied stress. Beyond this point, the reinforcement phase provides stiffness to the system. The response corresponding to the small strains assumption can also be observed. In this case, once the matrix's elastic limit is achieved, the material response decreases and the fibers do not participate in the response. This is because according to the small strain hypothesis the geometry is not updated and consequently the fibers cannot align themselves with the applied stress direction.

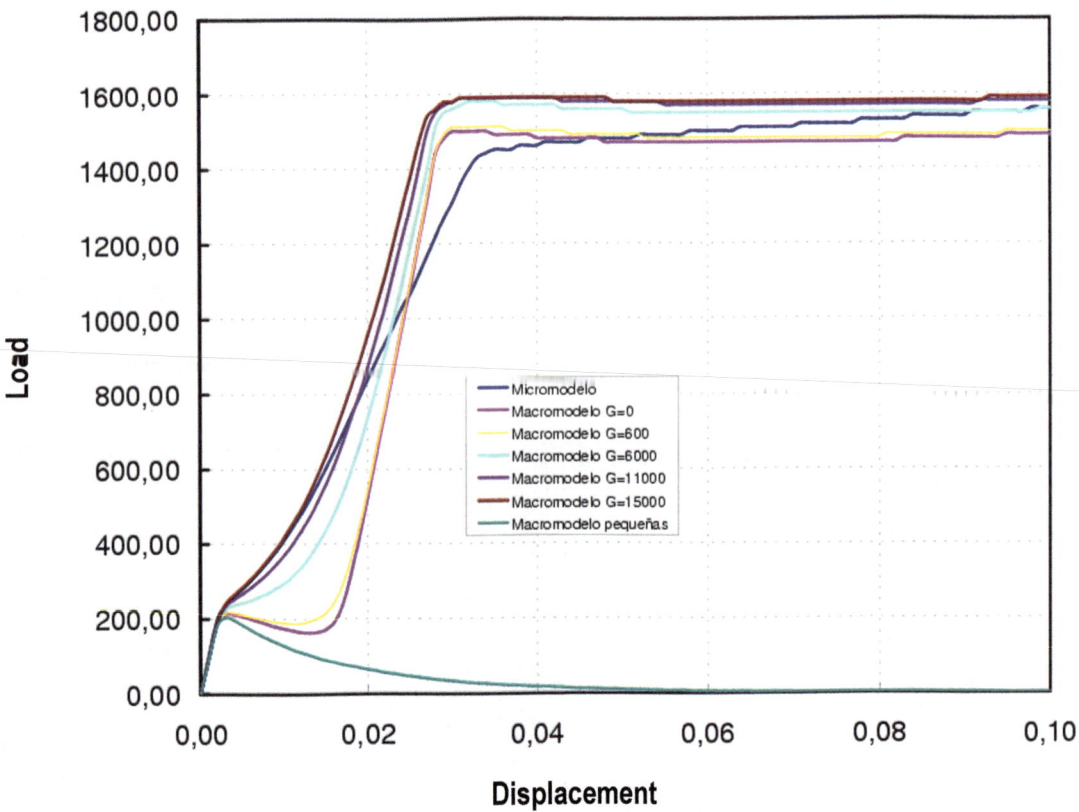

Figure 3.15 – Micro-macro model comparison. Load-displacement curves.

3.11 Behavior simulation of asphalt mixtures

3.11.1 Introduction

The classic mixing theory previously presented and modified for short length reinforcements is quite general and is suitable for the analysis of different types of composite materials such as reinforced concrete, asphalt mixtures and any other type of material made up by two or more phases.

The numerical simulation of a typical test used for asphalt mixtures characterization (Car et al., (1997)[27]) is presented.

A Marshall[32] test has been reproduced numerically for the behavior simulation of asphalt mixtures. This test can determine the strength to the "plastic strain of bituminous mixtures". The lab test procedure can be used for both the lab mixture project and the post field follow-up of the mixture manufacturing.

The Marshall test consists in determining the strength of a cylindrical specimen of 4" inches of diameter (101,6 mm) and 2,5" inches of height (63,5 mm), subjected to a load applied along the diameter at a temperature of approximately 60 ^{0}C. Two values can be determined with this test:

-    Stability: load required to produce the specimen failure.

-    Strain: Specimen diameter reduction expressed in [mm] from the beginning to the failure moment.

The load-displacement curves can also be carried out in the test which can be used to determine the stability and strain values previously defined. For details, see the standard NLT-159/86.

3.11.2 Problem motivation and description

The purpose of this section is to show numerically the behavior of a composite material (bituminous mixture) subjected to service loads using a constitutive model previously developed. Moreover, it has been assumed that the bituminous mixture undergoes an exogenous damage to the mechanical process caused by a non-mechanical potential (chemical problems, hygrometric, etc.). This bituminous mixture is made up by one reinforcement phase (arids) and one matrix (bitumen). The capacity evolution of the bituminous mixture is non-linear due to the exogenous agent, producing degradation in the mechanical characteristics of one or more phases of the composite, is coupled itself to the mechanical problem.

Three bituminous mixtures have been considered in this study (Mixture 1, Mixture 2, Mixture 3), according to the granulometric curve and bitumen percentage being incorporated to the mixture (see Table 3.3, Figure 3.16 and Figure 3.17). The granulometric characteristics of the arids of each mixture are a function of its position in the road surface profile.

	% weight particles to pass through		
Diameter [mm]	Mixture 1	Mixture 2	Mixture 3
25	90	100	100
20	78	94	100
12,5	56	71	82
10	52	67	65
5	36	50	41
2,5	25	36	14

---

[32] Yoder E. and Witczak M., (1975). *Principles of Pavement Design* - John Wiley & Sons. USA.

0,63	13	19	8
0,32	8	12	7
0,16	6	8	6
0,08	4	6	5
Ciego	0	0	0

Table 3.3 – Arid granulometry of the different mixtures.

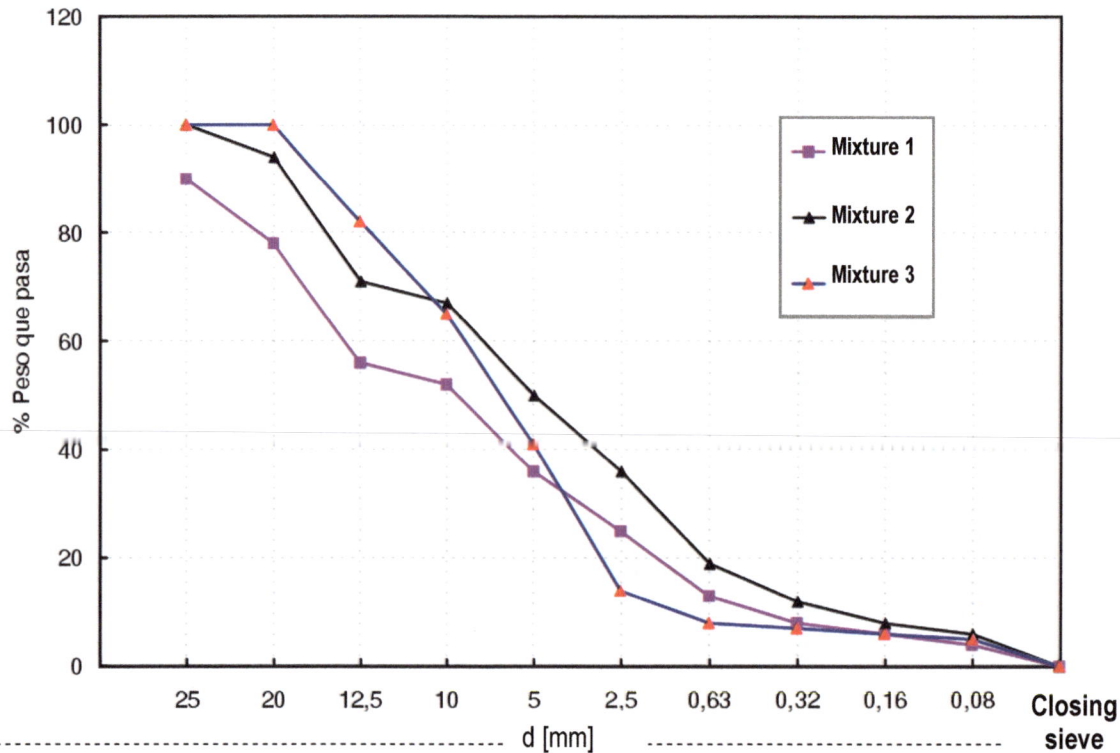

Figure 3.16 – Arid granulometric curve of different mixtures.

For the numerical simulation of the material's constitutive behavior the arids granulometric curve has been simplified and divided into two parts. The first part is comprised by arids with sizes ranging from 25 mm to 5 mm sieves (gravel and pebble) and the second part is comprised of the arids exceeding 5 mm sieve (sands). In each of these two subdivisions of the granulometric curve, different percentages of two types of arids have been considered, some of which are changeable against non-mechanical external agents (soft particles) and others non-changeable against external non-mechanical agents. Besides these two types of arids that make up each mixture, there is the bitumen part participating in different proportions in each case (see volume participation in Figure 3.16).

For the mechanical behavior simulation of an asphalt mixture through the developed model, it has been assumed that the different mixtures are made up in five phases:

-   **Reinforcement in the upper part of the granulometric curve** (gravel and pebble) likely to present degradation in their mechanical properties due to the presence of a non-mechanical external potential (soft particles).

- **Reinforcement in the upper part of the granulometric curve** (gravel and pebble) not presenting degradation in their mechanical properties due to the presence of an external non-mechanical potential (hard particles).

- **Reinforcement in the upper part of the granulometric curve** (gravel and pebble) likely to present degradation in their mechanical properties due to the presence of an external non-mechanical potential.

- **Reinforcement in the lower part of the granulometric curve** (sand) not presenting degradation in their mechanical properties due to the presence of an external non-mechanical potential.

- **Bitumen**

d [mm]	%weight particles to pass through			% retained accumulated weight			% partial accumulated weight		
	Mixture 1	Mixture 2	Mixture 3	Mixture 1	Mixture 2	Mixture 3	Mixture 1	Mixture 2	Mixture 3
20	78	94	100	22	6	0	22	6	0
12,5	56	71	82	44	29	18	22	23	18
5	36	50	41	64	50	59	20	21	41
0,32	8	12	7	92	88	93	28	38	34
0,08	4	6	5	96	94	95	4	6	2
closing sieve	0	0	0	100	100	100	4	6	5
bitumen	3,7	4,4	4,6						

Tabla 3.4 – Simplified granulometry of the arids of the different mixtures.

3.11.3 Material parameterization. Simplified granulometry and property correction by aspect relation

Mixture 1 has been called the granulometric curve used for the base layer, Mixture 2 for the intermediate layer and mixture 3 for the rolled asphalt layer. These bituminous mixtures can be considered as short reinforced composite materials. In other words, it is considered as a reinforcement of short length which cannot transfer the stresses from the matrix (bitumen) to the reinforcement (arids). The arids present in the bituminous mixtures have two types that can be clearly identified by their aspect. The first type corresponds to the soft particles likely to be affected by moisture producing a degradation of stiffness, which depends among other phenomena of evolution of the degradation variable by the humidity action. In the second type of arid, hard particles are not affected by the humidity action.

For the sake of simplicity, the diameters participating in each granulometric curve are summarized in Figure 3.16. According to their shape, separation, granulometric mechanical properties and their participation in the composite, the correction of the properties of the arids are derived. This correction follows the criteria described in Car et al., (1997)[27]. In Tabla 3.4 the mechanical properties and the correction of the reinforcement elastic modulus are shown, taking into account its length and distribution in the composite. For Young's modulus of the stone aggregated in natural state the value $\mathbb{C}_a = 300000$ kp/cm^2 is adopted and corresponds to a simple, homogenous and continuum material. In this case the material participates of its properties by adherence and stress transmission conditions between the matrix and the arid. Thus, the following $\mathbb{C}_a^{corr} = 1559,131$ kp/cm^2 is obtained, while Poisson's modulus is kept constant in $\nu = 0,15$. Young's modulus adopted for the bitumen is $\mathbb{C}_b = 300000 \ \text{kp/cm}^2$.

Correlation for short fiber $C_b = 80, 8529$									
L	r'	r	A	c_{mat}	ν_{mat}	G_{mat}	β	c_o	c_c
2	1,05	1,00	0,7854	80,375	0,4	28,705	0,12525	0,005197	1559,1311
1,25	0,65625	0,625	0,30679	80,375	0,4	28,705	0,20014	0,005197	1559,1311
0,5	0,2625	0,25	0,04908	80,375	0,4	28,705	0,50102	0,005197	1559,1311
0,032	0,0168	0,016	0,000201	80,375	0,4	28,705	7,82852	0,005197	1559,1311
0,008	0,0042	0,004	1,2566e-5	80,375	0,4	28,705	31,3140	0,005197	1559,1311
0,0001	0,0000525	0,00005	1,9635e-9	80,375	0,4	28,705	2505,12	0,005197	1559,1311

Table 3.5 – Mechanical parameters of the components and their correction.

The mechanical properties of the bitumen depend strongly on temperature. The numerical simulation has been obtained from a parameterization of the Marshall test of the mixture, taking into account its volumetric participation and the mixing theory (Car et al., (1998)[28]).

The threshold stress of the initial inelastic behavior is also obtained from a parameterization with the Marshall test. For example, for mixture 2, it is accepted as the initial point of the non linear behavior $\sigma_{Marshall}^{Y} - 20 \ \text{kg/cm}^2$. This threshold is strongly conditioned by the plasticity of the bitumen. After starting the test, a strong turning point is observed and immediately after the inelastic behavior of the arid starts. This occurs while the bitumen shows an excessive strain that leads to the contact between arids. Consequently, and based on the composite compatibility condition, the plastification threshold of the bitumen can be obtained for the moment when the non-linear behavior of the mixture starts.

The arid damage threshold is produced when the plastic behavior of the bitumen has progressed after the first threshold of the mixture inelastic behavior.

3.11.4 Numerical simulation

For the numerical simulation of the test a 101.6 mm-diameter circumference and two circumference sectors made of steel have been discretized to simulate the clamps of the machine used in the test. The finite element mesh used has 431 nodes and 776 triangular linear finite elements considering a plane stress state, therefore allowing the material to flow along the axial direction (see Figure 3.17).

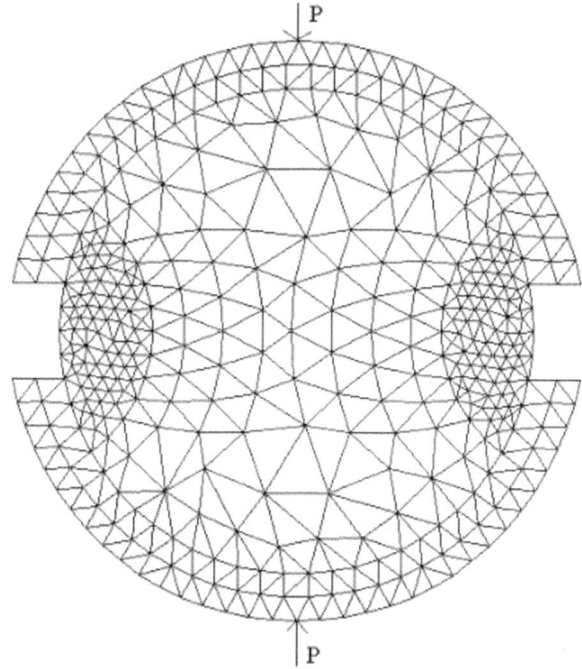

Figure 3.17 – Finite element mesh of the Marshall test for asphalt mixtures.

The outer cylinder cap is made of steel. Between the steel and the bituminous mixture, friction sliding is considered by controlling the shear stress between both materials.

From the test the load-displacement curves are obtained, by which the stability and strain values defined in the standard NLT-159/86 can be determined.

The basic substance mixing theory has been used in the numerical simulation. The intermediate layer C-2 and base C-1 of the road surface pavement has been modeled as composite materials with three basic components each:

Mixture 1:

- **Material M-1.a** corresponds to the arids ranging from 20 mm and 5 mm of the granulometric curve C-1. It participates in 58,62%, shows an initial damage as a function of humidity and is mechanically degradable.

- **Material M-1.b** corresponds to the arids ranging from 0.32 mm and the blind sieve of the granulometric curve C-1. It participates in 32.97% and is degradable by mechanical problems but not by humidity

- **Material M-1.c** corresponds to bitumen. This material participates in 8,41% of the mixture volume and has an elastoplastic behavior.

Mixture 2:

- **Material M-2.a** corresponds to the arids ranging from 20 to 5 mm of the granulometric C-2. It participates in 45.02%, shows an initial damage as a function of humidity and is mechanically degradable.

- **Material M-2.b** corresponds to the arids ranging from 0.32 mm to the blind sieve of the granulometric curve C-2. Its participation is 45.02% and it is degradable by mechanical problems but not by humidity.

-   **Material M-2.c** corresponds to the bitumen. This material's participation is 9.96% of the mixture volume and shows an elastoplastic behavior.

The finite element mesh and the materials used for the numerical simulation are shown in Figure 3.18.1. Figure 3.18-2, Figure 3.18-3 andFigure 3.18-4 show the evolution contours of the damage in the mixture M-2 for three different states of the loading process. Particularly Figure 3.18-2 shows the mechanical damage in mixture M-2 based on a null initial hygrometric damage. The scale observed shows the mechanical damage of all the material. Figure 3.18-4 shows the mechanical damage in mixture M-2 considering the volume of stony aggregate susceptible to damage, that not participate in the loading capacity of the mixture. In other words, it has an initial damage of 100%.

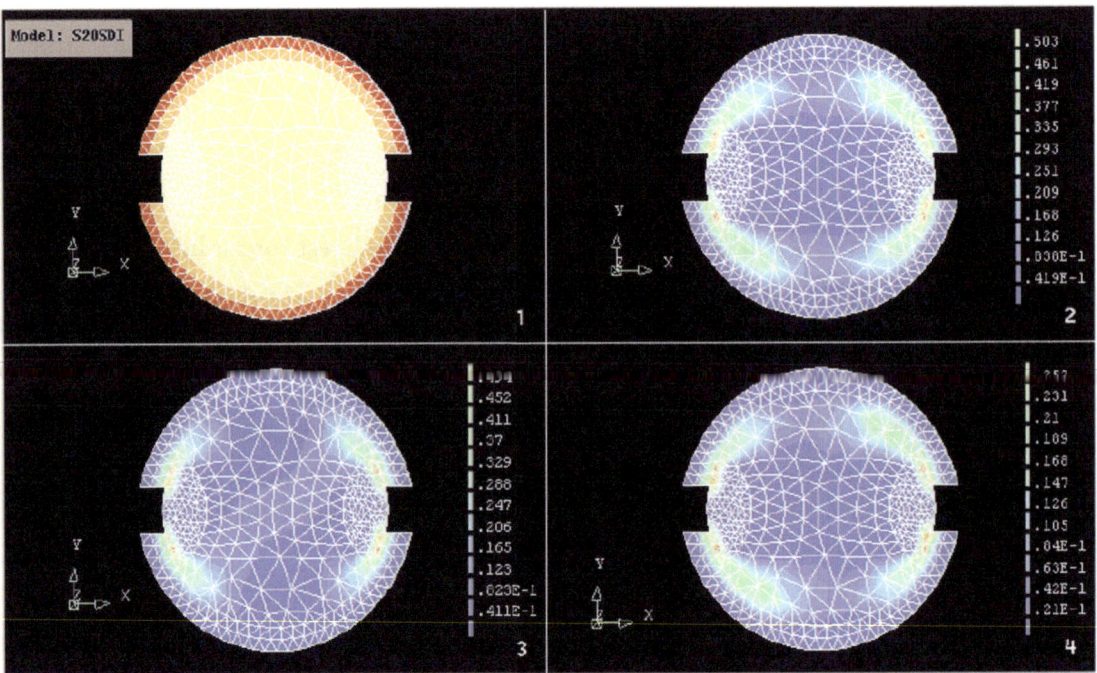

Figure 3.18 – Mapping of materials used – Internal variable of damage.

The problem is solved in plane stress, allowing the material to flow along the axial direction. The outer cylinder is made of steel and the inner part is made up of each of the three mixtures. There is a friction sliding between the steel and the bitumen due to the tangential stress control between the two materials. The results obtained in the numerical simulations are presented in Figure 3.19, Figure 3.20, Figure 3.21 and Figure 3.22. In all the figures it can be observed that the loading module $K = p/\delta$ decreases as the initial damage increases. However, any considerable strength loss is observed up to 40% of the initial damage in the mixture. This situation is evident for displacement over approximately 2 mm.

There is a loss of stiffness and strength as a function of the loading and the damage by hygrometric effects as observed in Figure 3.19 and Figure 3.20 for mixture M-2 and in Figure 3.21 andFigure 3.22 for mixture M-1. In both cases the loading capacity asymptomatically tends to a value close to 42% of the maximum for the undamaged material. It is important for the experimental response to show good agreement with the "undamaged" response as shown in Figures 3.19 and 3.21.

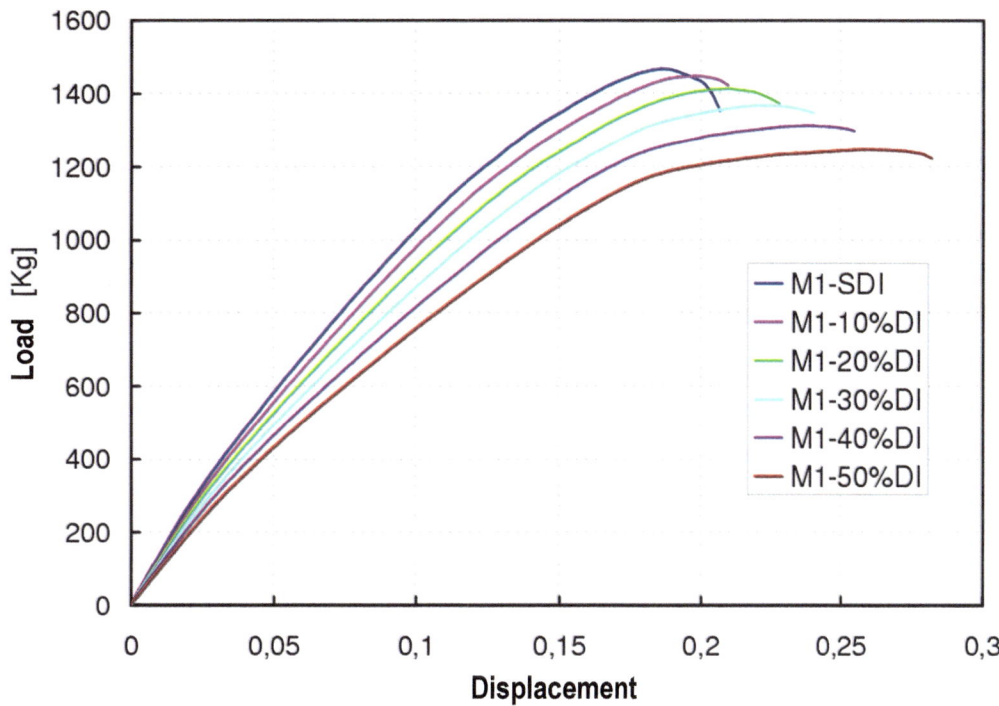

Figure 3.19 – Response curves of the Marshall test for **Mixture-1** $0 \le d \le 0,5$.

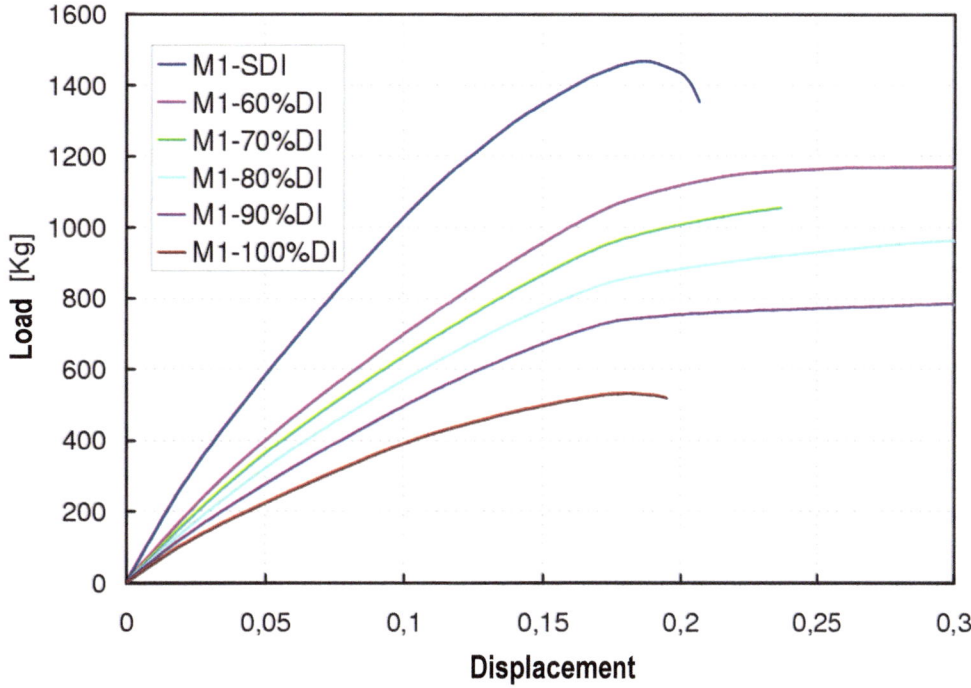

Figure 3.20 – Response curves of the Marshall test for **Mixture-1** $0,5 \le d \le 1$.

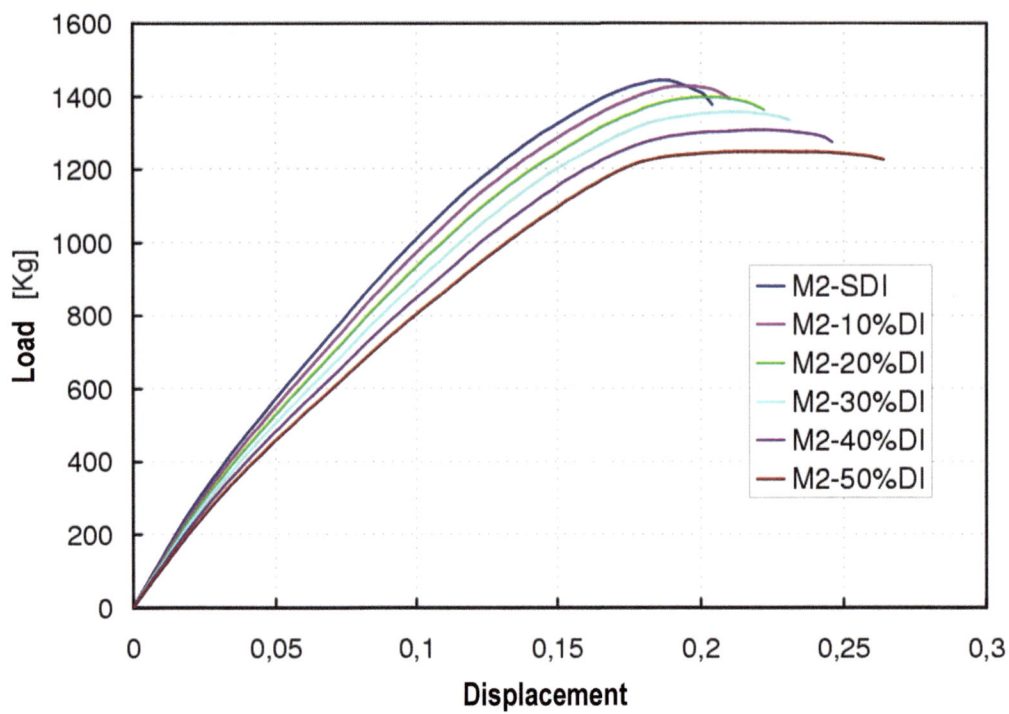

Figure 3.21 – Response curves of the Marshall test for **Mixture-2** $0 \leq d \leq 0,5$.

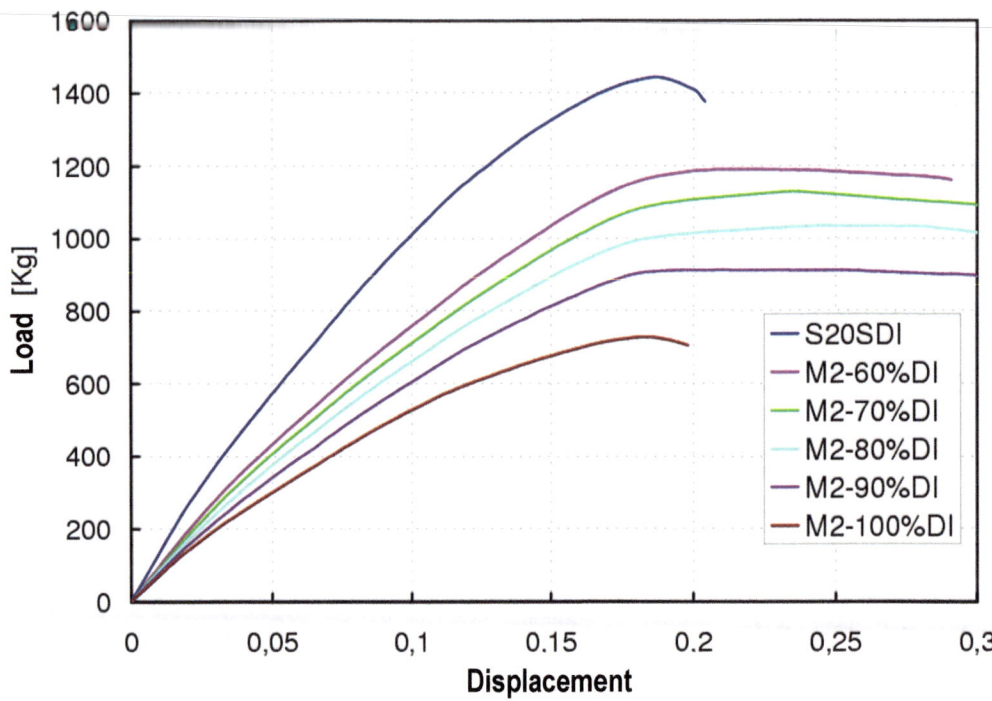

Figure 3.22 – Response curves of the Marshall test for **Mixture-1** $0,5 \leq d \leq 1$.

4 FIBER-MATRIX DISPLACEMENT (FMD) - Debounding

4.1 Introduction

This chapter focuses on the introduction of the relative movement phenomenon of the rigid body in *the mixing theory*. It takes place between the reinforcement and matrix phases when the debounding stress between both components is exceeded.

Among the causes of non-linear behavior for fiber reinforced composite materials are matrix crack formation and fiber-matrix displacement or relative movement. This phenomenon is known as "*debounding*" and is characterized by the matrix cracking and relative displacement between the fiber and the matrix. This loss of adherence involves the loss of stiffness of the composite material leading to inelastic or not recoverable strains between the fiber and the matrix. The aforementioned phenomenon will be called hereafter as "FMD" (*fiber-matrix displacement*) or "debounding".

As this phenomenon involves a fracture or relative movement between the fiber and the matrix, firstly, it is necessary to study the transfer mechanism of the fiber-matrix reinforcement. In composite material structures or pieces, the stresses are transferred from the matrix to the fiber because the loads are imposed on the composite matrix.

Generally, composite materials consist of a first low matrix strength phase and a reinforcement phase exceeding the strength of the matrix.

In fiber reinforced composite materials, the matrix fracture or cracks formation is accompanied by the fiber debounding phenomenon and subsequent displacement between both components (FMD). The matrix cracking process occurs at a considerably lower stress level than necessary to produce fiber fracture. Matrix fracture occurs at low stress values and is usually aligned with the principal stress direction producing a stiffness reduction and leads to inelastic strains and hysteresis loops (Beyerley *et al.* (1992)[1]), (Preyce and Smith (1992))[2].

---

[1] Beyerley D. and Spearing S. M. and Zok F. W. and Evans A. G. (1992). Damage, degradation and failure in a unidirectional ceramic-matrix composite. J. Am. Ceram. Soc., vol. 75, pp. 2719-2725.
[2] Pryce A. W. and Smith P. A. (1992). Modelling of the stress/strain behavior of unidirectional ceramic matrix composite laminates. J. Mater. Sci., vol. 27, pp. 2695-2704.

Over the last years, different models have been proposed for the micro mechanics modeling of the FMD phenomenon in reinforced fiber composite materials (Hild and Burr, (1996)[3]), (Owen and Lynnes (1972)[4]), (Agarwal and Bansal (1979)[5]).

4.2 Stress distribution along the reinforced fiber

The analysis of stress distribution along a fiber must consider the complex stress state developing at the fibers ends. As the relation between length l and diameter d decreases, the phenomena produced at the reinforcement ends affect considerably the stiffness of the whole set.

The stress transfer between the matrix and the fibers takes place between the two-phase interface. The stresses transfer mechanism between the phases is affected by different factors such as chemical factors of the interface and the fiber, superficial treatment of the fibers and the volume fraction of the fibers, as well as temperature conditions and humidity.

The fiber-matrix stresses of the transfer mechanism can be observed in Figure 4.1a. As observed, a length l-fiber is embedded in a matrix and oriented towards the loading direction. The stress applied to the matrix is transferred to the fiber through the interface. The matrix and the fiber undergo different tensile strains due to the difference between the matrix and fiber elastic modules. As shown in Figure 4.1a, the strains at the fiber ends are lower than in the matrix. As a result of the strain difference between both phases, the shear stresses around the fiber are induced in the direction of its longitudinal axis and the fiber is subjected to tensile state (Figure 4.1b., see also chapter 4, "Mixing Theory"). The fiber-matrix interface strength is relatively low. Figure 4.1b. shows the stress distribution in a fiber parallel to the loading direction. This distribution is based on the following hypotheses:

1. The fiber-matrix behavior is considered elastic.

2. The interface is thin and provides good stress transfer between the two components.

According to these assumptions, the resulting axial stress is zero at the fiber end and is maximum in its center. On the other hand, for long fibers, the resulting tangential stresses are maximum at the fiber ends and decreases until zero in its center. This has been confirmed by photoelasticity and laser spectroscopy analysis tests Raman (Hull (1987)[6]).

In the event that the maximum shear strength of the interface or the material surrounding the matrix is exceeded, the interface will break and consequently a relative displacement between the fiber and the matrix will occur. The shear failure does not necessarily mean that the matrix cannot transfer the stresses to the fiber due to the presence of friction forces between the fiber and the interface. The importance of this phenomenon depends on the properties of the fibers, the resin and the total volume of fibers in the composite material.

[3] Hild F. and Burr A. (1996). Matrix Cracking and Debounding of Ceramic-Matrix Composites. Int. J. Solids Structures Vol. 33 No. 8, pp. 1209-1220.

[4] Owen D.R.J. and Lyness J.F. (1972). Investigation of bond failure in fibre-reinforced materials by the finite element method. Fibre Sci. Technol., vol. 5, pp. 129-141.

[5] Agarwal B. D. and Bansal R. K. (1979). Effect of an interfacial layer on the properties of fibrouscomposites: a theoretical analysis. Fibre Sci. Technol. vol. 5, pp. 129-141.

[6] Hull D. (1987). *Materiales compuestos*. Editorial Reverté, España.

a)

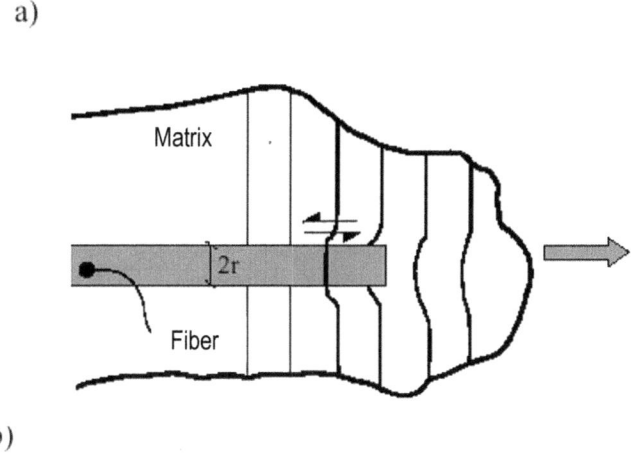

b)

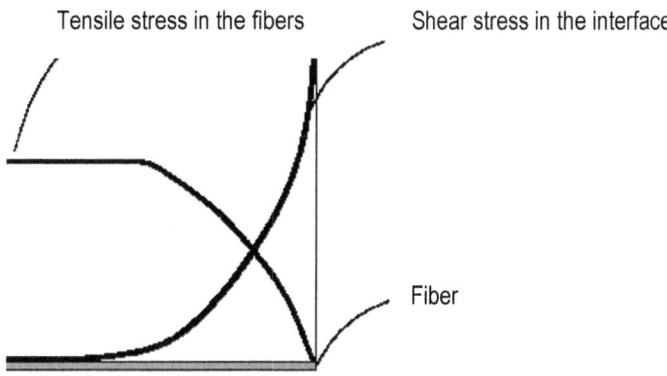

Figure 4.1 – a) Strain in a discontinuous fiber embedded in a matrix subjected to tension. b) Tensile and shear stress distribution in a fiber.

4.3 Crack and fiber interaction

Some of the important interaction during the cracking process can be understood by studying the mechanisms occurring when a sharp crack of the matrix and the fiber come into contact as shown in Figure 4.2. The stresses generated when a sharp and a brittle crack are exposed to a fiber are shown. A crack subjected to unidirectional tensile stress normal to its plane produces additional stresses.

The stress concentration around a crack is a function of the curvature radius of the crack end ρ and to the length l of the same and is given by the expression $(l/\rho)^{1/2}$. For a crack subjected to unidirectional tensile stress normal to its plane, additional tensile stresses parallel to the plane of the same are produced due the presence of the cracks.

In Figure 4.2, the stresses at the end of the elliptical crack are shown. The maximum tensile stress σ_1 is produced on the crack end and the tensile stress σ_2 parallel to the crack is produced just on its end. There is also a tangential stress τ in the plane normal to the crack plane.

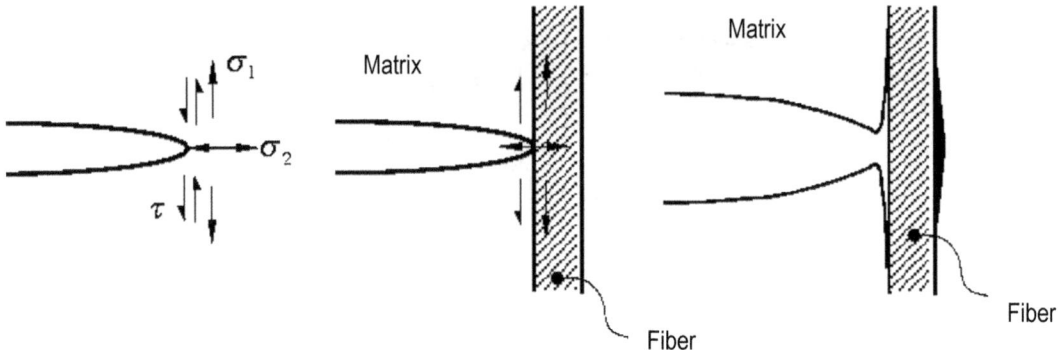

Figure 4.2 – a) Schematic representation of the stresses on a crack end. b) Crack end in the fiber interface. c) Interface fracture.

Stress σ_1 on the crack end tends to cause fiber fracture, stress σ_2 leads to bond fracture by traction in the interface and stress τ causes matrix shear fracture. In most composite materials the shear fracture in the interface is verified. However, a complete fiber unloading is not produced due to the presence of frictional forces between the fiber and the matrix.

The type of fracture is determined by the relative degree of fracture of the interface and the magnitude of the friction forces. It depends mainly on the composite material strength in parallel and perpendicular directions to the fiber and the shear strength. For most composite materials, the ratio between the final stress, parallel to the direction of the reinforcement, and the final shear stress is high. Therefore, the shear cracking in the interface occurs before the fiber fractures and a massive fiber-matrix cracking is observed. Since the matrix is fractured at values relatively low compared to the reinforcement strength, it is assumed that both mechanisms are not coupled. When this mechanism is produced, a complete loading is not produced due to frictional forces. Figure 4.3 shows the crack surface in an epoxy resin with Kevlar49 (Hull (1987)[6]). The crack surface looks very fibrous and with many extraction of the fibers.

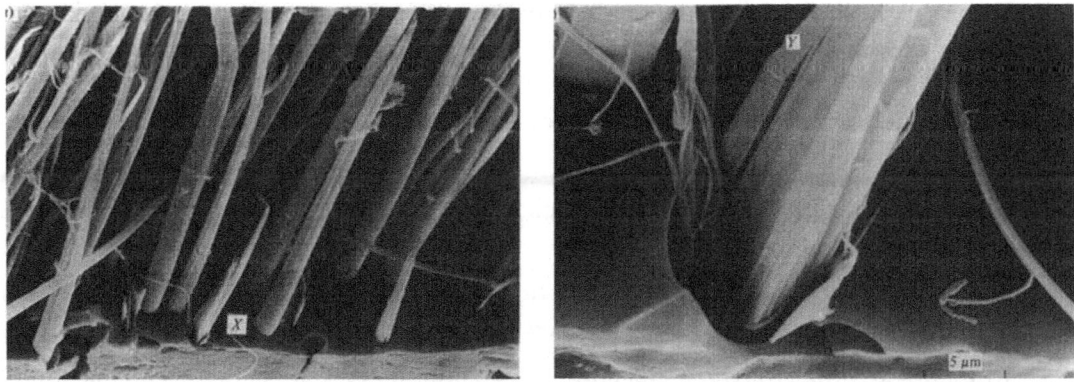

Figure 4.3 – a) Composite material fibrous crack. b) Fibrous crack detail.

4.4 Constitutive models for composite materials with FMD

Most of the research done for the analysis of the effective behavior of reinforced composite materials is based on the perfect fiber-matrix debounding. The cracking phenomenon between the phases or *"debounding"* is typically of this type of material and must be considered for the mechanical behavior analysis. Due to the lack of uniformity in the displacement field, this phenomenon must be incorporated to the mixing theory by the constitutive model.

Several authors have studied this phenomenon by means of the finite element method. Most of these research studies are based on the lack of contact between the fiber and the matrix and the use of numerical procedures to study this phenomenon (Owen and Lynnes (1972)[4]). Other studies introduced a layer between the components to simulate the interface zone between the fiber and the matrix (Agarwal and Bansal (1979)[5]). Lené and Leguillon proposed a model based on a tangential displacement between the fiber and the matrix and used the homogenization method along with the finite element method to model the composite material behavior.

The analysis of the behavior of composite materials disregarding the fiber-matrix debounding phenomenon was carried out by Aboudi (1982 and 1984)[7,8], who has used a Legendre development in a representative cell. Later, Beneviste and Aboudi (1984)[9] modified their model by taking into account the fiber-matrix debounding phenomenon. The *"debounding"* phenomenon is simulated by imposing the continuity condition into the normal displacements in the fiber-matrix interface zone. The tangential displacement concept has been used by Drumheller (1973)[10] to simulate the *"debounding"* phenomenon in bilaminated periodical composite materials. Drumheller studied the *"debounding"* effect in wave propagation in a laminated environment and is based on the use of elastodynamic equations. Beneviste and Aboudi used their model to study wave propagation velocity in fiber-reinforced composite materials.

Other authors proposed a model based on the micromechanics of this phenomenon (Cox (1952)[11]), (Aveston et alt. (1971)[12]), (Hsueh (1993)[13]). All these approaches use a model of length $2 \cdot l$ of the size of the separation between cracks and is made up by different materials designated by 1 and 2 (see Figure 4.4). The existence of a crack of size $2 \cdot a$ is assumed in the central zone as well as a friction length $2 \cdot l_f$. During the "FMD" phenomenon and due to the presence of friction between the fiber and the matrix there is a temperature difference between materials 1 and 2.

---

[7] Aboudi J. (1982). A continuum theory for fiber-reinforced elastic viscoplastic composites. Int. J. Engng. Sci., vol. 20, pp. 605-621.

[8] Aboudi J. (1984). Effective behaviour of inelastic fiber-reinforced composites. Int. J. Engng. Sci., vol. 22, pp. 439-449.

[9] Benveniste Y. and Aboudi J. (1984)A continuum model for fiber reinforced materials with debounding. Int. J. Solids Struct. vol. 20, pp. 935.

[10] Drumheller (1973). An effect of debounding on stress wave propagation in a composite material.
J. Appl. Mech., vol. 40, pp. 1146-1157.

[11] Cox H. L. (1952). The elasticity and the strength of paper and other fibrous materials. Br. J. Appl. Phys., vol. 3, pp. 72-79.

[12] Aveston J. and Cooper G. A. and Kelly A. (1971). Single and multiple fracture. Conference Proceedings of the National Physical Laboratory: Properties of Fiber Composites.

[13] Hsueh C. H. (1993). Evaluation of interfacial properties of fiber-reinforced ceramic composites using a mechanical properties microprobe. J. Am. Ceram. Soc., vol. 76, pp. 3041-3050.

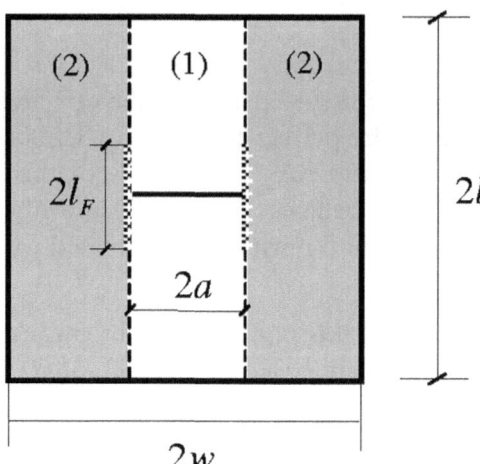

Figure 4.4 – Elemental cell of $(2l \times 2w)$ containing a crack size of $2a$.

Four variables are required to characterize the state of the composite material: total strain, friction length $2 \cdot l_f$, cell length $2 \cdot l$ and the crack opening in the matrix Δ (see Figure 4.4). The strain in zone 2 is a function that depends on the interface. The free energy density of the model is obtained by considering the superposition of two phenomena. First, the energy density is obtained when part 2 moves from part 1 a magnitude of Δ along the length l_f, without external loading. Then the second step is to load the cracked system preventing the develop friction phenomena. The presence of a crack in the matrix leads to a loss of stiffness defined through an internal variable D. This internal variable depends on the density of the crack and the elastic properties of both phases. The total free energy is the sum produced by both phenomena and is given by:

$$\psi = \frac{1}{2}(1-D)\mathbf{c}:(\bar{\mathbf{\varepsilon}} - \mathbf{\varepsilon}^i) + \frac{1}{2d}\mathbf{c}^*:(\mathbf{\varepsilon}^i)^2 \tag{4.1}$$

where $\mathbf{\varepsilon}^i$ represents the inelastic strains due to the displacements between the two phases.

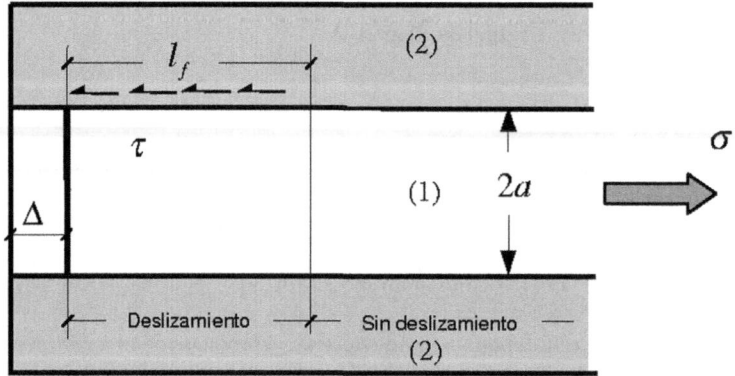

Figure 4.5 – Stresses due to the movement of phase (1) with respect to phase (2).

The evolution of the internal variables of the problem (D, d, ε^i) is established by considering the fragmentation process in a model with shear constant stresses along the friction zone (Curtin (1991)[14]). Hild et al. (1996)[15] proposed a practical method to determine the law of evolution of the internal variables through experimental results. However, this methodology has a disadvantage; it requires a number of tests to identify the laws of evolution of the internal variables for different types of material, or the same material, but with different matrix and reinforcement volumetric participations.

Later, Hutchinson and Jensen (1990)[16] presented different models for fibers embedded in a brittle matrix. These were limited to the analysis of composite materials subjected to compression stresses acting on the fiber-matrix interface and assuming a friction during the starting phenomenon after the cracking of fibers. The fiber-matrix interaction is modeled by a cylindrical cell with two types of boundary conditions: one isolated fiber-matrix and another matrix containing a set of unidirectional fibers. It is assumed that the fibers are transversally isotropic along their longitudinal axis and the matrix is isotropic

A few years later Hild, Burr and Leckie (1995)[17] presented a micro mechanical model based on the mechanics of the continuum medium to analyze the effects of the matrix fracture and the fiber-matrix relative displacement. This model uses four state variables and the free energy of the system is calculated following these variables.

4.5 A procedure proposed for FMD

The objective of this section is to modify the constitutive model presented in the previous chapters to incorporate the FMD phenomenon into the study of fiber-reinforced composite materials. The analysis of composite materials in this procedure is based on the mixing theory of basic substances (see Chapter 3). It is worth highlighting that this theory is based on the combination and interaction of the basic substances that make up the composite. It is assumed that all the component substances participate at the same time in each one of the material points under analysis and each one has its own constitutive law and volumetric proportion assigned. All the material components must satisfy the kinematic compatibility condition either of equal strains (classic mixing theory) or by the strain distribution according to the kinematics of the serial-parallel behavior formulated in the generalized mixing theory. Both kinematic hypotheses are only valid for long-fiber composite materials disregarding the phenomena occurring at the fiber ends. For short-fiber-reinforced composite materials, it is necessary to carry out a correction of the properties of each component and keep the kinematic condition of the classic or generalized mixing theory (see Chapter 3) (Car et alt. (1998)[18]).

---

[14] Curtin W. A. (1991).Exact theory of fiber fragmentation in single-filament composite. J. Mater. Sci. vol. 26, pp. 5239-5253.

[15] Hild F. and Burr A. (1996). Matrix Cracking and Debounding of Ceramic-Matrix Composites. Int. J. Solids Structures Vol. 33 No. 8, pp. 1209-1220.

[16] Hutchinson J. W. and Jensen H. M. (1990). Models of fiber debounding and pullout in brittle composites with friction. Mechanics of Materials, vol. 9, pp. 139-163.

[17] Hild F., Burr A. and Leckie A. (1994). Fiber breakage and fiber pull out of fiber-reinforced ceramic-matrix composites. Eur. J. Mech. A/Solids, vol. 13, pp. 731-749, No. 6.

[18] Car E. and Oller S. and Oñate E. (1998). Un modelo constitutivo elasto plástico acoplado con daño mecánico e higrométrico. Aplicación a pavimentos flexibles. Revista Internacional de Ingeniería de Estructuras, vol. 3(1), pp. 19-37.

Moreover, the compatibility equation is not valid for a relative displacement between the fiber and the matrix. This phenomenon occurs when the admissible maximum tangential stress of the interface between the fiber and the matrix is exceeded.

4.5.1 The constitutive model modification – Procedure for the fiber-matrix displacement phenomenon (FMD)

4.5.1.1 Introduction

Composite materials that consist of a matrix and reinforcement have a complex non-linear behavior due to the reinforcement displacement as a result of the loss of adherence between the matrix and the reinforcement. This relative movement between the reinforcement and the matrix causes a loss of stiffness in the whole set and a decrease of the composite mechanical parameters without fractures in the reinforcement phase is observed. (Hild et alt. (1994)[19], (Hild (1994)[20]).

The theory modification developed in previous chapters is based on the mechanics of the continuum medium to deal with the anisotropy and the mixing theory. It involves introducing an irrecoverable inelastic behavior in the constitutive equation to represent an approximation of the relative rigid movement of the body produced between the fiber and the matrix. The incorporation of the FMD into the constitutive equation must take into consideration two main characteristics: a) the global loss of stiffness due to the decrease of the fiber collaboration in the matrix and b) the irrecoverable relative displacement between the fiber and the matrix.

FMD composite materials subjected to tension do not satisfy the kinematic condition imposed by the basic theory of basic substances. A direct consequence of this phenomenon is the matrix limitation to transfer the stresses to the fiber. In other words, the fiber cannot increase its tensional state as a result of the limited adherence in the fiber-matrix interface zone.

The constitutive model is based on the assumption that the loading transfer from the matrix to the fiber varies when the matrix is under plastic strains. The relative movement between the fiber and the matrix can be represented by the mechanics of the continuum medium through an irrecoverable inelastic strain in the fiber. The starting point of this phenomenon is determined through a threshold condition of maximum strength which compares the effective stress on a point with respect to the fiber strength. Considering the fiber participation within the composite and the fiber-matrix stress transmission, its maximum strength or real strength and its collaboration capacity are determined depending on its own nominal strength $(f^\sigma)^N_{\text{fib}}$ (or fiber strength in isolated conditions), of the nominal strength of the matrix $(f^\sigma)^N_{\text{mat}}$ and the nominal strength of the fiber-matrix interface $(f^\tau)^N_{\text{fib-mat}}$, or stress transfer capacity from the matrix to the fiber. From another point of view, it can be stated that fiber participation in the composite depends on its own strength and on the stress transfer capacity of the fiber-matrix interface. Therefore, its strength is influenced by the medium containing it and its constitutive treatment might involve a non-local formulation. Then, the fiber strength contained in a matrix is defined as

[19] Hild F., Burr A. and Leckie A. (1994). Fiber breakage and fiber pull out of fiber-reinforced ceramic-matrix composites. Eur. J. Mech. A/Solids, vol. 13, pp. 731-749, No. 6
[20] Hild F. (1994). On the average pull-out length of the fibre-reinforced composites. C.R. Acad. Sci. Paris. vol. 319 (Serie II), pp. 1123-1128.

$$(f^{\sigma})_{\text{fib}} = \min\left\{(f^{\sigma})^{N}_{\text{fib}} , (f^{\sigma})^{N}_{\text{mat}} , \left[\frac{(f^{\tau})^{N}_{\text{fib-mat}} \cdot 2\pi r_{f}}{A_{f}}\right]\right\} \qquad (4.2)$$

in which r_{f} represents the radius of the fiber and A_{f} is the area of the fiber's transversal section. From equation (4.2) the following cases can be derived:

- When the matrix is more resistant than the fiber and the fiber-matrix adherence is perfect then the fiber participation capacity remains limited by its own nominal strength $(f^{\sigma})_{\text{fib}} \equiv (f^{\sigma})^{N}_{\text{fib}}$.

- When a failure occurs in the matrix due to microcrackings, while the fiber remains linear, the fiber strength is limited by the matrix strength as the stress transfer "mechanism" between the fiber and the matrix is broken. No more tension can be transferred except the tension allowed by the medium containing the fiber $(f^{\sigma})_{\text{fib}} \equiv (f^{\sigma})^{N}_{\text{mat}}$.

- When a failure occurs in the fiber-matrix interface, the fiber strength remains limited by the interface $(f^{\sigma})_{\text{fib}} \equiv \dfrac{2 \cdot (f^{\tau})^{N}_{\text{fib-mat}} \, 2\pi r_{f}}{A_{f}} = \dfrac{2 \cdot (f^{\tau})^{N}_{\text{fib-mat}}}{r_{f}}$.

In most composite materials the cracking due to shear in the interface is produced before fiber fracture and a massive fiber-matrix break off is observed and therefore the fiber strength is limited by the interface stress transmission capacity. The plastic phenomena arising in the composite matrix subjected to an increasingly monotonous loading prevents stress transfer from the matrix to the fibers. This produces irrecoverable strains as a result of the reinforcement phase displacement with respect to the matrix. From this moment, the load transfer from the fibers to the matrix is not zero due to the frictional phenomena between both composite phases. Consequently, the fibers' tensional state is increased based on an elastic module different from the initial one.

The model presented in this chapter sets the tensional state of the fiber when the composite matrix breaks off. From here, the irrecoverable strains due to displacement between both phases are considered in an elastoplastic model with hardening for the reinforcement phase because of the frictional forces between the fiber and the matrix. Therefore, this is *not a local material* model because the state of one of the materials (reinforcement) depends on the other (matrix).

4.5.1.2   Implementation

In order to consider the fiber-matrix interaction, the tensional state of the reinforcement phase must be determined when the fiber-matrix relative displacement occurs due to the plastic phenomena in the matrix. This tensional state is the maximum tensional state that can be transmitted from the matrix to the fiber under perfect adherence conditions between the reinforcement and the matrix. From this moment, the stresses are transmitted by frictional phenomena. Figure 4.6 shows the interaction between both phases for a composite material consisting of a matrix and a reinforced fiber for the reinforcement strength based on the aforementioned criteria.

The *"debounding"* phenomenon is included in the constitutive model formulation for composite materials, by using the relation,

$$r = \frac{\tau}{(f^\tau)_{\text{fib}}} = \frac{S}{(f^S)_{\text{fib}}} \qquad (4.3)$$

where $(f^\tau)_{\text{fib}}$ and $(f^S)_{\text{fib}}$ are the reinforced strengths in the referential and updated configurations, respectively, and τ and S are the stress in the reinforcement longitudinal direction in the referential and updated configurations when the fiber-matrix relative movement occurs due to plastic phenomena in the matrix.

For the simulation of the irrecoverable strains occurring as a result of the fiber-matrix relative movement, the yield criterion of the reinforcement phase must be redesigned and given by equation (4.1) in the updated configuration. The r factor set the starting of the mechanical process of the fiber-matrix displacement through the plastification mechanisms. This factor is considered constant after starting the plastification process.

Therefore, including the FMD phenomenon into the constitutive model requires the redefinition of the evolution function of the fiber strength. Thus, the yield criteria in the referential and updated configurations are then:

$$\mathbb{F}^\tau(\boldsymbol{\tau}, g, \boldsymbol{\alpha}) = f^\tau(\boldsymbol{\tau}, g) - r \cdot \mathcal{K}(\boldsymbol{\alpha}) = 0 \qquad (4.4)$$

where $\mathbb{F}^\tau(\boldsymbol{\tau}, g, \boldsymbol{\alpha}) = 0$ is an homogenous function in the stresses defined in the updated configuration called threshold function of plastic discontinuity (see Chapter 2), r is a factor limiting the fiber strength due to the FMD phenomenon and its expression is given by equation (4.3). The product $(r \mathcal{K}(\boldsymbol{\alpha}))$ can redefine the stress threshold in the referential and updated configurations of the reinforcement from the moment the plastic phenomena are produced in the matrix of the composite material. The yield condition modification requires the elastoplastic constitutive model redefinition of the reinforcement phase in the referential and updated configurations.

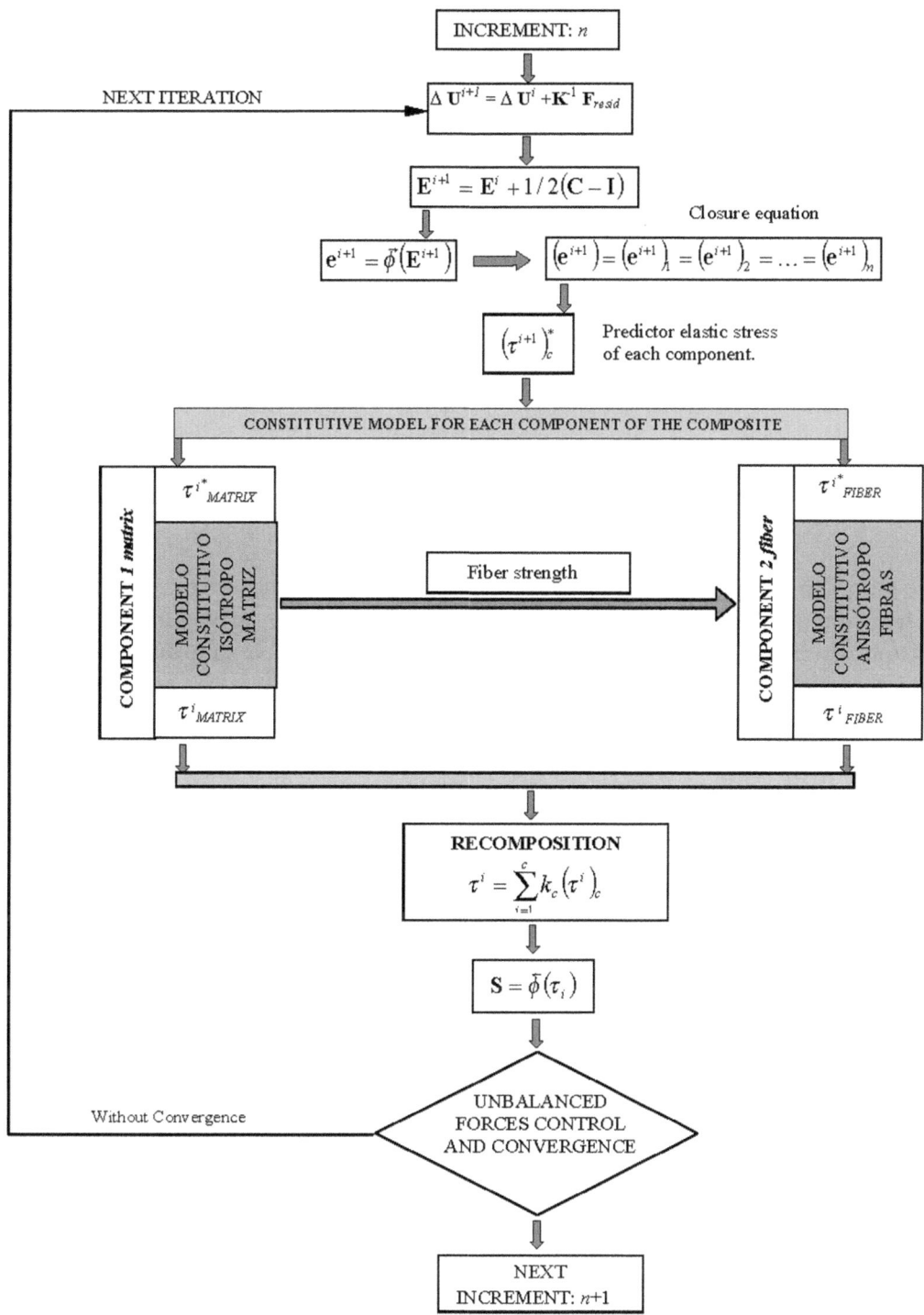

Figure 4.6 – FMD solution scheme of the non-linear biphasic problem.

4.6 Expression of the elastoplastic constitutive model of the reinforcement

4.6.1 Yield condition

The FMD phenomenon is introduced in the limit condition of the plasticity threshold through a modification of the yield function of the reinforcement phase. The reinforcement strength can be redefined from the moment the plastic phenomena are produced in the matrix of the composite material (see Section 4.5.1).

The factor limiting the strength from fiber r affects the material proportionality limit. This parameter is not introduced directly in the yield condition in order to modify very little the mathematical structure of the theory of plasticity. The yield condition expressed in equation (4.4) is written as

$$\mathbb{F}^\tau(\boldsymbol{\tau}, g, \boldsymbol{\alpha}) = \frac{1}{r} f^\tau(\boldsymbol{\tau}, g) - \mathcal{K}(\boldsymbol{\alpha}) = 0 \qquad (4.5)$$

The equation above represents the new yield criterion or discontinuity threshold behavior. The fiber-matrix displacement phenomenon can be considered by establishing the moment from which the matrix cannot transfer load to the fiber and simulating the fiber-matrix irrecoverable relative displacements through the plasticity.

4.6.2 Plastic flow rule

The flow rule establishes the evolution law of plastic strains and is defined in the updated configuration by Lee's objective derivative (Malvern (1969)[21])

$$L_v(\mathbf{e}^p) = \dot{\lambda} \frac{\partial g}{\partial \boldsymbol{\tau}} \qquad (4.6)$$

where $g = g(\boldsymbol{\tau}, g)$ is the function of plastic potential and λ is a non-negative scalar known as plastic consistency parameter . For the plasticity case associated to the evolution law of plastic strains it is expressed as

$$L_v(\mathbf{e}^p) = \dot{\lambda} \frac{1}{r} \frac{\partial f}{\partial \boldsymbol{\tau}} \qquad (4.7)$$

The theory of plasticity also requires satisfying the loading-unloading conditions or Kuhn-Tucker conditions (Crisfield (1991)[22])

$$\dot{\lambda} \geq 0 \quad , \quad \mathbb{F}^\tau(\boldsymbol{\tau}, g, \boldsymbol{\alpha}) \leq 0 \quad , \quad \dot{\lambda} \cdot \mathbb{F}^\tau(\boldsymbol{\tau}, g, \boldsymbol{\alpha}) \leq 0 \qquad (4.8)$$

Moreover, the persistence condition or plastic consistency condition must be satisfied, and it is expressed by the temporal variation of the yield function:

---

[21] Malvern L.E. (1969). Introduction to the Mechanics of a Continuous Medium. Prentice-Hall.
[22] Crisfield M.A. (1991). Non-linear finite element analysis of solids and structures. John Wiley & Sons Ltd.

$$\dot{\mathbb{F}}^{\tau}(\tau, g, \alpha) = \frac{1}{r} \frac{\partial f^{\tau}(\tau, g)}{\partial \tau} \dot{\tau} - \frac{\partial \mathcal{K}(\alpha)}{\partial \alpha} \dot{\alpha} = 0 \tag{4.9}$$

4.7 "Total" and "updated" Lagrangian formulation

The implementation of the constitutive model presented in this chapter completes the mechanical formulation detailed in chapters 2 and 3 and has been developed for its incorporation within the finite element method. Nevertheless, for the sake of efficiency, improvement and implementation simplicity, special attention must be paid to the strategy to follow and the kinematic space for the formulation. In the mechanics of the continuum medium the problems can be formulated using the material coordinates $\mathbf{X}$ or the spatial coordinates $\mathbf{x}$ as a reference, obtaining the *material* or *spatial* kinematic descriptions. The material description —*Referential configuration* — is characterized by the fact that the properties of all the body particles always refer to their original position and their kinematic description is called *total Lagrangian*. The spatial description —*updated configuration*— is characterized by the fact that the properties of all the body particles are tracked down from their original to their current position and their kinematic description is called *updated Lagrangian*.

Given a solid in the space and a Cartesian orthogonal coordinate system, the equilibrium in time $(t + \Delta t)$ is expressed through the principle of *virtual works* establishing the equality of the internal and external virtual works.

In the updated configuration the principle of virtual works is given by the following expression

$$\underbrace{\int_{V^{t+\Delta t}} \boldsymbol{\tau}^{t+\Delta t} : \delta \mathbf{e}^{t+\Delta t} \ dV^{t+\Delta t}}_{\mathbf{f}^{\text{int}}} = \underbrace{\int_{V^{t+\Delta t}} \rho \, \mathbf{b}^{t+\Delta t} \cdot \delta \boldsymbol{u}^{t+\Delta t} \ dV^{t+\Delta t} + \oint_{S^{t+\Delta t}} \mathbf{t}^{t+\Delta t} \cdot \delta \boldsymbol{u}^{t+\Delta t} \ dS^{t+\Delta t}}_{\mathbf{f}^{\text{exr}}} \tag{4.10}$$

where $\boldsymbol{\tau}^{t+\Delta t}$ is the Kirchhoff stress tensor , $\delta \mathbf{e}^{t+\Delta t}$ is the Almansi virtual strain tensor compatible with the virtual displacement $\delta \boldsymbol{u}^{t\Delta t}$, $\mathbf{b}^{t+\Delta t}$ represents the forces per unit mass and volume, $\mathbf{t}^{t+\Delta t}$ represents the surface forces per unit surface, $V^{t+\Delta t}$ is the volume of the solid, and $S^{t+\Delta t}$ the surface loading. All variables are defined at the instant of time $t + \Delta t$ in the *updated configuration*.

In the referential configuration, the formulation is carried out at the reference instant, in other words, when $t = 0$. In this configuration the principle of virtual works is formulated through the following expression

$$\underbrace{\int_{V^0} \mathbf{S}^{t+\Delta t} : \delta \mathbf{E}^{t+\Delta t} \ dV}_{\mathbf{F}^{\text{int}}} = \underbrace{\int_{V^0} \rho_0 \, \mathbf{b}_0^{t+\Delta t} \cdot \delta \boldsymbol{u}^{t+\Delta t} \ dV + \oint_{S^0} \mathbf{t}_0^{t+\Delta t} \cdot \delta \boldsymbol{u}^{t+\Delta t} \ dS}_{\mathbf{F}^{\text{exr}}} \tag{4.11}$$

where $\mathbf{S}^{t+\Delta t}$ is the Piola Kirchhoff second stress tensor, $\delta \mathbf{E}^{t+\Delta t}$ is the Green-Lagrange virtual strain tensor compatible with the virtual displacement $\delta \boldsymbol{u}^{t\Delta t}$, $\mathbf{b}_0^{t+\Delta t}$ represents the forces per unit mass, and $\mathbf{t}_0^{t+\Delta t}$ represents the surface forces per unit surface, V^0 is the volume of the solid, and S^0 the surface loading. All variables are defined at the instant of time $t + \Delta t$ in the *referential configuration*.

Both, the formulation at the referential configuration −total Lagrangian− and the one resulting from the equilibrium in the updated configuration −updated Lagrangian− include a kinematic formulation of large displacements, strains and rotations. The total and updated Lagrangian formulations resulting from equations (4.10) and (4.11) give equivalent results. The only difference is how the constitutive model must be written in each of them because it may not possible to find such a simple solution for both formulations. In other words, both formulations lead to identical results if they are appropriately used. (Bathe (1982)[23]).

Having the stress and constitutive tensors in the referential configuration, the corresponding tensors in the updated configurations are obtained through the *"push-forward"* tensorial transport operations. The transport back to the referential configuration is carried out through the *"pull-back"* operation.

The implementation of the *"push-forward"* and *"pull-back"* tensorial transport operations between the referential and updated configurations in a finite element code considering large strains allows either of the formulations to deal with the problem. The results are identical with either the total or the updated Lagrangian. Choosing one or the other depends mainly on the configuration in which the constitutive tensor expression is known because although this tensor can be transferred from one configuration to the other by the *"push-forward"* and *"pull-back"* operations, these transfers have a high computational cost as well as non-linear problems. These operations must be carried out in each iteration of the linearization process until the equilibrium is reached.

---

[23] Bathe K. J. (1982). Finite Element Procedures in Engineering Analysis. Prentice-Hall, Inc.

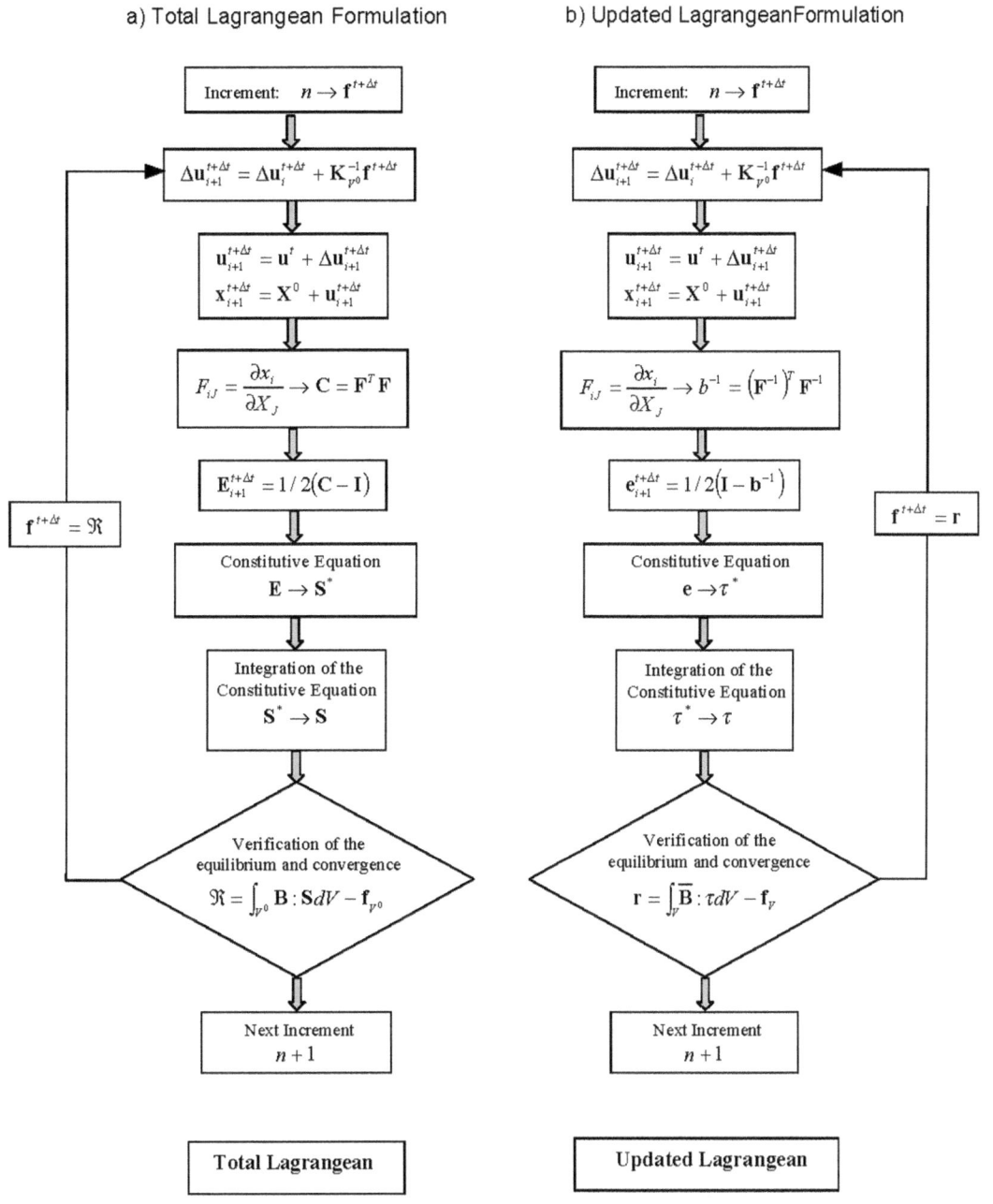

Figure 4.7 – Resolution of a non-linear problem through Lagrangean formulations a) Total formulation and b) updated formulations

Figure 4.7 shows a graphic that compares the steps to follow for the implementations of a total Lagrangian and an updated formulation in a finite element code.

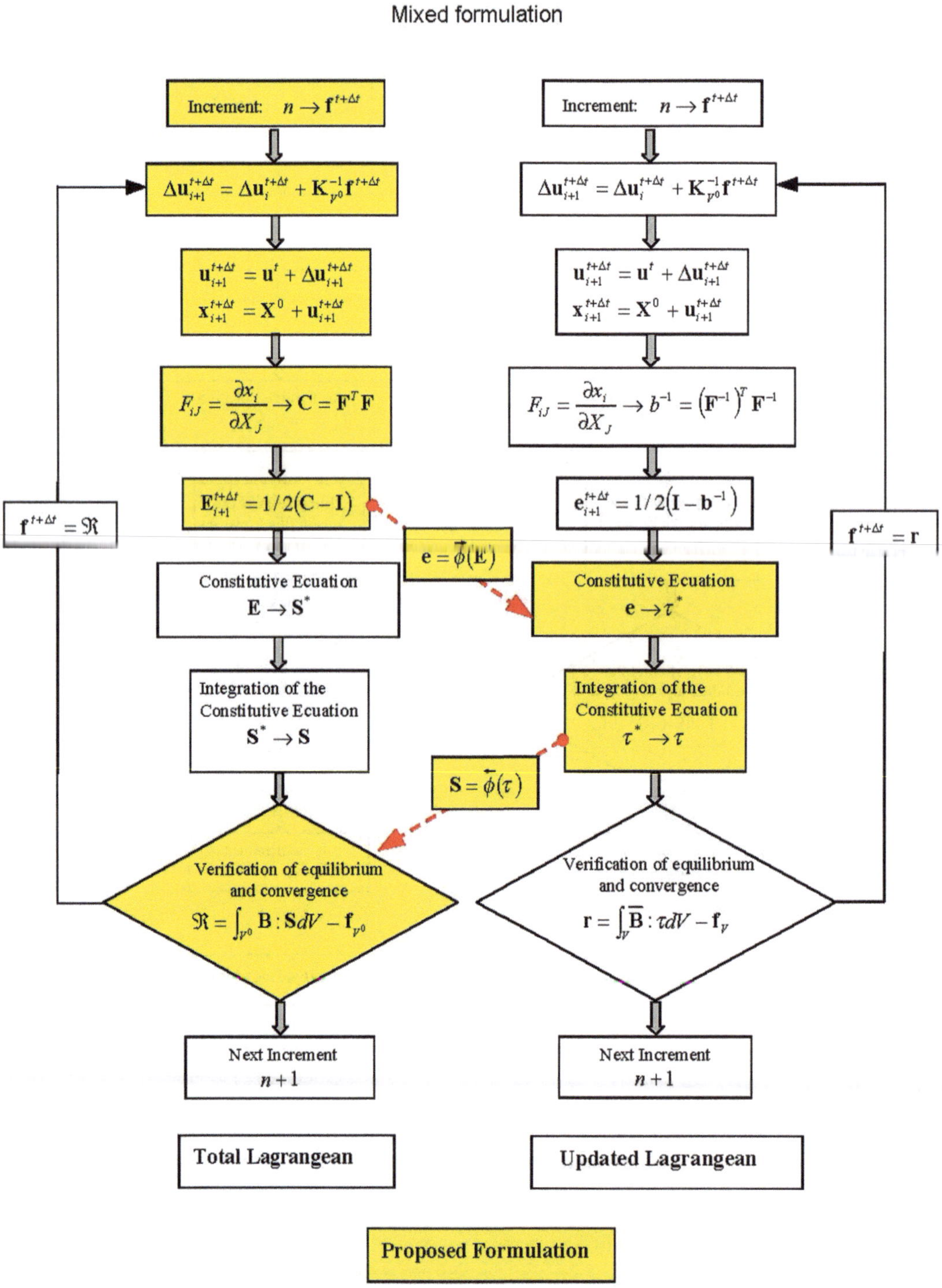

Mixed formulation

Figure 4.8 – Resolution of a non-linear problem through a mixed formulation.

Figure 4.8 presents the algorithm used in the constitutive model for composite materials proposed in previous chapters. This algorithm has a special feature: it combines the total Lagrangian formulation and the updated one. Its main characteristic is that it only carries out the constitutive equation integration in the updated configuration while the rest of the operations are carried out in the referential configuration. This formulation is chosen because of several reasons, such as equilibrium verification. This is carried out in the referential configuration as the volume there is constant and loading follow-ups are not necessary. The integration of the constitutive equation is carried out in the updated configuration as here is where the material constitutive equation and the material yield or discontinuity threshold are known more accurately. The strain tensor of the total Lagrangian formulation is passed to the updated one through the "*push-forward*" operation and this strain tensor obtained from the integration of the constitutive model is returned to the total formulation through the "*pull-back*" operation.

4.8 Implementation of the mixing and anisotropy theory in the FEM context

Implementing the anisotropy and the mixing theory within the FEM context involves adopting a transformation algorithm of the stress tensors, strains and constitutive model from the referential to the updated configuration. Nevertheless, strictly speaking, not all the tensors can be transferred. Transporting the *space transportation tensors* $-\boldsymbol{a}^{\tau} = \vec{\phi}(\mathbf{A}^{S})$ and $\boldsymbol{a}^{e} = \vec{\phi}(\mathbf{A}^{E})-$ for the anisotropy description of each of the material points (see Chapter 2) can be a very expensive process. It is more convenient to use these fourth-order tensors in the referential configuration $-\mathbf{A}^{S}$ and $\mathbf{A}^{E}-$ and then transfer the stress and strain tensors resulting from the transformation of the anisotropic space into an isotropic one $-\overline{\mathbf{S}} = \mathbf{A}^{S} : \mathbf{S} \Rightarrow \overline{\tau} = \vec{\phi}(\overline{\mathbf{S}})$ and $\overline{\mathbf{E}} = \mathbf{A}^{E} : \mathbf{E} \Rightarrow \overline{e} = \vec{\phi}(\overline{\mathbf{E}})-$. Both tensors are functions of the gradient of the strains expressing the movement of the referential system and of the transformations tensor of the stresses and strains spaces in the referential configuration (see Chapter 2). Transporting the space transformation tensors leads to tensors which are not suitable for their representation as a matrix. Therefore, all the tensorial operations must be carried out with this algorithm, which involves a high computational cost.

The proposed algorithm has a disadvantage for the transformation of any anisotropic real or fictitious space variable or vice versa: it has to first transfer these magnitudes to the referential configuration.

Graph 1 show the transportation operations between *the isotropic fictitious spaces* and *the anisotropic real spaces* carried out for the constitutive equation integration (Lubliner (1990)[24]).

For the mixing theory implementation within the FEM context, it is necessary to take into account the calculation of the predictive stress and the integration of the constitutive equation must be carried out for each phase making up the composite material. It is important to highlight that composite components presenting no reversible phenomena such as plasticity, damage, etc., require the existence of internal variables controlling the irreversible processes for each composite phases. The existence of these internal variables for each composite phase involves a high computational cost due to the memory required to store these variables, which take into account the loading history of the component. Graph

---

[24] Lubliner J. (1990). *Plasticity Theory*. Macmillan Publishing, U.S.A.

1hows the calculation of the predicting stress and the integration of the constitutive equation. These operations must be carried out for each of the component phases of the composite material.

Graph 1. Numerical implementation

- Definition of the constitutive tensor in the anisotropic real space, choice of the isotropic fictitious constitutive tensor and rotation tensor for each phase of the composite material.

$$(\overline{\mathbb{C}}^S)_{c_{m,r}} \; ; \; (\mathbb{C}^S_{\text{loc}})_{c_{m,r}} \; ; \; (\mathbb{R})_{c_{m,r}}$$

- Calculation of the anisotropic constitutive tensor in the global coordinate system for each phase of the composite material. Hyper elastic linear referential model.

$$(\mathbb{C}^S)_{c_{m,r}} = (\mathbb{R})_{c_{m,r}} : (\mathbb{C}^S_{\text{loc}})_{c_{m,r}} : (\mathbb{R})_{c_{m,r}}$$

- Definition of the space mapping for each phase of the composite material.

$$(\mathbf{A}^E)_{c_{m,r}} = (\overline{\mathbb{C}}^S)^{-1}_{c_{m,r}} : (\mathbf{A}^S)_{c_{m,r}} : (\mathbb{C}^S)_{c_{m,r}}$$

- Calculation of the anisotropic and isotropic constitutive tensors of each phase to be considered in the generalized mixing theory (see chapter 3)

$$(\hat{\mathbb{C}}^S)_{c_{m,r}} = \left[(1 - \chi_{c_{m,r}}) \cdot \mathbf{I}_4 + \chi_{c_{m,r}} \cdot (\boldsymbol{\Phi})_{c_{m,r}}\right]^T : (\mathbb{C}^S)_{c_{m,r}} : \left[(1 - \chi_{c_{m,r}}) \cdot \mathbf{I}_4 + \chi_{c_{m,r}} \cdot (\boldsymbol{\Phi})_{c_{m,r}}\right]$$

$$(\hat{\overline{\mathbb{C}}}^S)_{c_{m,r}} = \left[(1 - \chi_{c_{m,r}}) \cdot \mathbf{I}_4 + \chi_{c_{m,r}} \cdot (\boldsymbol{\Phi})_{c_{m,r}}\right]^T : (\overline{\mathbb{C}}^S)_{c_{m,r}} : \left[(1 - \chi_{c_{m,r}}) \cdot \mathbf{I}_4 + \chi_{c_{m,r}} \cdot (\boldsymbol{\Phi})_{c_{m,r}}\right]$$

- Calculation of the anisotropic and isotropic constitutive tensors of the composite according to the generalized mixing theory for long and short reinforcements (see Chapter 3).

$$\mathbb{C}^S = \underbrace{\sum_{c_m=1}^{n_m} k_{c_m} \frac{m^0}{m_{c_m}} (\hat{\mathbb{C}}^S)_{c_m}}_{\text{Matrix components}} + \underbrace{\sum_{c_r=1}^{n_r} k_{c_r} \varsigma_{c_r} \frac{m^0}{m_{c_r}} (\hat{\mathbb{C}}^S)_{c_r}}_{\text{Fiber components}}$$

$$\overline{\mathbb{C}}^S = \underbrace{\sum_{c_m=1}^{n_m} k_{c_m} \frac{m^0}{m_{c_m}} (\hat{\overline{\mathbb{C}}}^S)_{c_m}}_{\text{Matrix components}} + \underbrace{\sum_{c_r=1}^{n_r} k_{c_r} \varsigma_{c_r} \frac{m^0}{m_{c_r}} (\hat{\overline{\mathbb{C}}}^S)_{c_r}}_{\text{Fiber components}}$$

❖ **Loading incremental loop :** $n = 1$, to Maximum number of Increments.

$$\textbf{For} \quad n=1, \; i=1 \;\Rightarrow\; \mathbb{C}^T = \mathbb{C}^S$$

- **Equilibrium iteration loop:** $i = 1$, to Maximum number of Iterations.

  - (1) Calculation of the global stiffness matrix I

$$\left[\mathbf{K}^e\right]^i_n = \int_{V^e} (\nabla^S \mathbf{N}) : \mathbb{C}^T : (\nabla^S \mathbf{N}) dV \;\Rightarrow\; (\mathbf{K})^i_n = \mathbf{A}_{e=1}^{\text{Nelem}} \left[\mathbf{K}^e\right]^i_n$$

  - Calculation of the strain in the referential configuration.

$$(\Delta \mathbf{U})_n^i = \left[(\mathbb{K})_n^{i-1} \right]^{-1} \cdot \underbrace{\left(\mathbf{F}^{\mathrm{mas}} + \mathbf{F}^{\mathrm{int}} - \mathbf{F}^{\mathrm{ext}} \right)_n^{i-1}}_{\Delta \mathbf{F}}$$

$$(\mathbf{U})_n^i = (\mathbf{U})_n^{i-1} + (\Delta \mathbf{U})_n^i$$

$$(\mathbf{E})_n^i = \frac{1}{2} \left(\mathbb{F}^T : \mathbb{F} - \mathbf{I} \right)_n^i$$

➢ **COMPOSITE PHASE LOOP:** $j = 1$, to Maximum number of Components.

- Calculation of the predicting stress. Referential hyper elastic model.

$$(\mathbf{E}_{c_{m,r}})_n^i = \left[\left(1 - \chi_{c_{m,r}} \right) \cdot \mathbf{I}_4 + \chi_{c_{m,r}} \cdot (\mathbf{\Phi})_{c_{m,r}} \right] : (\mathbf{E})_n^i$$

$$(\mathbf{S}^*_{c_{m,r}})_n^i = (\hat{\mathbf{C}}^S_{c_{m,r}}) : \left[(\mathbf{E}_{c_{m,r}})_n^i - (\mathbf{E}^P_{c_{m,r}})_n^{i-1} \right]$$

- Transformation from the real space to the isotropic fictitious space.

$$(\overline{\mathbf{S}}^*_{c_{m,r}})_n^i = \mathbf{A}^S : (\mathbf{S}^*_{c_{m,r}})_n^i$$

- Transportation of the predicting stress to the updated configuration.

$$(\boldsymbol{\tau}^*_{c_{m,r}})_n^i = \vec{\phi} \left[(\overline{\mathbf{S}}^*_{c_{m,r}})_n^i \right]$$

- Integration of the constitutive equation.

- Transportation of the stress, plastic strain and tangential constitutive tensor to the referential configuration.

$$(\overline{\mathbf{S}}_{c_{m,r}})_n^i = \overleftarrow{\phi} \left[(\boldsymbol{\tau}_{c_{m,r}})_n^i \right]$$

$$(\overline{\mathbf{E}}^P_{c_{m,r}})_n^i = \overleftarrow{\phi} \left[(\overline{\mathbf{e}}^P_{c_{m,r}})_n^i \right]$$

$$(\overline{\mathbb{C}}^T_{c_{m,r}})_n^i = \overleftarrow{\phi} \left[(\overline{\mathbf{c}}^T_{c_{m,r}})_n^i \right]$$

- Transformation of the stress, plastic strain and tangential constitutive tensor from the fictitious isotropic space to the real space.

$$(\mathbf{S}_{c_{m,r}})_n^i = \left[\mathbf{A}^S \right]^{-1} : (\overline{\mathbf{S}}_{c_{m,r}})_n^i$$

$$(\mathbf{E}^P_{c_{m,r}})_n^i = \left[\mathbf{A}^E \right]^{-1} : (\overline{\mathbf{E}}^P_{c_{m,r}})_n^i$$

$$(\mathbb{C}^T_{c_{m,r}})_n^i = \left[\mathbf{A}^S \right]^{-1} : (\overline{\mathbb{C}}^T_{c_{m,r}})_n^i : \left[\mathbf{A}^E \right]$$

➢ **END OF THE COMPOSITE PHASE LOOP**

- Calculation of the stress and the constitutive tensor of the composite material.

$$(\mathbf{S})_n^i = \sum_{c=1}^n k_{c_{m,r}} \frac{m}{m_{c_{m,r}}} \left[\left(1 - \chi_{c_{m,r}} \right) \cdot \mathbf{I}_4 + \chi_{c_{m,r}} \cdot (\mathbf{\phi})_{c_{m,r}} \right]^T : (\mathbf{S}_{c_{m,r}})_n^i$$

$$(\mathbb{C}^T)_n^i = \sum_{c=1}^n k_{c_{m,r}} \frac{m}{m_{c_{m,r}}} \left[\left(1 - \chi_{c_{m,r}} \right) \cdot \mathbf{I}_4 + \chi_{c_{m,r}} \cdot (\mathbf{\phi})_{c_{m,r}} \right]^T : \left\{ (\mathbb{C}^T_{c_{m,r}})_n^i : \left[\left(1 - \chi_{c_{m,r}} \right) \cdot \mathbf{I}_4 + \chi_{c_{m,r}} \cdot (\mathbf{\phi})_{c_{m,r}} \right] \right\}$$

- Calculation of the residual load in the referential configuration.

$$\Delta \mathbf{F} = \left(\mathbf{F}^{mas} + \mathbf{F}^{int} - \mathbf{F}^{ext} \right)_n^i$$

If $\left\| \Delta \mathbf{F} \right\| > 0 \;\Rightarrow\; i = i + 1$ Return to **(1)**

Otherwise "Converged Solution" $\Rightarrow\; n = n + 1$ "New load increment"

- **END OF THE EQUILIBRIUM ITERATION LOOP**
- ❖ **END OF THE LOADING INCREMENT**

END

4.9 FMD phenomenon: micro model and mixing theory with anisotropy

An example of the formulation application combining the mixing theory, the anisotropic model in large strains and the theory that includes the FMD phenomenon analysis (Fiber-Matrix Displacement) is described below. This example compares the numerical simulation of the composite material specimen (reinforced concrete) with a central notch subjected to traction where the reinforced and matrix phases have been discretized (micro model), with a similar specimen in which only a composite material made up by a reinforced phase and the matrix (macro model) exists.

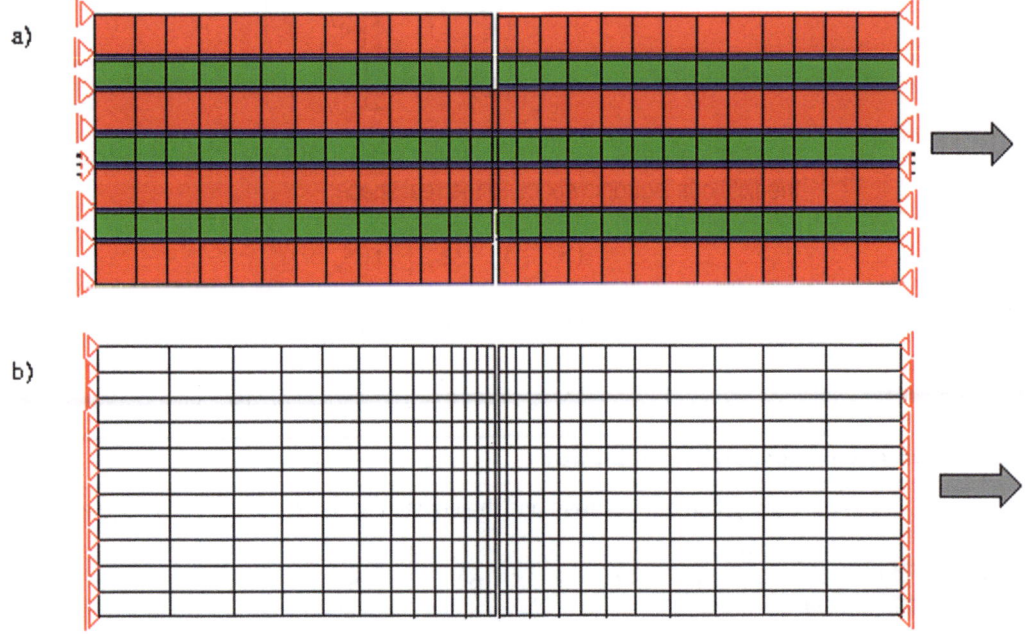

a)

b)

Figure 4.9 – Finite element mesh: a) micro model. b) macro model.

The numerical simulations have been carried out using a rectangular finite element mesh of 4 nodes with a total of 343 elements, 392 nodes and 766 degrees of freedom for the micro model and 291 elements, 336 nodes and 644 degrees of freedom for the macro model. Figure 4.9 shows the meshes and boundary conditions used for each case.

	Material 1 Matrix of the Concrete	Material 2 Steel Reinforcement	Material 3 Matrix-Reinforcement Interface
Type behavior	Mohr-Coulomb Isotropic Elasto-plastic Model	Isotropic Elastic Model	Kachanov Damage Model
Young Modulus [kp/cm²]	$3{,}5{\times}10^5$	$2{,}1{\times}10^6$	$3{,}5{\times}10^5$
Poisson Coefficient	0,2	0,0	0,0
Internal Friction	30°	-	30°
Compression Strength [kp/cm²]	200	2000	20
Tension Strength [kp/cm²]	20	2000	20
G_f, G_c [kp/cm]	0,25, 26,0	-	2,0, 2,0
Behavior law after the Yield point	Line function with softening		Exponential function with softening

Table 4.1 – Mechanical properties of the micro model.

The micro model consists of three materials: matrix, fiber-matrix interface zone and reinforcement. The macro model is made up of one composite material consisting of two phases: reinforced fiber and matrix. Table 4.1 shows the mechanical properties of the materials used in the micro model. The mechanical properties of the composite material phases in the macro model are identical to the corresponding ones in the matrix and reinforcement of the micro model.

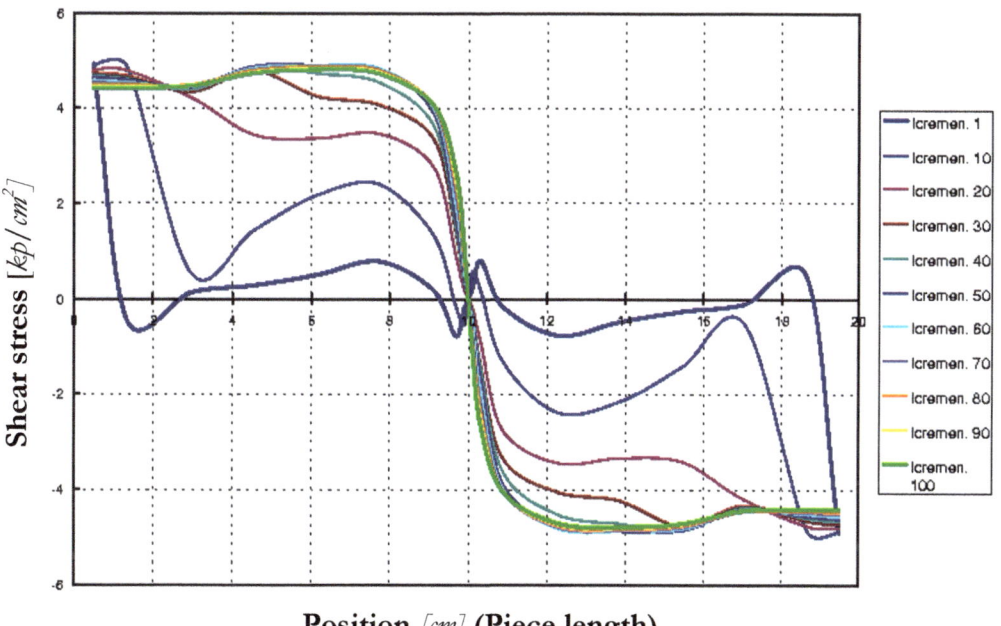

Position *[cm]* **(Piece length)**

Figure 4.10 – Shear stresses in the fiber-matrix interface. Increments 1-100.

The purpose of this example is to show the loading transfer phenomenon from the matrix to the reinforcement phase. This is achieved first by comparing the "load-displacement" curve obtained by the micro model and then by the macro model made up of composite material, the components of which are not possible to identify physically (mixing theory). Figures 4.10 and 4.11 show the shear stress evolution in the fiber-matrix interface zone for different loading increments. As shown, in the first loading increment where the irreversible processes are not verified, the shear distribution along the reinforcement is similar to the theoretical curve shown in Figure 4.1. The change of sign of the stresses, which are mainly due to the presence of the notch, can be observed in the central zone. Figures 4.12 and 4.13 show the longitudinal stress evolution in the reinforcement phase for different loading increments. It can be observed that for the first loading increment at the reinforcement end zone the maximum shear stresses are obtained at the reinforcement end zone, while the longitudinal stresses increase from zero at the end zone to a constant value along the reinforcement. Moreover, a variation of the longitudinal stress in the central zone due to the presence of the notch is observed. Additionally, in those figures an increment of the stresses applied leads to irreversible phenomena in the fiber-matrix interface zone at the reinforcement ends. Thus, a decrease of the stress transfer capacity from the matrix to the fibers is observed. This phenomenon also causes a modification of the stress state and, as observed, the stress distribution curve along the reinforcement is no longer constant. Figure 4.14 shows the interface zones exceeding the material proportionality limit for different loading stages. It can be noted that the fiber-matrix relative sliding starts at the fiber's end zone and moves towards the specimen's center.

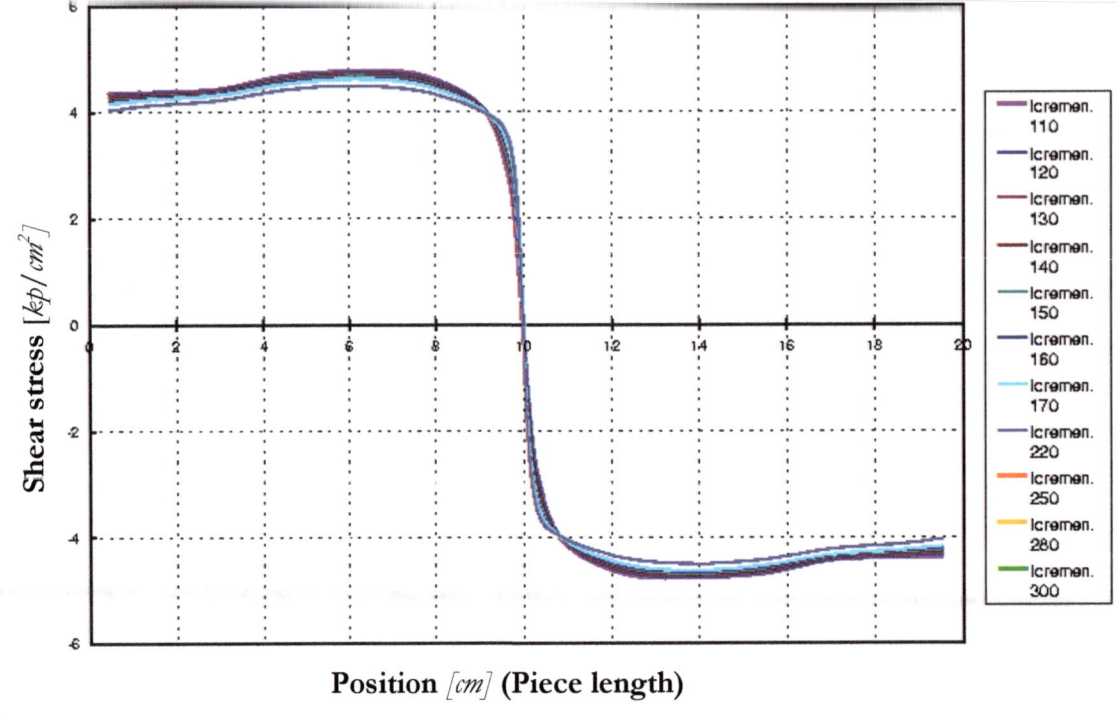

Figure 4.11 – Shear stresses in the fiber-matrix interface. Increments 110-300.

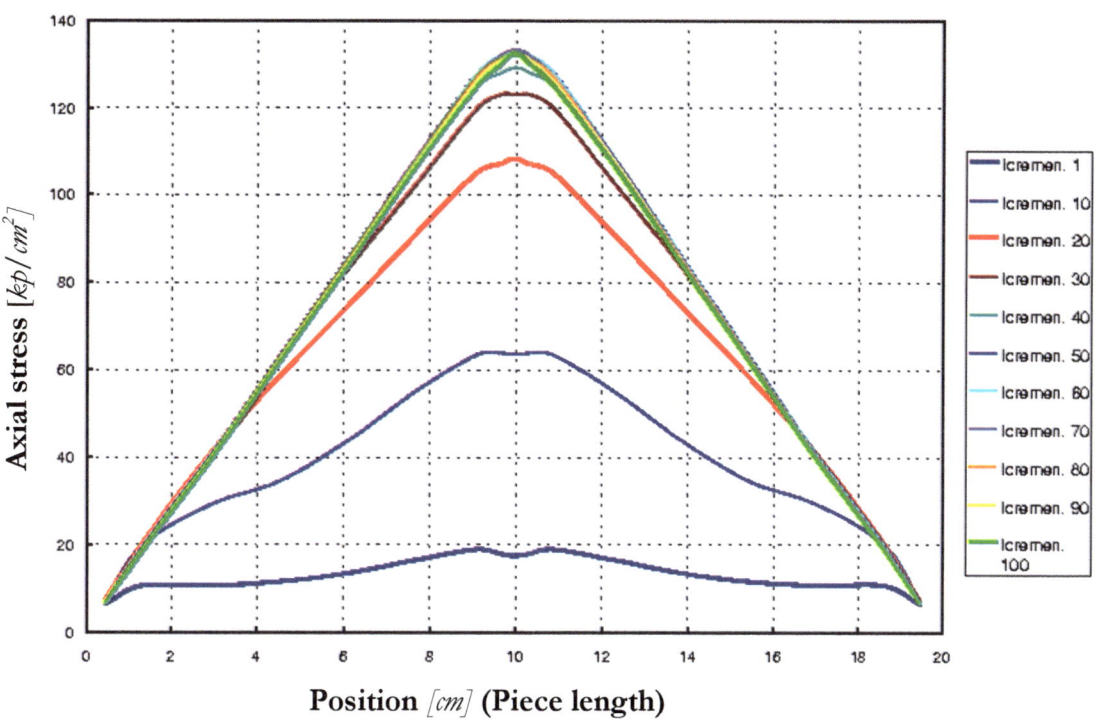

Figure 4.12 – Longitudinal stresses in the reinforcement. Increments 1-100.

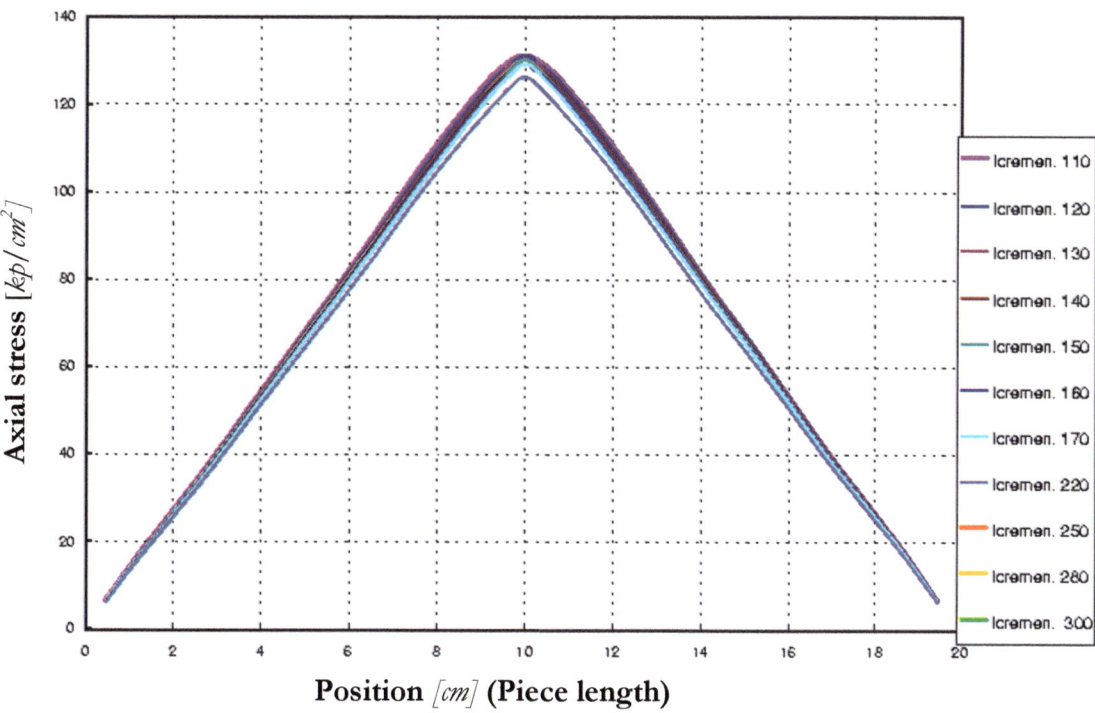

Figure 4.13 – Longitudinal stresses in the reinforcement. Increments 110-300.

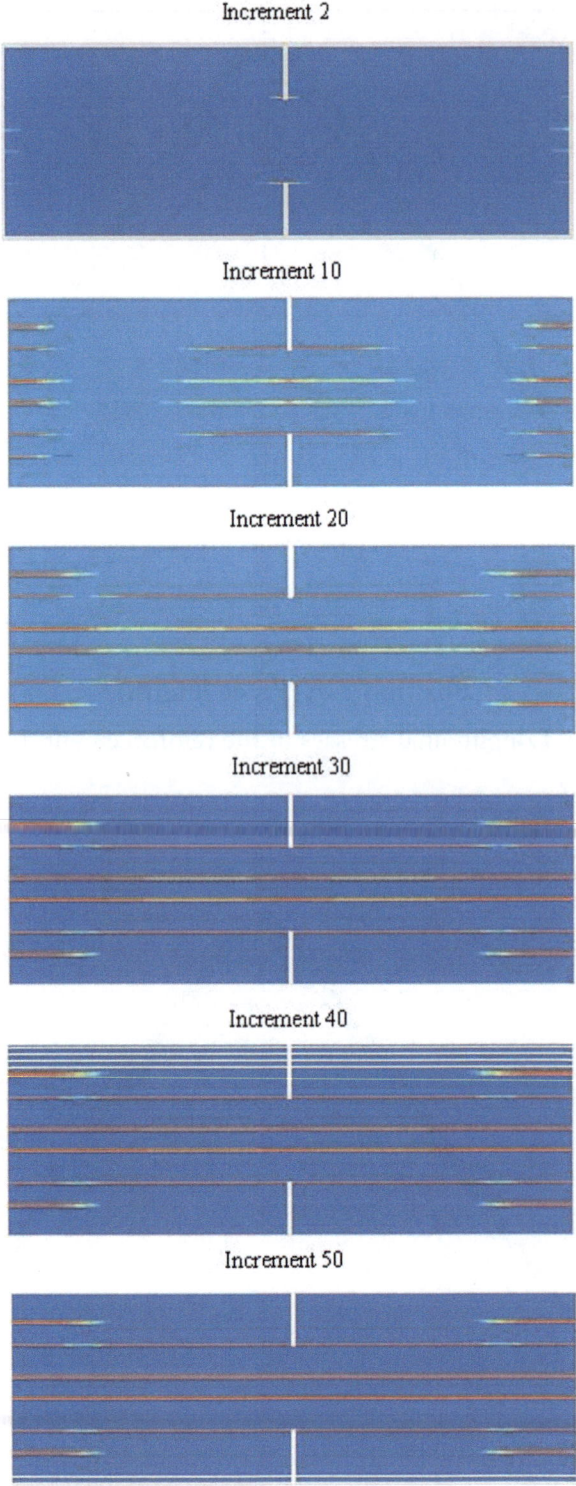

Figure 4.14 – Plastic strains in the fiber-matrix interface for different loading increments.

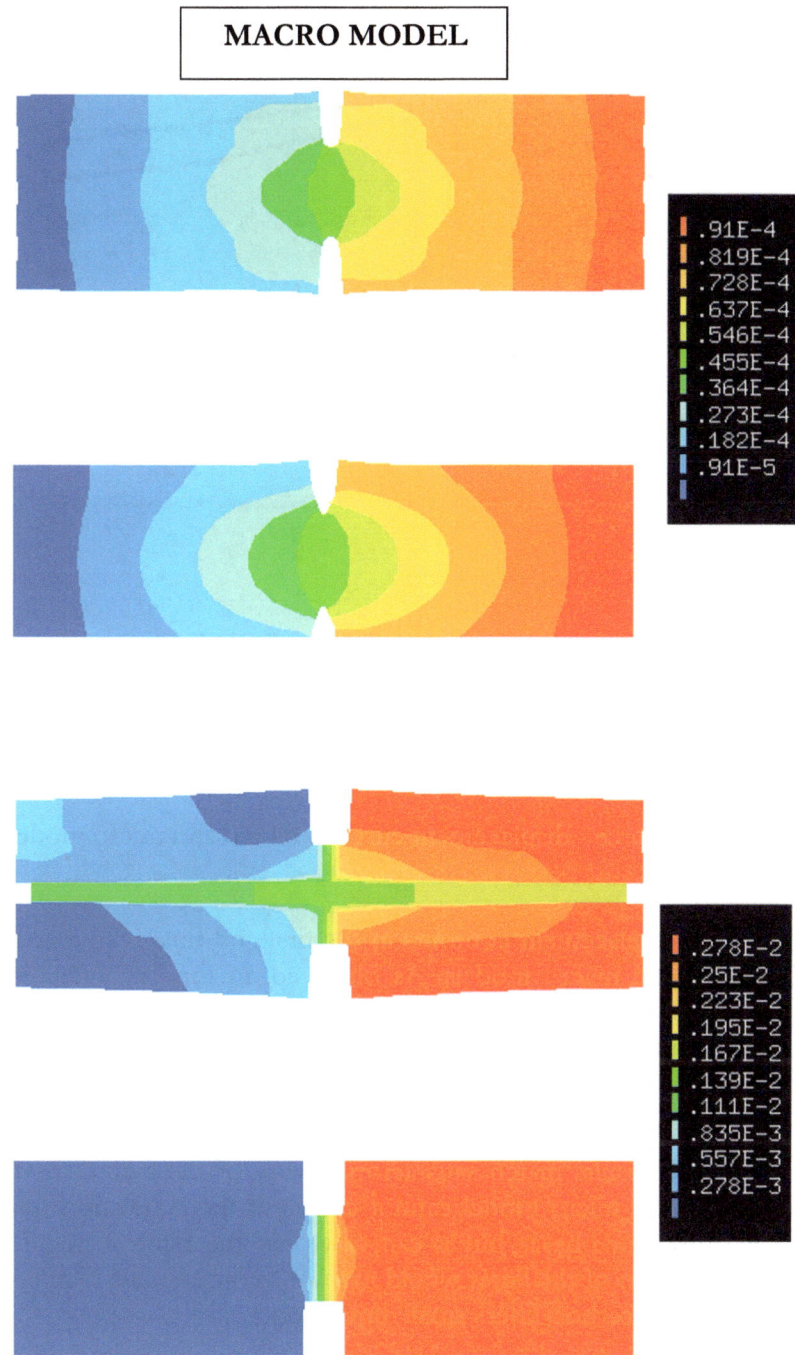

Figure 4.15 – Displacement contours of the macro and micro models in the first and final converged loading step.

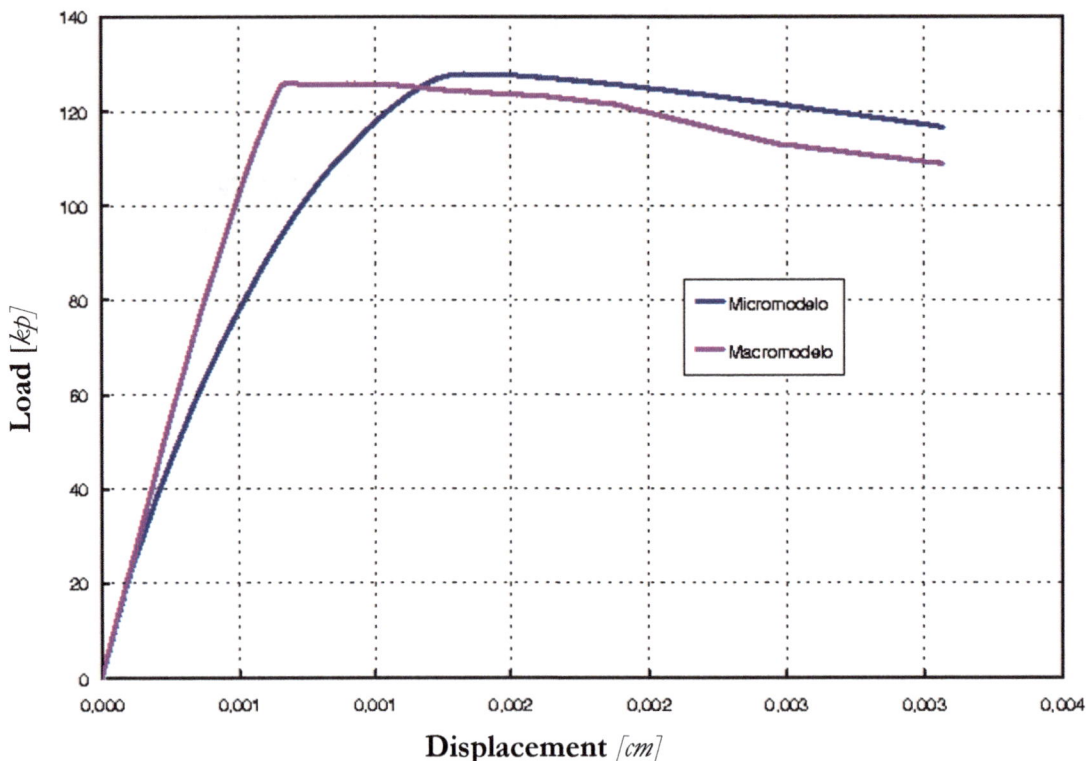

Figure 4.16 – Force - displacement curve in macro and micro models.

Figure 4.15 shows the displacement contours in the first and final converged loading in-crements for the micro and macro models. As observed, in this last increment the dis-placements are basically in the specimen's central zone and along the central reinforcement. A displacement is observed between the fiber and the reinforcement at the specimen's ends.

Figure 4.16 shows the total forces' response for micro and macro models. It can be not-ed that the micro model's results match satisfactorily those of the macro model. It is im-portant to highlight that the micro model cannot carry out the simulation of the relative movements between different phases but it can carry out the reinforcement simulation. However, the characterization of the latter would involve a considerable high computation-al cost of analysis due to the carbon fibers small dimension.

5 HOMOGENIZATION THEORY

5.1 Introduction and state of the art

The representation of the behavior of homogeneous materials is carried out through mathematical laws or equations formulated at macroscopic level in the context of the *mechanics of continuous media*. Most formulations extrapolate this concept and regard the behavior of composite materials from a macroscopic point of view but it is disregarded from the compounding materials' perspective. However, during the last decades several formulations have been developed to obtain the global behavior of composite materials through the stress and strain fields generated in the compounding materials. Consequently, new "multiscale approaches" have been proposed for the representation of the behavior of heterogeneous materials which use a representative elemental volume for the composite material modeling. The sphere assembly model proposed by Hashin (1962)[1], (1983)[2], is among these representation approaches, in which a domain is filled up by spheres of different sizes respecting the volumetric relation among the phases. Another method called "self-consistent method" was proposed by Hill (1965)[3], Budiansky (1965)[4], Hashin (1970)[5], and Christensen (1979)[6], where the heterogeneities of a medium are represented by an ellipsoidal or cylindrical inclusion within an infinite matrix of unknown elastic properties. There are also models based on the "Mori-Tanaka method" by Mori-Tanaka (1973)[7], which considers cylindrical, ellipsoidal or plane fibers or fractures embedded in an isotropic matrix transversely isotropic or orthotropic. These methods follow an "eigenstrain"[*]-based formulation, which considers an elastic, linear, homogeneous and infinite solid and the eigenstrain is also admitted. This idea was originally proposed by Eshelby (1958)[8].

Moreover, according to Sánchez-Palencia (1987)[9], the heterogeneous media can be studied from two different points of view:

---

[1] Hashin Z. (1962). The elastic moduli of heterogeneous materials. *J. Appl. Mech.*, Vol. 29, pp. 143-150.

[2] Hashin Z. (1983). Analysis of composite materials: a survey. *J. Appl. Mech.*, Vol. 50, pp. 481-505.

[3] Hill R. (1965). A self-consistent mechanics of composite materials. *J. Mech. Phys. Solids*, Vol. 13, pp. 213-222.

[4] Budiansky B. (1965). On the elastic moduli of heterogeneous materials. *J. Mech. Phys. Solids*, Vol. 13, pp. 223-227.

[5] Hashin Z. (1970). *Mechanics of composite materials. Theory of composite materials.* Pergamon. Oxford.

[6] Christensen R. M. (1979). *Mechanics of composites materials.* Wiley. New York.

[7] Mori T. and Tanaka K. (1973). Average stress in matrix and average elastic energy of materials with misfitting inclusions. *Acta Metall.* Vol. 21, pp. 571-574.

[*] It is admitted that a *total strain* consists of a pre-written elastic and inelastic part, $\varepsilon_{ij} = \varepsilon_{ij}^e + \varepsilon_{ij}^*$.

[8] Eshelby J. D. (1958). The determination of the elastic field of an ellipsoidal inclusion and related problems. *Proc. R. Soc. London.* Vol. 241, pp. 376-396.

[9] Sanchez-Palencia E. (1987). *Boundary Layers and Edge Efects in Composites. Homogenization Techniques for Composite Media.* Ed. E. Sanchez-Palencia and A. Zaoui. Spring-Verlag, Berlin.

- A *global* point of view or *macroscopic* level, in which the heterogeneities of the medium are very small and therefore ignored. Thus, the composite is considered as a homogeneous material represented at *global scale* through the x_i coordinates, as shown in Figure 5.1.

- A *local* point of view or *microscopic* level, in which the composite's internal structure is analyzed. Consequently, at *local scale* y_i is used and a part of the domain is represented. It is called *representative elemental volume*. This domain consists of small components or phases. Obviously, it is assumed that the dimensions of these phases are big enough to satisfy the hypothesis of the mechanics of the continuous medium, in other words, they are much bigger than the intermolecular distance.

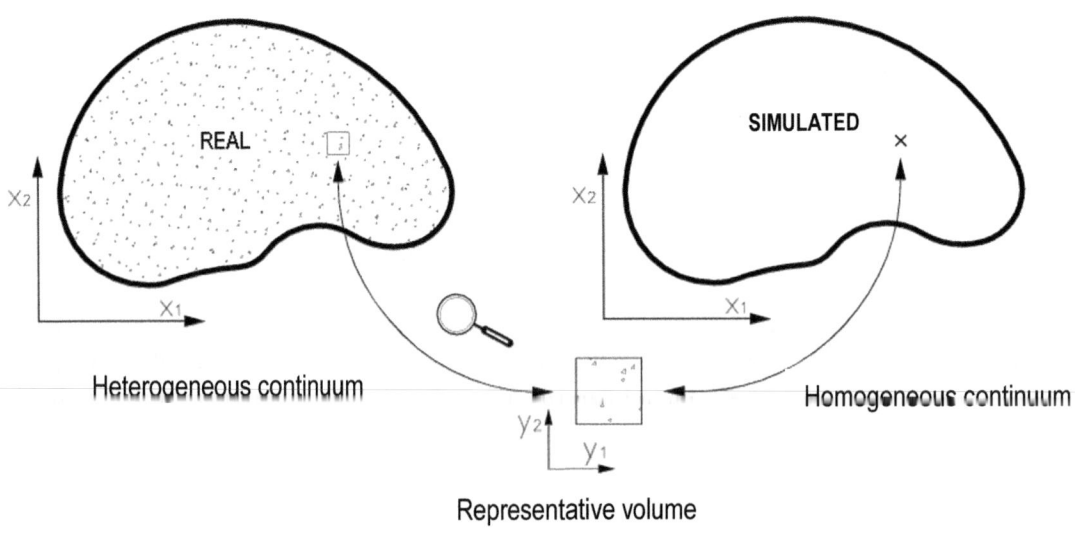

Figure 5.1 – Homogenization theory, two-scale representations.

Moreover, the hypothesis that the heterogeneous medium is *statistically homogeneous* is also assumed. Although experience shows that finding a statistically homogeneous domain is not an easy task, it is assumed that a representative elemental volume of this medium can be determined (Suquet (1987)[10]). Based on these general concepts, there are different formulations such as the *average methods*, the *asymptotic expansion theory*, etc. A brief description of some of these developments will be presented. References to the original sources (Zalamea (2001)[11] and Badillo (2012)[12]) are recommended for further information.

5.1.1 Average methods

The *average method*, also known as the *heuristic method* was the starting point of Suquet's research work (1982)[13]. He proposed various contributions to the homogenization theory.

[10] Suquet P. M. (1987). *Elements of homogenization for inelastic solid mechanics. Homogenization Techniques for Composite Media.* Ed. E. Sanchez-Palencia and A. Zaoui. Spring-Verlag, Berlin.
[11] Zalamea F. (2001). Tratamiento numérico de materiales compuestos mediante la teoría de homogeneización- Ph.D. thesis, Universidad Politécnica de Cataluña (UPC), Barcelona- España.
[12] Badillo H. (2012). Numerical modelling based on the multiscale homogenization theory. Application in composites materials and structures- Ph.D. thesis, Universidad Politécnica de Cataluña (UPC), Barcelona-España.
[13] Suquet P. M. (1982). *Plasticité et Homogénéisation.* Université Pierre et Marie Curie. Paris.

The author defined the homogenization theory as: "the procedure of replacing a strongly heterogeneous material by a homogeneous material similar to the previous one within the usual loading range". A *strongly heterogeneous* material or a composite material is the material containing a large number of heterogeneities (grains, crystals, fibers, holes, cracks, etc.) in a manageable-sized sample embedded in a matrix that has different properties (the matrix and the heterogeneities are the composite components).

In order to obtain a macroscopic behavior law of some composite materials, Hill (1967)[14] and Mandel (1972)[15] proposed that the characteristic parameters of the material should result from an *average* of the microscopic magnitudes defining the system's state. In other words, assuming V is the volume of a representative domain of the heterogeneous media, the stresses and strains of the macroscopic variables can be obtained through the average of their respective microscopic values within such domain. Therefore,

$$\boldsymbol{\sigma}^x = \langle \boldsymbol{\sigma} \rangle_V \qquad , \qquad or \qquad \boldsymbol{\varepsilon}^x = \langle \boldsymbol{\varepsilon} \rangle_V \qquad (5.1)$$

where $\boldsymbol{\sigma}^x$ and $\boldsymbol{\varepsilon}^x$ are the stress and strain tensors at macroscopic level and $\boldsymbol{\sigma} = \boldsymbol{\sigma}^y$ and $\boldsymbol{\varepsilon} = \boldsymbol{\varepsilon}^y$ are the corresponding stress and strain fields at microscopic level. The transition from these microscopic fields to their respective macroscopic ones is justified through the *average operator* $(\langle \cdot \rangle)$, which is defined as follows:

$$\boldsymbol{f}^x(x) = \frac{1}{V} \int_V \boldsymbol{f}^y(x,y) dV := \langle \boldsymbol{f}^y(x,y) \rangle_V \qquad (5.2)$$

x being the coordinates of a point at a macroscopic level, V the representative volume of the domain and $\boldsymbol{f}$ a state function of such domain depending on the macroscopic x_i and microscopic level y_i.

By assuming the stress and strain tensors as state variables, then,

$$\boldsymbol{\sigma}^x(x) = \langle \boldsymbol{\sigma}^y(x,y) \rangle_V \qquad or \qquad \boldsymbol{\varepsilon}^x(x) = \langle \boldsymbol{\varepsilon}^y(x,y) \rangle_V \qquad (5.3)$$

In this case, the problem at microscopic level shows two important differences regarding the classic boundary value problem.

1. The loads consist of field averages, not in displacements or mass or surface forces.

2. The boundary conditions in the representative domain do not exist or are not clear.

In any case, the elastic problem of composite materials is tackled as follows: finding an approximation of the *elastic homogenized constitutive tensor* $\mathbb{C}^x$ (or its inverse $\mathbb{D}^x = (\mathbb{C}^x)^{-1}$). Thus, it is possible to decouple the problem and establish a behavior law of the elastic heterogeneous media as a function the variables of which depend only on the macroscopic level x_i. Suquet generalized this average method and applied it to periodic media.

There are different procedures that can be considered as subgroups of the average method, such as:

1. *Average method formulated in strains.* Some works related to this procedure can be found in references by Hill (1967)[14], Mandel (1972)[15] and Suquet (1987, 1982)[10,13].

---

[14] Hill R. (1967). The Essential Structure of Constitutive Laws for Metal Composites and Polycristals. *J. Mech. Phys. Solid.* Vol. 15, pp. 79-95.

[15] Mandel J. (1972). *Plasticié classique et viscoplasticité.* Springer-Verlag, Nr. 97, series CISM Lecture Notes.

2. *Average method formulated in stresses.* Some works related to this procedure can be found in references by Mandel (1972)[15] and Suquet (1987, 1982)[10,13].

5.1.2 The asymptotic expansion theory

This mathematical formulation carried out through the *asymptotic developments* splits the heterogeneous media problem into scales of different magnitudes. Consequently, at each of the scales, the equations governing the material stress-strain behavior are obtained based on a rigorous mathematical work. This theory was proposed and developed by Sánchez-Palencia (1974, 1980)[16,17], Bensoussan (1978)[18], Duvaut (1976)[19] and Lene (1981)[20], among others. The interested reader can enhance these mathematical concepts of the asymptotic methods through the following references: Sánchez-Huber (1992)[21], Cole (1980)[22].

As mentioned at the beginning of this chapter, two different points of view can be adopted for composite material problems. If a microscopic point of view is chosen, then the material's small components can be observed as well as the respective fields of the state variables. Flows or oscillations occur in these microscopic fields and their wavelengths are related to the component's dimensions. However, if a macroscopic point of view is adopted, neither the medium's heterogeneities nor the rapid flows of the variables fields will be perceived. This difference in magnitude between the wavelengths of the microscopic and the macroscopic fields leads to the division of the reference space into two different magnitude spaces. Consequently, a global or macroscopic space called x_i is used, as well as a local or microscopic space called y_i, where *the internal structure of the composite or microstructure* is represented. The scales of these two spaces are related by the scale relationship ε, representing the magnitude difference between the wavelength of the two scales,

$$y = \frac{x}{\varepsilon} \tag{5.4}$$

Thus, a two-scale differential operator applied on a function $f^\varepsilon = f(x_i, y_i)$ based on two scales can be obtained,

$$\frac{\partial f^\varepsilon}{\partial x_i} = \frac{1}{\varepsilon}\frac{\partial f^\varepsilon}{\partial y_i} + \frac{\partial f^\varepsilon}{\partial x_i} \tag{5.5}$$

and the displacement field can be written within the multiscale context as,

$$u^\varepsilon(x) = u(x,y) = u^0(x,y) + \varepsilon u^1(x,y) + \varepsilon^2 u^2(x,y) + \cdots \tag{5.6}$$

[16] Sanchez-Palencia E. (1974). Comportement local et macroscopique d'un type de milieux physiques hétérogènes. *Int. J. Eng. Sc.,* Vol. 12, pp. 331-351.
[17] Sanchez-Palencia E. (1980). *Non-homogeneous media and vibration theory. Lecture Notes in Physics* Vol. 127. Springer-Verlag, Berlin.
[18] Bensoussan A. and Lions J. L. and Papanicolaou G. (1978). *Asymptotic Analysis for Periodic Structures.* North-Holland, Amsterdam.
[19] Duvaut G. (1976). *Analyse Fontionnelle et Mécanique des Milieux Continus. Th. Appl. Mech.* Ed. W. Koiter. pp. 119-132, Nord Holland.
[20] Lene G. and Duvaut G. (1981). *Résultats d'isotropie pour des milieux homogénésiés. Comptes Rendus,* Vol. 293, pp. 477-480, Acad. Sci. Paris, Series II.
[21] Sanchez-Hubert J. and Sanchez-Palencia E. (1992). *Introduction aux méthodes asymptotiques et á l'homogénéisation.* Masson.
[22] Cole J. D. and Kevorkian (1980). *Perturbation Methods in Applied Mathematics.* Springer. New York.

where $\boldsymbol{u}^{\varepsilon}(\boldsymbol{x})$ is the strain field of the heterogeneous medium which is decomposed into

the sum of different functions $(\boldsymbol{u}^{0}, \boldsymbol{u}^{1}, \cdots)$ that coexist under different magnitude orders $(\varepsilon^{-1}, \varepsilon^{0}, \varepsilon^{1}, \cdots)$. Generally, it is accepted that these functions change slowly in each of the scales.

The principle is known as *local periodicity hypothesis* (Bensoussan et al. (1978)[18]), (Sánchez-Palencia (1987)[9]), (Levi (1987)[23]) if the composite is a periodic medium. For example, consider a medium divided into unit cells of domain Y, as shown in Figure 5.2. Also, consider that two points *P1* and *P2* are identified in neighboring cells which are homologous by periodicity. The principle states that the function value f^{ε}, which represents a variable field of the state of the problem, turns out to be approximately the same in these two points as they are equivalent with respect to the local scale and are very close to each other's position at a global scale. On the other hand, the function value is generally different for any point *P3* that is homologous by periodicity but far from *P1*, due to the big difference with respect to scale x_i. Moreover, the function value will be different for any point *P4* inside the same cell but far from the initial point *P1*, as the distance between these points is big with respect to scale y_i.

To obtain the governing equations in each of the two scales, most research works use the first three terms of the displacement field decomposition, as shown in equation (5.6), and the stresses are decomposed, too. However, for the sake of simplicity and clarity, the same expressions can be obtained by using only the first two terms of the displacement field. (Lene (1986)[20]).

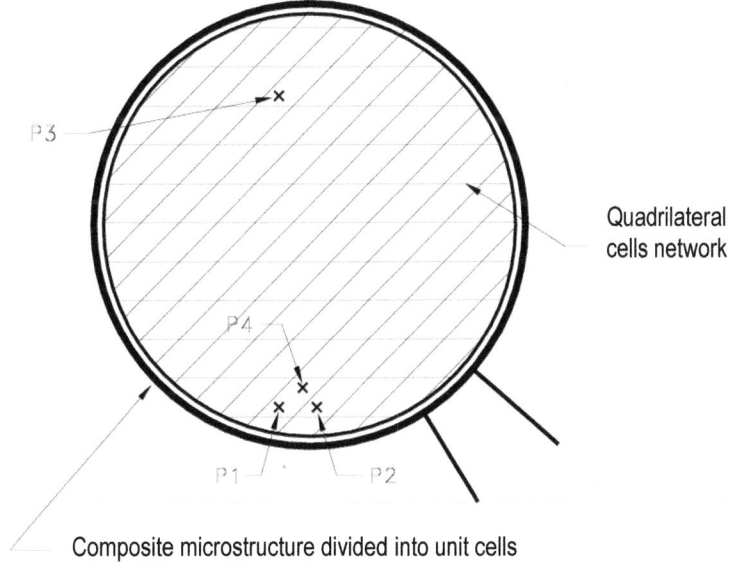

Composite microstructure divided into unit cells

Figure 5.2 – Continuous medium divided into unit cells.

[23] Sanchez-Palencia E. (1987). *Fluids in Porous Media and Suspensions. Homogenization Techniques for Composite Media.* Ed. E. Sanchez-Palencia and A. Zaoui. Spring-Verlag, Berlin.

5.1.3 Extension of the "average method" and the "asymptotic expansion method" to the non-linear problem

The introduction of the aforementioned average and asymptotic expansion methods focuses on the study of composites in linear elastic behavior. Extending these formulations to the inelastic non-linear field is not simple and it is an issue still open for discussion.

Among the first developments in the homogenization theory in the non-linear field are Suquet's works (1982, 1987)[10,13]. Due to its high complexity, in his work he takes into consideration the behavior of component materials, in other words, he studies composite materials which are viscoelastic, elastoplastic, etc. One of the main results of his research is that *the macroscopic variables of the problem depend on the microscopic variables*. A detailed description of the mathematical development of these original ideas can be found in the reference by Suquet (1987)[10] and a general idea of the state of the art about this subject can also be found in the references by Zalamea (2000)[11] and Badillo (2012)[12].

5.1.4 Other homogenization-related subjects

The homogenized elastic parameters of the composite material are obtained through a representative elemental volume or unit cell. Once the homogenized parameters are known the elastic problem can be uncoupled as a simple material. Some of the representative works dealing with the elastic problem in each scale will be mentioned further. The use of adaptive methods and error estimators to guarantee the reliability of the results will also be mentioned. On the other hand, reference will be made to works questioning the homogenization theory. Moreover, attention will be paid to the solution of the homogenization problem in the non-linear behavior of composite materials such as the imposition of boundary conditions on the unit cell, as decomposing the fields into one uniform and another periodic part may not be suitable for non-linear problems. Finally, some non-conventional techniques will be presented to deal with non-linear problems such as the finite element method in multiple scales.

5.1.4.1 Boundary conditions and implementation

One of the basic problems of the homogenization theory that researchers have not agreed on yet is how to impose boundary conditions on the unit cell. As known in solid mechanics, the stresses σ and the displacements fields u are obtained by the standard problem of the boundary values within a domain Ω, produced by some determined stresses. Thus, the domain boundary conditions, as the prescribed values of the stress vector $\bar{t}(n)$ or displacements $\bar{u}$ in the domain boundary $\partial\Omega$, are known. The existing boundary conditions in these cells are:

- Dirichlet conditions: these establish the part or the total of the boundary subjected to imposed displacements $\partial\Omega_u$

- Neumann conditions: they establish the part or the total of the boundary subjected to imposed loads $\partial\Omega_t$.

- Mixed condition: a combination of the two above-mentioned conditions $\partial\Omega_u \cup \partial\Omega_t$.

Obviously, the boundary conditions at microscopic level try to reproduce the interactions produced inside the composite internal structure.

The average theory assumes that the macroscopic stresses and strains correspond to the average of their respective microscopic fields. The Dirichlet condition is used to force the compliance of at least one of the two fields by imposing a uniform strain field (Hill (1967)[14]) or the Neumann condition by imposing a uniform stress field (Maldel (1972)[15]). The error introduced by such boundary condition is negligible when the dimensions of the heterogeneities are very small with respect to the representative elemental volume. However, the numerical solution of the volume containing many of the heterogeneities requires a considerable computational effort. Consequently, in later research works materials with periodic structures were considered with an elemental and relatively simple volume containing the information of the whole composite microstructure. Thus, the homogenization theory proposals with periodic average, based on the average theory (Suquet (1982)[13]), and the asymptotic developments by Sánchez-Palencia (1980)[17]), Duvaut (1976)[19], (Bensoussan(1978)[18]), and Lene (1981)[20], reproduced the periodicity conditions through the periodic functions. But these methods, by decomposing the microscopic variables fields into a uniform and periodic part, use the superposition of effects condition which limits them to linear problems. Therefore, some recent works are exploring different approaches to impose the boundary conditions for non-linear problems.

Swan (1994)[24] proposed a strain control technique ε^x for the homogenization of inelastic periodic composites based on the additive decomposition of the displacement fields $u = \bar{u} + u_p$, of an imposed linear contribution $\bar{u}$, plus one unknown periodic contribution u_p. In this chapter a brief presentation of the concepts is given, which also shows that the opposite sides of the unit cells have the same shape. The solution is obtained through an incremental form based on a *prediction-correction* technique in which the predictor is the linear part of the displacement field and from there periodic corrections are applied until the equilibrium of the domain is reached. Moreover, a very complex stress control technique σ^x is used as it must reach the equilibrium of the domain and then add strain and displacement restrictions. This problem is usually solved by implementing the mixed formulation of the finite elements through the penalization method. For more information about the principles of the mixed formulations see the references by Zienkiewicz (1984)[25], SIMO (1998)[26] and Hughes (1987)[27].

An interesting alternative approach to impose the boundary conditions was proposed by Anthoine (1995)[28] based on the average theory. For the masonry particular case, it is modelled in two dimensions and subdivided into homogeneous quadrilateral and hexagonal cells. Their sides are related by a vector base known as *periodicity vectors* in F. Zalamea's thesis (2000)[11]. *Lagrange multipliers* are used to solve the imposition problems of boundary conditions on the cell (Badillo 2012[12]).

In a later publication, Anthoine (1996)[29] analyzed a masonry case again, but this time component materials (brick and mortar) with non-linear behavior. The materials' "*softening*" produced convergence problems solved through "*arc-length*" numerical techniques.

---

[24] Swan C. (1994). Techniques for Stress- and Strain-Controlled Homogenization of Inelastic Periodic Composites. *Computer Methods in Applied Mechanics and Engineering*, Vol. 117, pp. 249-267.

[25] Zienkiewicz O. C. and Taylor R. L. (1994). *El método de los Elementos Finitos: formulación básica y problemas lineales, Vol 1*. CIMNE, McGraw-Hill. Barcelona.

[26] Simo J. C. and Hughes T. J. R. (1998). *Computational Inelasticity*. Springer-Verlag. New York.

[27] Hughes T. (1987). *The Finite Element Method*. Prentice-Hall.

[28] Anthoine A. (1995). Derivation of the in-Plane Characteristics of Masonry Through Homogenization Theory. *Int. J. Solids Structures*, Vol. 32, No. 2, pp. 137-163.

[29] Anthoine A. and Pegon P. (1996). Numerical analysis and modelling of the damage and softening of brick masonry. *Numerical Analysis and Modelling of Composite Materials*. Blackie academic and professional pp. 152-184.

In an article written by Michel (1999), he proposed another strain and stress control method for the determination of the composite behavior. The formulation is based on the decomposition of the displacement field into a uniform and a periodic part $u = \overline{u} + u_p$. Its application is restricted to elastic materials and plastic-stiff materials (elastic-perfectly plastic). The concept *of macroscopic degree of freedom* is proposed in this formulation. Thus, in each of the elements of the discretized unit cell an additional node called *macroscopic node* is introduced to add more degrees of freedom in the points where the macroscopic restrictions are imposed or introduced ε^x y σ^x.

5.1.4.2 Two-scale solution for the elastic problem

In this section reference is made to the work presented by Guedes (1990)[30], in which the composite is considered as a periodic material and the homogenization theory is used through the asymptotic developments. The homogenized elastic coefficients of the constitutive tensor for the composite material are determined from here.

The solution to the elastic problem in composite materials as follows can be found in the references by (Lene (1986)[31], Devries (1989)[32]):

- Pre-processing: this consists of obtaining the homogenized coefficients through the unit cell.

- The macroscopic problem is solved based on the assumption that the material is homogeneous and its behavior is established in the previous step.

- Post-processing: then the stress and strain fields are obtained in the macrostructure points of interest.

Finally, an adaptive method is introduced to improve the accuracy of the homogenized coefficients. This method consists of carrying out a finer discretization in the unit cell parts where the biggest approximation errors associated with the numerical solution (the finite element method) are found. These areas are determined by the developed error estimator.

5.1.4.3 Questioning the homogenized theory and the use of "multi-grid" and adaptive methods

As known, the adaptive methods are strategies to improve the quality of the solution obtained by the finite element method. These methods vary depending on the improvement to be achieved. Among these are:

- Techniques using a mesh refinement are known as "*h-methods*",

- Techniques to increment the polynomial order of the approximation functions are known as "*p-methods*".

- Techniques using the node relocation are known as "*r-methods*".

- Techniques that result from the combination of the previous ones, for example, are known as "*hp-methods*".

---

[30] Guedes J. M. and Kikuchi N. (1990). Preprocessing and Postprocessing for Materials Based on the Homogenization Method with Adaptive Finite Element Method. *Computer Methods in Applied Mechanics and Engineering*. Vol. 83, pp. 143-198.

[31] Lene F. (1986). Damage Constitutive Relations for Composite Materials. *Engineering Fracture Mechanic* Vol. 25, pp. 713-728.

[32] Devries F. and Dumontet H. and Duvaut G. and Lene F. (1989). Homogenization and Damage for Composite Structures. *International Journal for Numerical Methods in Engineering*. Vol. 27, pp. 285-298.

- Moreover, the "*multi-grid methods*" are techniques aimed at approximating the solution to the problem or to speed up the convergence by the iterative solution of the discretized structure in different levels. In other words, the transmission of the solution from a thicker mesh to a finer one or vice versa.

The works presented by Fish and coauthors (1993-a)[33], (1993-b)[34], (1994-a)[35], (1994-b)[36], (1995-a)[37], (1995-b)[38] are mentioned in this section, who have questioned the homogenization theory and have also made some contributions to the refinement of the solution to the elastic problem of heterogeneous media in various scales through the methods before mentioned. For example, quoting (Fish (1993-a)[33]): "The asymptotic developments follow a rigorous mathematical deduction in the heterogeneous media when the following assumptions are satisfied:

- The microstructure is periodic, in other words, the composite is formed by the spatial repetition of a very small structure or unit cell.

- The terms of the displacement decomposition $u^k(x,y)$ are periodic inside y with the same microstructure period".

However, if the material is locally non-periodic or the material is periodic but the solution in y is not periodic due to the presence of local effects (macrostructure boundaries, components break off, etc.), then the asymptotic developments give a poor approximation of the local fields. As stated in an article published by Fish (1994-a)[35]: "Unfortunately, in areas with high concentration of stresses, either at macro mechanic level ("macro-crack") or at meso mechanic level (free sides, components break off), the assumption that the macroscopic or microscopic solution is uniform in the domain of the representative elemental volume is not valid ".

Finally, as mentioned in another reference (Fish (1895-a))[37]: "It is widely known that in the limit when $\varepsilon \to 0$ the solution to the heterogeneous media problem is close to the boundary values problem with homogenized coefficients". Unfortunately, in many practical situations when the value of $\varepsilon \to \infty$ and the solution to the homogenized problem has a high gradient, the solution obtained can be considerably different from the initial solution the problem. The main sources of error are located in the problem domain areas where the solution has high gradients. Ironically, from a practical point of view, these are precisely the regions of greater interest. Moreover, the researchers vision can be summarized as follows (Fish (1995-b))[38]: "The mathematical theory of the homogenization is used to capture the low frequency of the heterogeneous medium while in the high gradient areas the response oscillates, therefore a disturbing term is introduced, the solution is determined by applying relaxation techniques until the convergence is achieved".

[33] Fish J. and Wagiman A. (1993-a). Multiscale finite element method for a locally nonperiodic heterogeneous medium. *Computational Mechanics*. Vol. 12, pp. 164-180.

[34] Fish J. and Markolefas F. (1993-b). Adaptive s-method for linear elastostatics. *Computer Methods in Applied Mechanics and Engineering*. Vol. 104, pp. 363-396.

[35] Fish J., Nayak P. and Holmes M. H. (1994-a). Microscale reduction error indicators and estimators for a periodic heterogeneous medium. *Computation Mechanics*. Vol. 14, 323-338.

[36] Fish J. ,Markolefas S., R. Guttal and P. Nayak (1994-b). On adaptive multilevel superposition of finite element meshes for linear elastostatics. *Applied Numerical Mathematics*. Vol. 14, pp. 135-164.

[37] Fish J. and Belsky V. (1995-a). Multi-grid method for periodic heterogeneous media, Par 1: Convergence studies for one-dimensional case. *Comput. Methods Appl. Mech Engrg*. Vol. 126, pp. 1-16.

[38] Fish J. and Belsky V. (1995-b). Multi-grid method for periodic heterogeneous media, Par 2: Multiscale modeling and quality control in multidimensional case. *Comput. Methods Appl. Mech Engrg*. Vol. 126, pp. 17-38.

These ideas from publications help to deal with the problem. In an article entitled "Multiscale finite element method for a locally non periodic heterogeneous medium" (Fish 1993-a))[33], the decomposition of the global domain in a periodic part is proposed, where the homogenization theory solution is used. The rest of the domain, where the local effects are supposed to be dominant, is considered as a non-periodic portion (for example: the boundary). In this part of the problem a fine mesh is superimposed and its boundary conditions are the displacement obtained in the solution of the global problem. In this case, the displacements correspond to the macroscopic field u^0 enriched by a microstructure contribution through the second term of the asymptotic decomposition u^1; the solution is obtained in a coupled form (see details in the reference). However, it is not clear whether dividing or introducing this disturbance will lead to better results.

A more detailed work by Fish (1993-b)[34] proposed an adaptive "*s-method*" for elastic problems based on a fine mesh superimposition where an error estimator indicates what is required. Its operation is as follows:

- An error estimator is proposed to be applied node by node.

- Critical regions are identified according to the density boundaries of the estimated error.

- A fine mesh is superimposed in places where unacceptable errors are determined.

- The solution of the adaptive refinement is obtained.

- The quality of the solution is assessed through a global and local norm.

Later Fish (Fish (1994-a)[35] developed a "Microscale error reduction indicator and estimator", which is based on the estimation of high-order terms that are neglected in the double-scale classic formulation in the asymptotic expansion theory.

A further work compiles previous works (Fish (1994-b))[36]. The "*s-method*" is generalized by superimposing meshes at various levels. Two versions are presented, one for structured meshes and another for non-structured ones. The structure mesh version is solved with an iterative "solver" for positive defined symmetric systems. Moreover, two procedures are analyzed: the use of a preconditioned conjugated gradient and a hierarchical "*multi-grid*" algorithm. The adaptive process based on an error estimator and a refinement strategy is discussed. Also a strategy is proposed to simulate discontinuity formation and its propagation through a hierarchical adaptive method using discontinuous fields.

Furthermore, a change of strategy is found in the publication by (Fish (1995-a))[37] and (Fish (1995-b))[38] entitled "Multi-grid method for periodic heterogeneous media" (parts 1 and 2). In this work the uniformity hypothesis of the variable field at macroscopic level is abandoned. Some transfer operators are proposed to pass the information from the thick mesh to the fine one and vice versa "intergrid transfer operators", using the asymptotic developments terms. The solution is carried out iteratively in the two scales, one at macroscopic level and another at the components level. Again the error reduction estimator and indicator are used at microscale level.

Because the response depends on the solution obtained by the finite element method, it is obvious that the use of adaptive methods (in each of the scales) improves the result. However, it is not clear whether the introduction of such disturbances and their corresponding relaxations is appropriate. In this case the solution seems to depend on both the disturbance value and the introduction level or scale. Also the use of all these refinement techniques increases considerably the elastic problem complexity of composite materials.

Finally, a different approach has been proposed for heterogeneous materials, but based on a research line called *Homogenized Dirichlet Projection Method* (HPDM). In this case, the microstructure effects are obtained at different scales over the response at macroscopic level of the heterogeneous medium. As in the previous case, an error estimator is introduced and it is refined repeatedly at various levels or scales by a hierarchical method. For further details, see the following references (Zohdi (1996)[39], Moes (1998)[40], Oden (1999)[41].

5.1.4.4 Homogenization by the Voronoi finite element method

This method is an innovative approach proposed for non-linear problems to represent a heterogeneous material through the *Voronoi finite element method* (Ghosh (1995)[42], Ghosh (1996)[43]). This method was created to reproduce the behavior of the randomly distributed heterogeneities of materials. Thus, a medium volume with arbitrary dispersed heterogeneities is represented by a domain partition in convex polygons of various sides called *Voronoi elements*. As shown in Figure 5.3, in order to represent heterogeneities, each of the elements contains a second phase or inclusion in the domain.

In this case, each one of the Voronoi elements can be considered as a base cell so that a number of these elements inside the representative domain can characterize a volume of a medium with randomly dispersed heterogeneities. In order to make its use easier, the authors have developed mesh generators which create these polygons based on their shape, size and heterogeneities location (Ghosh (1991)[44], Ghosh (1993)[45]).

This formulation is intended to reduce the computational cost involved in determining the microstructural fields in the composite because the exact determination of the microscopic fields in random materials is not as important as in the fields' determination at macroscopic level.

Through this procedure a finite element is represented by each Voronoi element and its formulation is carried out through a *stresses hybrid method* introduced by Pian (1964)[46]. Later, heterogeneity is incorporated inside the matrix of the Veronoi element for the composite representation (Ghosh (1993). Its effect is introduced by the continuity restriction of the traction vector in the matrix-inclusion interface. In the interface, the stress and strain fields

[39] Zohdi T. I., Oden J. T. and Rodin G. J. (196). Hierarchical modeling of heterogeneous bodies. *Comput. Methods Appl, Mech. Engrg.* Vol. 138, pp. 273-298.

[40] Moes N., Oden J. T. and Zohdi T. I. (1998). Investigation of the interaction between the numerical and the modeling errors in the Homogenized Dirichlet Projection Method. *Comput. Methods Appl. Mech. Engrg.* Vol. 159, pp. 79-101.

[41] Oden J. T., Vemaganti K. and Moes N. (1999). Hierarchical modeling of heterogeneous solids. *Comput Methods Appl. Mech. Engrg.* Vol. 172, pp. 3-25.

[42] Ghosh S., Lee K. and Moorthy S. (1995). Multiple Scale Analysis of Heterogeneous Elastic Structures Using Homogenization Theory and Voronoi Cell Finite Element Method. *Solids Structures.* Volume 32, No. 1, pp. 27-62.

[43] Ghosh S., Lee K. and Moorthy S. (1996). Two Scale Analysis of Heterogeneous Elastic-plastic Materials with Asymptotic Homogenization and Voronoi Cell Finite Element Model. *Comput. Methods Appl. Mech. Engrg.* Volume 132, pp. 63-116.

[44] Ghosh S. and Mukhopadhyays s. N. (1991). A two-dimensional automatic mesh generator for finite element analysis of randomly dispersed composites. *Computers Struct.* Vol. 41, pp. 245-256.

[45] Ghosh S. and Mukhopadhyays s. N. (1993). A materials based finite element analysis of heterogeneous media involving Dirichlet tessellations. *Comp. Meth. Appl. Mech. Engng.* Vol. 104, pp. 211-247.

[46] Pian T. (1964). Derivation of element stiffeness matrices by assumed stress distribution. *AIAA J.* Vol. 2. pp. 1333-1336.

can be discontinuous whereas the displacement fields can be continuous. The discontinuities in the stress field are achieved by generating hikes in the coefficients in the polynomic interpolation of the stresses. Therefore, the functional of the *complementary energy* of the element satisfies the continuity restriction of the traction vector on the imposed matrix-inclusion interface through Lagrange multipliers.

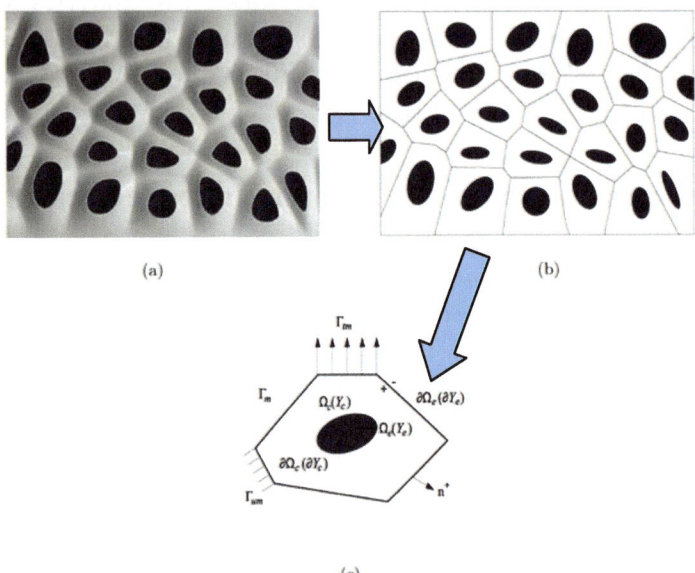

(a)   (b)

(c)

Figure 5.3   Representation of a heterogeneous structure with the VCFEM. (a) Composite material formed by two randomly distributed composites; (b) heterogeneous structure discretized by Dirichlet tessellation; (c) basic structural element represented by a Voronoi cell.

Later, Ghosh (1995)[42] proposed a two-scale method to deal with the elastic problem of composite materials. In the proposed method the governing equations in each scale are obtained by the asymptotic development theory and the analysis of the representative volume is carried out by the Voronoi finite element method

In (Ghosh 1996)[43]) the method is extended to non-linear problems. Moreover, interesting examples are shown comparing the proposed method to the homogenization theory through asymptotic developments. The results are obtained from lab tests and the constitutive equations are determined by certain composites. The results were quite good and the computational cost was lower than the unit cell methods. In a later publication (Lee 1999)[47]) a fracture criterion inside the element and a macroscopic scale adaptive method is introduced.

This innovative method is highly effective for the study of randomly distributed composites. A periodic medium unit cell can be represented even with only one Voronoi finite element, which is a great advantage for its fast results. However, due to its formulation assumptions, the precision of the microscopic fields provided by one single Voronoi finite element is lower than the precision obtained by the unit cell (with various finite elements), although the global average value of these fields is approximated. On the other hand, due to its formulation it may be difficult to extend this methodology to non-periodic media problems presenting: component decomposition, complex heterogeneity forms, composites with more than two phases, etc.

---

[47] Lee K., Moorthy S. and Ghosh S. (1999). Multiple scale computational model for damage in composite materials. *Cumput. Methods Appl. Mech. Engrg.* Vol. 172, pp. 175-201.

5.2 Homogenization theory based on "local periodicity"

5.2.1 Introduction

Various *multiscale methods* have been developed to solve the composite material problem. Some of these methods have already been presented in previous sections and are included in the *homogenization theory* where composite materials are divided into two different scales. *A* composite material is accepted as a homogenous material *at macroscopic scale* (x_i) and its behavior can be studied from a unit volume –elemental volume, represented by a second or a microscopic scale (y_i). When the material's internal structure is periodic, the representative unitary volume of this periodicity is called *cell* or *unit cell*. This two-scale technique is equivalent to determining the properties of the composite materials when the unit cell dimensions tend to zero.

The homogenization theory proposed by Sanchez-Palencia (1980)[17], Bensoussan (1978)[18] and Duvaut (1976)[19] is among the most relevant methods. Its formulation is based on the asymptotic expansion theory. Due to these developments, the two-scale technique to deal with composite material problems is based on rigorous theoretical analysis. On the other hand, as mentioned in the previous section, Suquet (1982[13], 1987)[10] used an *average method* and extended it to the non-linear case by coupling macroscopic variables with the microscopic ones. Coupling the microscopic variables with the macroscopic ones involves an additional difficulty to the two-scale treatment problem because there are an infinite number of variables to solve. In order to obtain a simple constitutive law for composite materials some simplifications have been proposed which are only applicable to simple structural problems. Ghosh (1996)[43] was also involved with non-linear structures. He found a solution by applying the Veronoi finite elements and an asymptotic expansion as a link between the two scales.

A new alternative for the *homogenization theory* for periodic media is proposed by F. Zalamea (2000)[11] as a result of his thesis. The formulation developed uses the standard *mechanics of the continuum media* and it coincides with the main ideas presented in the "state of the art" summarized in the previous section. In other words, it is admitted that the effective values at the stress and strain macroscopic level are associated with the average of the corresponding microscopic fields and the two-scale problem decomposition as presented by the asymptotic expansion theory. However, new mechanisms or concepts that might have gone unnoticed are sought. The analysis starts from the consequences deriving from the periodicity of the medium and its division into unit cells. Through a number of concepts provided in this analysis along with the *local periodicity hypothesis* (Sanchez-Palencia (1987)[9]) some of the macroscopic variables from the fields of the microscopic variables can be deduced. Moreover, the equations governing the problem at each scale are obtained without using asymptotic developments. Then, a boundary value problem at microstructure level is posed through the appropriate boundary restrictions for the unit cell at both linear and non-linear range. Its solution is determined by the *finite element method* where the boundary restrictions are imposed by the Lagrande (Anthoine (1997)[48]) multipliers.

The composite material's global problem occurs within the double-scale context. To solve the non-linear problem it is admitted that microstructures state variables cannot be

---

[48] Anthoine A. and Pegon P. (1996). *Numerical Analysis and Modelling of Composite Materials. Chapter: Numerical analysis and modelling of the damage and softening of brick masonry.* Blackie academic and professional. pp. 152-184.

decoupled (Suquet (1982)[13], Suquet (1987)[10]). Therefore, a numerical method through the *finite element method* is proposed to obtain the composite material behavior. In other words, solving the problem by the natural way indicating the macrostructure variables depend on an infinite number of internal variables. This approach contradicts the opinion of other researchers about the subject. They say it is not possible to obtain the exact solution (or numerical) of the microstructural fields due to the enormous computational work required (Fish (1997)[49]). Nevertheless, despite this opinion and due to the computational advances, this approach has been used successfully for the solution to the problem in this work. It has set a path forward to be certainly improved.

This formulation is fully justified by the geometrical aspects of the microstructure involved. Moreover, no assumptions about the nature of the material's components or dubious simplifications of the problem are required. The result is a general method that does not require the explicit development of a constitutive equation because the computer determines the material's behavior from the microstructure information through an algorithm 1. Field computation at microstructural level involves a high cost both in calculation processing and information management. However, the complexity of the problem justifies it. A rigorous and consistent formulation is presented in this work which due to the current computational capacity can solve composite material problems through variable field determination at microscopic level. Likewise, the micromechanical phenomena present in these materials are also analyzed.

5.2.2 Periodic structure concepts

The periodic distribution of the composite material components shows some symmetries that can divide the composite material in unit structural cells. This virtual division is carried out through polygons that have their sides in contact. Thus, a plane section of the composite material represented in the two-dimensional space can be divided in four-sided cells (two pairs of periodic sides) called periodic cells or in six-sided cells (three pairs of periodic sides) called hexagonal periodic cells. The points in which the material coordinates X_i have the same relative position in the cell neighborhood are called periodic points (see Figure 5.4). Each periodic point in a cell domain is indicated as P. The relative position between two points determines the vector base D which will be called periodic vector (also known as cell base-vector). The latter establish the cell domain dimensions and directions.

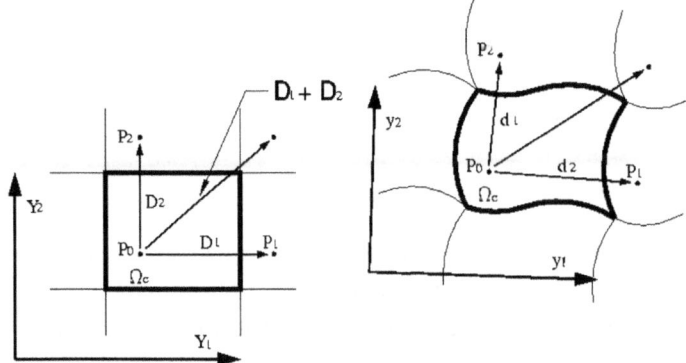

Figure 5.4 – Periodic vectors in the referential configuration D and periodicity vector d in the updated configuration of the cell domain Ω_c.

[49] Fish J., Shek K., Pandheeradi M. and Shephard M. S. (1997). Computational Plasticity for Composite Structures Based on Mathematical Homogenization: Theory and Practice. *Computer methods in applied mechanics and engineering.* Vol. 148, pp. 53-73.

In heterogeneous materials, magnitudes such as the stress or the strain depend on the scale. Thus, from a macroscopic point of view, composite materials can be considered as homogenous, so each point of the solid has a magnitude of its homogenized stress or strain. The information passed from the microscopic to the macroscopic variables is carried out by the periodicity hypothesis introduced by Sánchez-Palencia (1987)[23]. This hypothesis comes from the equation minimization stating that the accumulated energy by the microscopic variables inside a cell is equal to the one accumulated by the forces on the cell contour, but it may be very different to the one developed by distant cells.

5.2.3 Local periodicity of variables

In a continuum medium the strain and stress concepts are assumed for a point, in other words, it is not linked to the dimensions of a domain. It is also assumed that these variables set up continuum fields which vary smoothly in their neighborhood. In this same context, if the composite material is only considered at macroscopic level it can be assumed that the global stress and strain fields meet these requirements. However, at the components' level, these variables change abruptly due to their different properties. At microstructure level these fields have a great number of irregularities or fluctuations that make it extremely difficult to handle. Consequently, new mechanisms must be sought to deal with the problem variables.

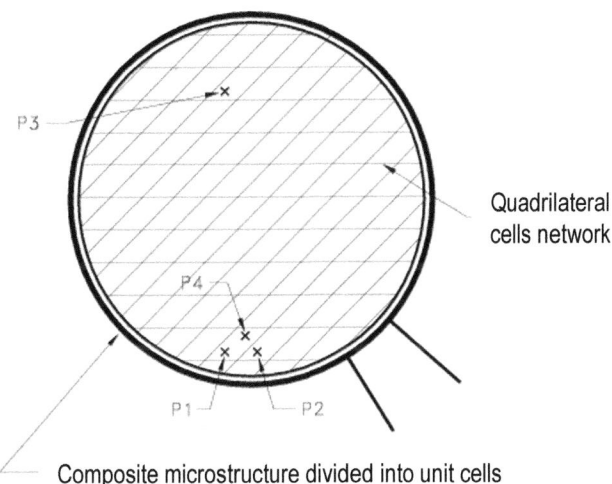

Figure 5.5 – The points P1, P2, and P3 are considered homologous due to their periodicity.

The alternative approach proposed in this work is based on the *local periodicity* hypothesis (Sánchez-Palencia (1987)[23], Levi (1987)[50]) formulated by the asymptotic developments. This principle says that in a periodic medium the fields of the state variables have local periodicity. Based on the energy minimization, in periodic media the state variable fields such as displacements and forces or strains and stresses tend to be periodic as their geometry in the space (in small strains). In other words, the variables' value in two close periodic points P_1 and P_2 is practically the same. In contrast, if a third point P_3 is identified as homologous by its periodicity to P_1 and P_2, but far from the former, then the

[50] Levi T., Sanchez-Palencia E. and Zaoui A. (1987). *Homogenization Techniques for Composite Media. Chapter: Fluids in Porous Media and Suspensions.* Springer-Verlag.

field value will be different. Moreover, it will also be different from any point P_4 even if it is inside the same cell as P_1 but far from the macroscopic scale.

Variable periodicity arises naturally when considering the problem of this type of media. Assume that a body Ω is made up by a periodic composite material which is divided into very small cells Y (Figure 5.5). Also assume that the body has a displacement or a strain produced by the stresses on its boundary. At microscale level, consider any cell located not exactly at its boundary. Obviously, such cell will undergo a transformation due to the forces generated around its boundary as will the neighboring cells. The energy minimization principle states that each domain reaches its internal equilibrium with the minimum energy consumption. Because, by definition, these domains are the same both in shape and properties and are distributed generally one after the other, the stress and displacement field that is generated in these domains is the same (it is also possible to justify this result by *Sain-Venant's principle* in elasticity (Oleinik (1992)[51]). In other words, if the cell boundary is taken as a reference, the periodic faces of the neighboring cells keep "parallel" to one another, even if their shape changes (see Figure 5.6). This ensures the displacement compatibility; otherwise overlapping would occur and form hollows.

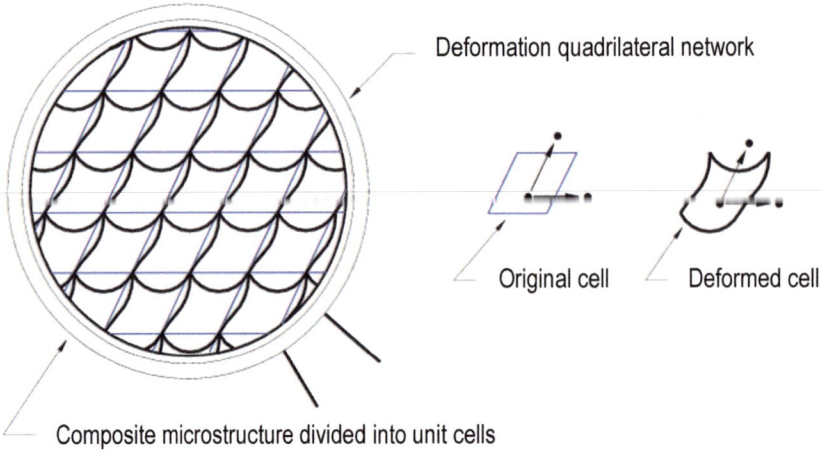

Figure 5.6 – Deformation of a quadrilateral mesh, under a periodic displacement field without periodicity vectors change.

It should be noted that this displacement field has the peculiarity of keeping the local periodicity of the medium. Something similar occurs with the forces generated on the cell boundary. The stresses on a cell face are transmitted with the same magnitude and opposite direction to the neighboring cell due to the principle of action and reaction. Thus, stresses of the same magnitude and opposite direction on its periodic faces are produced in these cells. In the homogenization literature this principle is called the antiperiodic stress field in the formulation for periodic media.

5.2.3.1 Periodic displacement field effect

As mentioned before, in periodic media displacement fields are produced which keep the local periodicity of the medium. The field's analysis reveals that these displacements have two simultaneous consequences:

[51] Oleinik O. A., Shamaev A. S. and Yosifian G. A. (1992). *Mathematical Problems in Elasticity and Homonenization. Elsevier Science Publishers* B. V. Vol. 26, Series:Studies in mathematics and its applications.

1. The first is a differential displacement of particles within the cell which causes face deformation (undulations). If this displacement does not alter the periodicity vectors, such displacement can be considered as a perturbation as it only occurs at microstructural level. Figure 5.6 shows a domain divided into cells through a mesh formed by quadrilaterals; a periodic displacement field is imposed on the domain without affecting the distance relation of the cell faces but affecting their shape. It should be noted that despite the cell domain perturbation the global dimension of the *quadrilateral mesh* does not change. Moreover, the distance among vertices (and in periodic points in general) is not altered.

2. The second consequence of a periodic displacement field is the periodic vector's base modification. For example, a cell is deformed but the distance among the periodic faces is affected (see Figure 5.7).

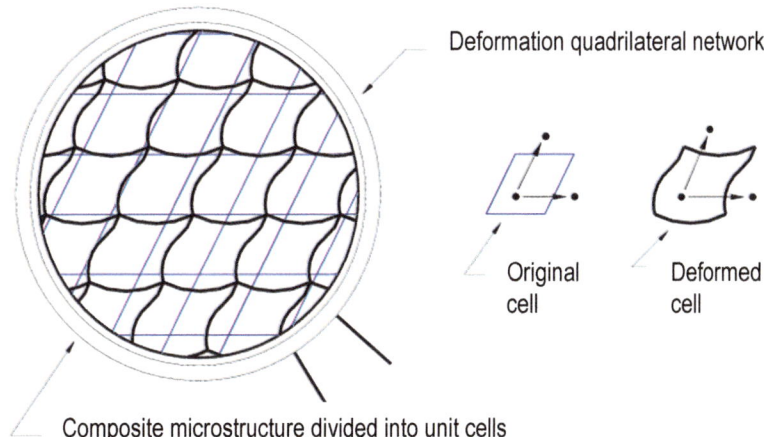

Figure 5.7 – Deformation of a mesh under a periodic displacement field that modifies the cell periodicity's vectors base.

As the neighboring cells undergo the same transformation (periodic condition) and the *mesh's* global dimension are altered, the effect is amplified from the microscopic to the macroscopic level. Each of these two effects is relevant and is produced simultaneously:

1. The first one can be considered as the differential displacement of particles. It is important for the cell's internal equilibrium.

2. The second one alters the periodicity vectors and is related to the strain of the medium.

In hexagonal cells, perturbations can occur from shape changes in the faces as well as displacements of three of its periodic vertices without affecting the global level as shown in Figure 5.8.

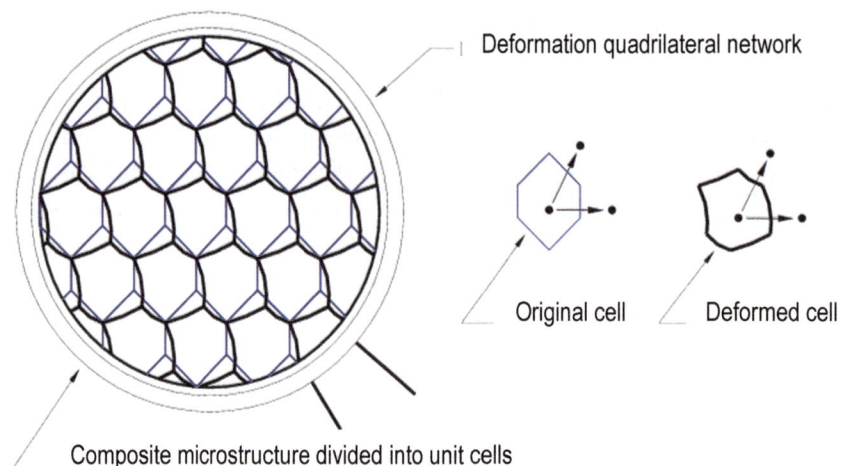

Figure 5.8 – Periodic displacement field showing a relative displacement of three of the periodic vertices.

This occurs because the periodicity vectors within a cell are defined by three periodic vertices of the cell. The other three vertices can be moved with the same displacement without altering the periodicity vectors and, due to the very small size of the cell, this domain perturbation cannot be noticed at macroscopic level. However, if the relative distance among the periodic vertices is increased or decreased, the other vertices must undergo the same transformation so that the medium keeps the local periodicity and the effect is amplified from the microscopic to the macroscopic scale.

It should be highlighted that the strain at macroscopic level is related to the transformation of the periodicity vectors or, in other words, the relative displacement among periodic points. This relation is obtained in the next section.

5.2.4 Strain tensor homogenization

Assume that Ω_c represents the cell domain of a composite material (represented in the referential space Y_i – see Figure 5.4), and that the domain of this cell is characterized by the periodicity vector $\boldsymbol{D} = Y_p - Y_{p0}$. The local periodicity hypothesis states that the composite material subjected to displacements keep the periodicity relation with its neighbors despite the strains experienced –the deformed cell along with its neighbors undergo the same transformation. Thus, the new periodic vector $\boldsymbol{d}$ can be written as,

$$\boldsymbol{d} = \boldsymbol{y}_p - \boldsymbol{y}_{p0} = \boldsymbol{D} + (\boldsymbol{u}_P - \boldsymbol{u}_{P0}) \tag{5.7}$$

where $\boldsymbol{u}_P - \boldsymbol{u}_{P0}$ is the displacement difference between periodic points. The cell space transformation is associated with the periodicity's vector change. The vector's partial derivative with respect to its reference coordinate is obtained as follows,

$$\frac{\partial \boldsymbol{d}}{\partial \boldsymbol{D}} = \frac{\partial (\boldsymbol{y}_p - \boldsymbol{y}_{p0})}{\partial (\boldsymbol{Y}_p - \boldsymbol{Y}_{p0})} \tag{5.8}$$

At macroscopic level, it can be assumed that the periodicity vectors are infinitesimally small ($|\boldsymbol{D}| \to 0$). Consequently, at macroscopic scale these tend to the following limit,

$$\lim_{D \to 0} \left[\frac{\partial d}{\partial D} \right] = \lim_{D \to 0} \left[\frac{\partial (y_p - y_{p0})}{\partial (Y_p - Y_{p0})} \right] = \frac{\partial x}{\partial X} = F \tag{5.9}$$

Thus, the periodicity vector in the updated configuration can be written as,

$$d = F\ D \tag{5.10}$$

where F is the *homogenized strain gradient tensor*. With this simple scale change the *homogenized macroscopic strain tensor* can be obtained based on the mechanics of the continuum media. The square of the length of this new periodicity vector is represented as follows,

$$|d|^2 = D^T\ F^T\ F\ D \tag{5.11}$$

The difference between the square of the length between the periodicity vectors located in the updated and referential configurations, respectively, provides the following relation for determining the *homogenized strain or macroscopic strain.*

$$|d|^2 - |D|^2 = \left[D^T\ F^T \right] \left[F\ D \right] D^T D$$
$$|d|^2 - |D|^2 = 2\ D^T\ \tilde{E}\ D \tag{5.12}$$

Thus, to obtain the *homogenized Green Lagrange strain tensor,*

$$\tilde{E} = \frac{1}{2} \left[F^T\ F - I \right] = \frac{1}{2} \left[\frac{\partial F}{\partial D}^T \frac{\partial F}{\partial D} - I \right] \tag{5.13}$$

This tensor $\tilde{E}$ is associated with the change of the periodicity vectors. It also coincides with the classic definition given by the *average theory* (Suquet (1982)[13]),

$$\tilde{E} = \left\langle E(y) \right\rangle_{\Omega_c} = \frac{1}{V_c} \int_{V_c} E(y)\, dV_c \tag{5.14}$$

Where $\left\langle E(y) \right\rangle_{\Omega_c}$ is the macroscopic strain field, Ω_c is the cell domain and V_c is the volume contained in Ω_c.

5.2.5 Homogenized stress and the equilibrium equation

Cauchy' s equilibrium equation at macroscopic scale can be represented in the classic form as,

$$\int_{S_c} \sigma_{ij}\ n_j\ dS = \int_{V_c} \rho\ a_i\ dV - \int_{V_c} \rho\ b_i\ dV \tag{5.15}$$

From a macroscopic scale point of view the cell is very small ($V_c \to 0$). Consequently, the magnitudes of the volume and inertia forces are also small and tend to zero. The equilibrium of the cell domain is guaranteed by this equation as follows,

$$\lim_{V_c \to 0} \left(\int_{S_c} \sigma_{ij}\ n_j\ dS \right) = \lim_{V_c \to 0} \left(\int_{V_c} (\rho\ a_i - \rho\ b_i)\, dV \right) = 0 \tag{5.16}$$

where S_c represents the cell boundary surface, σ_{ij} is the macroscopic stress field and n_j is the unit normal vector of the surface element dS_c. Note that the direction of two surface elements located at periodic points (Figure 5.9) has unit normal vectors ($\mathbf{n}_1$ and $\mathbf{n}_2$) in opposite directions. The action and reaction principle guarantees that the surface forces $\mathbf{f} = \mathbf{t}(\mathbf{n})\,dS_c$ in these two surface elements are equal and of opposite direction. In the homogenization literature this is known as antiperiodic field forces on the cell sides (Lene (1986)[31]).

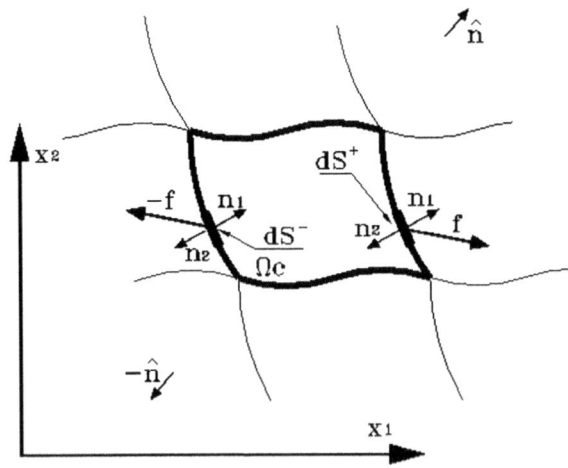

Figure 5.9 – The resulting forces on the cell periodic boundary are equal and of opposite direction.

The stress tensor at macroscopic scale or *the homogenized stress tensor* $\widetilde{\sigma}_{ij}$ is defined by the average of the forces acting on the cell sides,

$$\widetilde{\sigma}_{ij} = \frac{\displaystyle\int_{S_C} y_k \sigma_{ij}\, n_j\, dS}{\displaystyle\int_{S_C} y_k\, n_j\, dS} \tag{5.17}$$

Taking into account Cauchy's equilibrium equation or *the momentum balance* –disregarding influence of the volume forces– at microscopic level $\sigma_{ij,j} = 0$, and using the divergence theorem, this stress can also be written as,

$$\widetilde{\sigma}_{ij} = \frac{\displaystyle\int_{V_C} y_{k,j}\, \sigma_{ij}\, dV}{\displaystyle\int_{V_C} y_{k,j}\, dV} = \frac{\displaystyle\int_{V_C} \delta_{k,j}\, \sigma_{ij}\, dV}{\displaystyle\int_{V_C} y_{k,j}\, dV} = \frac{1}{V_c}\int_{V_C} \sigma_{ij}\, dV \tag{5.18}$$

The *surface forces* $\widetilde{\mathbf{t}}(\hat{\mathbf{n}})$ are defined as the forces average on the cell sides S_c determined by the unit vector direction at macroscopic level $\hat{\mathbf{n}}$,

$$\tilde{t}_i(\hat{n}) = \left[\frac{\int_{S_c} y_k \sigma_{ij} n_j \, dS}{\int_{S_c} y_k n_j \, dS}\right] \hat{n}_k = \left[\frac{1}{V_c} \int_{V_c} \sigma_{ij} \, dV\right] \hat{n}_k \qquad (5.19)$$

Since the term on the right hand-side of this equation is a linear function of the vector direction $\hat{n}$ ($\tilde{t}_i(\hat{n}) = \tilde{\sigma}_{ik} \hat{n}_k$) and as the tensor $\tilde{\sigma}$ meets the same requirements of the stress tensor at macroscopic level $\tilde{\sigma}$, hereafter it will be known as *homogenized stresses tensor*. Following this concept and by considering the whole domain Ω of the composite material which is formed by a big number of cells, the following static equilibrium equation is obtained,

$$\int_S \left[\frac{1}{V_c} \int_{V_c} \sigma_{ij} \, dV\right] \hat{n}_j \, dS + \int_V \left[\frac{1}{V_c} \int_{V_c} \rho b_i \, dV\right] dV = 0 \qquad (5.20)$$

As observed, in contrast to the equilibrium equation at micro scale level, the forces by unit volume have been considered, as their magnitude is important at macro scale level. This force's magnitude is also obtained by the volume force's average inside the cell, as follows,

$$\tilde{b}_i = \frac{1}{V_c} \int_{V_c} \rho b_i \, dV = \text{constant}; \qquad (5.21)$$

By substituting equations (5.17) and (5.20) into equation (5.19) and transforming the surface integral by the divergence theorem, the Cauchy's equilibrium equation is obtained in its weak form and corresponding to the whole composite solid at macro scale level,

$$\int_{V_c} \tilde{\sigma}_{ij,j} \, dV + \int_{V_c} \tilde{b}_i \, dV = 0 \qquad (5.22)$$

This equation is valid for any arbitrary volume V at macro scale level and therefore it is also valid in case the volume is very small, assuming its inferior limit will be the unitary cell domain ($\Omega \to \Omega_c$; $\Omega_c \to 0$). This inferior limit gets the weak form (differential),

$$\tilde{\sigma}_{ij,j} + \tilde{b}_i = 0 \qquad (5.23)$$

The above equation is the homogenized local equation of static equilibrium.

5.2.6 Elastic problem basis at micro-macro scales

Assume a body in a domain Ω made up by a composite material with a fine periodic structure. The domain of the boundary Ω is represented by $\partial\Omega$, in which $\partial\Omega_u$ is the part of the boundary where the known displacements are imposed (Dirichlet's conditions) and $\partial\Omega_t$ is the part of the boundary where the known surface forces are imposed (Newman's conditions). The structure of the material can also divide itself into very small unit cells. The sub-domains of these cells will be represented as Ω_c, so that their orderly repetition makes up the composite material Ω. Moreover, the existence of two other scales of different magnitude order are assumed, such that one particle position x_i at macro scale level Ω can be identified as the position y_i of a cell at microscopic scale level.

Consequently, the composite material's problem "at macroscopic level" becomes a "boundary-value problem" of homogeneous solids in which stress and displacement fields $(u(x), \tilde{\sigma}(x))$ satisfy the following equations,

$$
\text{Macroscopic scale:}
\begin{cases}
\dfrac{\partial \tilde{\sigma}(x)}{\partial x} + \tilde{b} = 0 & \text{constitutive equation on } \Omega \\[2mm]
\tilde{\sigma}(x) = \dfrac{1}{V_c} \displaystyle\int_{V_c} \sigma(x,y)\, dV_c & \text{constitutive equation on } \Omega \\[2mm]
u(x) = \overline{u}(x) & \text{displacements on } \partial\Omega_u \\[2mm]
\tilde{\sigma}(x){:}n = \overline{t}(x) & \text{load on } \partial\Omega_t
\end{cases}
\tag{5.24}
$$

On the other hand, at "microscopic level" the cell boundary conditions must be able to reproduce the "material microstructure conditions". The periodicity vectors obtained at microscopic level are then formulated based on the local periodicity hypothesis.

$$
d - D = u_p - u_{p0}
\tag{5.25}
$$

where u_p and u_{p0} represent the displacements of two periodic points, D is the local periodicity vector defined between such points in the "microscopic referential configuration" X_i and d is the new periodicity vector resulting from the strain field. By considering small strains, it can be demonstrated that the relative displacement between the cell boundary's periodic points is equivalent to (F. Zalamea *et al.* (1999[a])[52]),

$$
u_p - u_{p0} \cong E\,D
\tag{5.26}
$$

This relative displacement between periodic points of the cell boundary will be henceforth be called *displacement periodic field*. Furthermore, the forces generated on the cell boundaries have the same magnitude and direction but opposite direction. Then, the problem at "microscopic level" y_i focuses on solving the *boundary values problem* in the cell domain Ω_c,

$$
\text{Microscopic scale:}
\begin{cases}
\dfrac{\partial \sigma(y)}{\partial y} = 0 & \text{Constitutive equation on } \Omega_c \\[2mm]
\dot{\sigma}(y) = \mathbb{C}^s(y) : \dot{E}(y) & \text{Constitutive equation on } \Omega_c \\[2mm]
u_p - u_{p0} = \tilde{E}\,D & \text{Periodic displacent on } \partial\Omega_c \\[2mm]
t_p = -t_{p0} & \text{Periodic load on } \partial\Omega_c
\end{cases}
\tag{5.27}
$$

These equations represent the static equilibrium at microstructural level, disregarding the volume forces, and must be applied in each cell point during the evolution of the mechanic process to which the composite solid is subjected. The solution to the boundary values problem leads to the forces' equilibrium at microscopic level.

---

[52] Zalamea F., Miquel Canet J., and Oller S., 1999a. Teoría de homogeneización para el análisis de materiales compuestos con estructura interna periódica. *Proceedings IV Congreso de Métodos Numéricos en Ingeniería*, Ed. R. Abascal, J. Domínguez, G. Bugeda. SEMNI CD-ROM.

The constitutive behavior of the material components is represented by the second equation of (5.27), in which $\mathbf{C}^S$ is the local constitutive tensor. Thus, the constitutive equation of the material component can represent any type of mechanic behavior (elastic, plastic, viscous, etc.) In the same equation (5.27) the displacement condition and the periodic forces imposed on the cell boundary $\partial\Omega_c$ can also be observed, which are simultaneously related to the homogenized strain tensor $\widetilde{E}$ at microscopic level. Consequently, the micro-macro problems are coupled and therefore the solution at macro scale level is obtained by following the classic equilibrium procedure for homogeneous solids[*] (see Zienkiewicz and Taylor (1994)[25]), to satisfy the boundary values conditions at each point of the domain Ω. This task involves a considerable high number of problems to solve at microscopic level. Nevertheless, being aware of this difficulty and considering the current advances in numerical and computational techniques, the discrete solution to the elastic equilibrium equations at microscopic level can be tackled through the finite element method (Zienkiewicz and R. Taylor (1994)[25]),

$$f^{\text{int}}(\boldsymbol{\sigma}) = f^{\text{ext}} = f \quad \Rightarrow \quad \boldsymbol{K} \cdot \boldsymbol{U} = f \tag{5.28}$$

where $\boldsymbol{K}$ is the stiffness matrix assembled in a cell, $\boldsymbol{U}$ is the nodal displacement vector and f is the force's vector in the cell boundary. There are different approaches to solve this equilibrium equation subjected to periodic boundary conditions, but in this approach the method of Lagrange multipliers is used (λ). However, this method has some drawbacks such as the increase in the number of equations as well as the cell stiffness matrix bandwidth which cause computational difficulties. This drawback can be solved by considering the boundary conditions imposed by Anthoine (1995)[28], who divided Lagrange multipliers into two groups, λ_1 and λ_2. The first of them contains the boundary forces in the restricted nodes (λ_1) and the second one contains the corresponding periodic magnitudes (λ_2). The stationarity of the functional increased by the Lagrange multipliers,

$$\Pi = \frac{1}{2}\boldsymbol{U}^T \cdot \boldsymbol{K} \cdot \boldsymbol{U} - \boldsymbol{U}^T \cdot f + \lambda_1^T \cdot (\boldsymbol{k}_p \cdot \boldsymbol{U} - \boldsymbol{U}_r) +$$
$$+ \lambda_2^T \cdot (\boldsymbol{k}_p \cdot \boldsymbol{U} - \boldsymbol{U}_r) + \frac{1}{2}(\lambda_1 - \lambda_2)^T \cdot (\lambda_1 - \lambda_2) \tag{5.29}$$

leads to the following well-conditioned system of linear equations,

$$\begin{bmatrix} \boldsymbol{K} & \boldsymbol{k}_p^T & \boldsymbol{k}_p^T \\ \boldsymbol{k}_p & \boldsymbol{I} & -\boldsymbol{I} \\ \boldsymbol{k}_p & -\boldsymbol{I} & \boldsymbol{I} \end{bmatrix} \cdot \begin{bmatrix} \boldsymbol{U} \\ \lambda_1 \\ \lambda_2 \end{bmatrix} = \begin{bmatrix} f \\ \boldsymbol{U}_r \\ \boldsymbol{U}_r \end{bmatrix} \tag{5.30}$$

where $\boldsymbol{K}$ is the cell stiffness matrix and $\boldsymbol{k}_p$ is the matrix of the cell boundary degrees of freedom. This matrix has a code -1 in the component's restricted degree of freedom. Code $+1$ represents the position of the current degree of freedom and 0 occupies the remaining matrix positions. $\boldsymbol{I}$ is the identity matrix, $\boldsymbol{U}$ is the nodal displacement vector, λ_1 and λ_2 are Lagrange multipliers representing the forces in the cell boundary, f is the vector of the nodal forces in the boundary and $\boldsymbol{U}_r$ is the relative nodal displacement between the boundary nodes.

---

[*] Note that the classic constitutive equation used for homogeneous materials is replaced by the problem's solution at macro scale level.

5.2.7 Basis of the inelastic problem at micro-macro scales

The behavior of the composite material is characterized by a wide scope ranging from the linear to the non-linear field. The non-linear behavior occurs when at least one of the composite materials components of a point of a solid at microstructural level has reached its elastic limit and develops an inelastic behavior such as plastic, damage, viscous, plastic, fracture, etc. The fact that at least a single point of one of the components has a non-linear behavior makes the whole solid have a non-linear behavior. The first papers on non-linear behavior in the homogenization theory were presented by Suquet (Suquet (1982)[13], (1987)[10]). This author concludes that "the composite material's behavior depends on an infinite number of internal variables", which is very difficult due to the high computational cost involved. In later papers the author introduced some simplifications based on the knowledge and nature of each composite material (see also Lene (1986)[31], Devries *et alt.* (1989)[32]). In each case the proposals were carried out within the non-linear field. They were approximated and only applicable to some particular composites.

Another interesting approach to tackle this problem was proposed by Fish (Fish et al. (1997)[49], Fish & Shek (1999)[53]), who used the asymptotic expansion theory to obtain the governing equations in each of the two scales. Consequently, the composite behavior is obtained by solving a unit cell for each integration points at macro scale level. However, the solution for all the cells is not carried out by the non-linear behavior procedure due to the computational cost. Thus, the cells to be solved by the non-linear field are selected through a search algorithm of integration points at macro scale taking into consideration the stress and strain levels of each of them and only the cells associated with these points are solved. Additionally, only the internal variables average is stored in the data base for each cell showing non-linear behavior.

Another approach based on the asymptotic expansion theory was developed by Ghosh (Ghosh et al. (1996)[43], Lee et al. (1999)[47]), in which the composite behavior is obtained by the Veronoi finite element method by representing the whole cell with only one of them. These elements provide very good results at micro and macro scales, but they cannot describe internally the microstructure's details.

An alternative to Fish and Ghosh formulations previously mentioned is presented here. The conceptual basis has been presented in different publications (Zalamea *et al.* (1998)[54], Zalamea *et al.* (1999[a])[52], Zalamea *et al.* (1999[b])[55], Zalamea *et al.* (2000)[56], Zalamea (2001)[11]). This is based on a direct formulation of the governing equations in each of the two scales disregarding the use of the asymptotic expansion theory. Another characteristic is the use of the finite element method for the analysis of the two scales by imposing the boundary conditions at micro scale level by the "Lagrange Multipliers" (equations (5.29) and (5.30)). Also, during the linearization of non-linear equilibrium equations at macro scale, a cell is solved for each of the integration points and in order to partly reduce the high computational cost a parallelization strategy is used through a "PVM" (Parallel Virtual

[53] Fish J. and Shek K. (1999). Finite deformation for composite structures: Computational models and adaptive strategies. *Comput Methods Appl. Mech. Engrg.* Vol. 172, pp. 145-174.

[54] Zalamea F., Miquel J. and Oller S. (1998). Treatment of Composite Materials based on the Homogenization Method. *Proceedings of the Fourth World Congress on Computational Mechanics.* CIMNE.

[55] Zalamea F., Miquel J., Oller S., (1999b). Un Método en doble escala para la simulación de materiales compuestos. *III Congreso Nacional de Materiales Compuestos. "MATCOMP 99".* Ed. A. Corz, J. M. Pintado. Materiales Compuestos 99 - AEMAC, pp 381-393 - Málaga. España. 1999.

[56] Zalamea F., Miquel J., Oller S. (2000). A double scale method for simulating of periodic composite materials. *European Congress on Computational Methods in Applied Sciences and Engineering,* ECOMAS 2000 - COMPLAS VI. Ed. E. Oñate, R. Owen. CD-ROM - CIMNE Barcelona.

Machine[57]) system. Thus, a parallel process is solved for each of the cells and additionally one for the whole structure at macro scale. All the information distribution in each of the sub-processes opened for each cell unit is controlled by this latter process. The internal variables are stored in each integration point at micro scale for non-linear behavior cells.

The solution of the non-linear problem through the Newton-Raphson linearization approach is carried out by keeping the stress and strain definitions valid during such linearization, as shown in the previous section, as well as the governing equations formulated at micro and macro scales. However, the definition of the constitutive model at micro scale level is no longer elastic and adopts a non-linear form such as the year, plasticity, viscous-elasticity, fracture, etc. The basic contents of a non-linear constitutive equation which must replace the second equation of (5.27) is detailed as follows,

$$
\left\{
\begin{array}{l}
\text{Free energy for a single component of the composite,} \\[4pt]
\quad \Psi = \Psi(\boldsymbol{E}, \alpha) \\[4pt]
\text{Free variable of the problem: the strain tensor,} \\[4pt]
\quad \boldsymbol{\varepsilon} = \nabla^{S}\mathbf{u} \\[4pt]
\text{Internal variables,} \\[4pt]
\quad \boldsymbol{\alpha} = \{\alpha_{k}\}, \qquad k = 1, \ldots, n \\[4pt]
\text{Dependent variable: stress tensor,} \\[4pt]
\quad \dot{\boldsymbol{\sigma}} = \dot{\boldsymbol{\sigma}}(\dot{E}, \alpha)
\end{array}
\right. \tag{5.31}
$$

The constitutive equation for the whole composite in the current time $t+\Delta t$ is expressed in the incremental form and its expression coincides strictly with equation (5.18),

$$
\widetilde{\boldsymbol{\sigma}}^{t+\Delta t} = \frac{1}{V_{c}} \int_{V_{C}} \boldsymbol{\sigma}^{t+\Delta t}\, dV = \frac{1}{V_{c}} \int_{V_{C}} \left(\boldsymbol{\sigma}^{t} + \dot{\boldsymbol{\sigma}}^{t+\Delta t}\Delta t\right) dV = \widetilde{\boldsymbol{\sigma}}^{t} + \left\{\widetilde{\mathbb{C}}^{T}\right\}^{t+\Delta t} : \dot{\widetilde{\boldsymbol{E}}}^{t+\Delta t}\Delta t \tag{5.32}
$$

where $\left\{\widetilde{\mathbb{C}}^{T}\right\}^{t+\Delta t}$ represents the homogenized tangent constitutive tensor in time $(t+\Delta t)$, from which the stress $\dot{\widetilde{\boldsymbol{\sigma}}}^{t+\Delta t} = \left\{\widetilde{\mathbb{C}}^{T}\right\}^{t+\Delta t} : \dot{\widetilde{\boldsymbol{E}}}^{t+\Delta t}$ is obtained, in an incremental step during the internal equilibrium equation linearization of the composite solid. Once the homogenized stresses tensor $\widetilde{\boldsymbol{\sigma}}^{t+\Delta t}$ is obtained in each point of the composite solid, the traditional path is followed for the solid equilibrium solution as if it were a homogeneous material.

The non-linear equilibrium equation for a discretized solid by the finite elements is written by deleting the unbalanced (residual) forces resulting from the difference between the internal forces f_{k}^{int} and the external forces f_{k}^{ext},

$$
0 = \mathop{\mathbf{A}}_{\Omega^{e}}\left[f_{k}^{\text{int}} - f_{k}^{\text{ext}}\right]_{\Omega^{e}} = \Delta f_{k}\big|_{\Omega} \tag{5.33}
$$

In this case $\mathop{\mathbf{A}}_{\Omega^{e}}$ represents the "assembly" operator of the elemental forces in a finite element to obtain the forces' state in all the structure (Zienkiewicz and Taylor (1994)[25]). These unbalanced forces are eliminated by a linearization process established in the

---

[57] Geist Al, Beguelin A., Dongarra J., Jiang W., Manchek R. and Sunderam V. (1994). *Paralle Virtual Machine - A Users' Guide and Tutorial for Networked Parallel Computing.* Ed. Jasnusz Kowalik. Massachusetts Institute Technology.

iteration $(i+1)$ of time $t+\Delta t$, close to the equilibrium state between the internal and external forces. Thus, it is necessary to force the equilibrium state in the current instant $(i+1)$ and express this condition based on the Taylor-series expansion truncated in the first variation,

$$0 = \underset{\Omega^e}{\mathbf{A}}^{i+1}\left[\Delta f_k\right]_{\Omega^e}^{t+\Delta t} \cong \underset{\Omega^e}{\mathbf{A}}^i\left[\Delta f_k\right]_{\Omega^e}^{t+\Delta t} + \underset{\Omega^e}{\mathbf{A}}\left[\left[\frac{\partial(\Delta f_k)}{\partial U_r}\right]_{\Omega^e}^{i}\right]^{t+\Delta t} \cdot^{i+1}\left[\Delta U_r\right]_{\Omega^e}^{t+\Delta t}$$

$$0 = \underset{\Omega^e}{\mathbf{A}}^i\left[\Delta f_k\right]_{\Omega^e}^{t+\Delta t} + \underset{\Omega^e}{\mathbf{A}}\left[\left[\frac{\partial(f_k^{\text{int}})}{\partial U_r}\right]_{\Omega^e}^{i}\right]^{t+\Delta t} \cdot^{i+1}\left[\Delta U_r\right]_{\Omega^e}^{t+\Delta t}$$

(5.34)

By substituting the unbalanced or residual force into this equation by the expression $\Delta f_k\big|_\Omega = \underset{\Omega^e}{\mathbf{A}}\left[f_k^{\text{int}} - f_k^{\text{ext}}\right]_{\Omega^e} = \underset{\Omega^e}{\mathbf{A}}\left[\int_{V^e} \tilde\sigma_{ij}\nabla_i^S N_{jk}dV - f_k^{\text{ext}}\right]_{\Omega^e}$, in which the internal forces are expressed as a function of the homogenized stress $\tilde\sigma_{ij}$, and taking into account the displacement field approximation in each finite element through the following polynomial form based on the shape functions, $u_i(x,y,z) = N_{ij}(x,y,z)\cdot U_j$, (Zienkiewicz and Taylor (1994)[25]), then,

$$0 \cong \underset{\Omega^e}{\mathbf{A}}^i\left[\int_{V^e}\tilde\sigma_{ij}\nabla_i^S N_{jk}\,dV - f_k^{\text{ext}}\right]_{\Omega^e}^{t+\Delta t} + \underset{\Omega^e}{\mathbf{A}}^i\left[\frac{\partial}{\partial U_r}\left(\int_{V^e}\tilde\sigma_{ij}\nabla_i^S N_{jk}\,dV\right)\right]_{\Omega^e}^{t+\Delta t} \cdot \underset{\Omega^e}{\mathbf{A}}^{i+1}\left[\Delta U_r\right]_{\Omega^e}^{t+\Delta t}$$

$$0 \cong \underset{\Omega^e}{\mathbf{A}}^i\left[\int_{V^e}\tilde\sigma_{ij}\nabla_i^S N_{jk}\,dV - f_k^{\text{ext}}\right]_{\Omega^e}^{t+\Delta t} + \underset{\Omega^e}{\mathbf{A}}^i\left[\int_{V^e}\frac{\partial\tilde\sigma_{ij}}{\partial\tilde E_{st}}\frac{\partial\tilde E_{st}}{\partial U_r}\nabla_i^S N_{jk}\,dV\right]_{\Omega^e}^{t+\Delta t} \cdot \underset{\Omega^e}{\mathbf{A}}^{i+1}\left[\Delta U_r\right]_{\Omega^e}^{t+\Delta t}$$

(5.35)

By calculating and substituting the stress and strain continuous field from the displacements into the previous equation, the following expression is obtained,

$$0 \cong \underset{\Omega^e}{\mathbf{A}}^i\left[\int_{V^e}\tilde\sigma_{ij}\nabla_i^S N_{jk}\,dV - f_k^{\text{ext}}\right]_{\Omega^e}^{t+\Delta t} + \underset{\Omega^e}{\mathbf{A}}^i\left[\int_{V^e}\tilde C_{ijst}^T\frac{\partial(\nabla_s^S N_{tr}U_r)}{\partial U_r}\nabla_i^S N_{jk}\,dV\right]_{\Omega^e}^{t+\Delta t} \cdot \underset{\Omega^e}{\mathbf{A}}^{i+1}\left[\Delta U_r\right]_{\Omega^e}^{t+\Delta t}$$

$$0 \cong \underset{\Omega^e}{\mathbf{A}}^i\left[\int_{V^e}\tilde\sigma_{ij}\nabla_i^S N_{jk}\,dV - f_k^{\text{ext}}\right]_{\Omega^e}^{t+\Delta t} + \underset{\Omega^e}{\mathbf{A}}^i\underbrace{\left[\int_{V^e}(\nabla_s^S N_{tr})\tilde C_{ijst}^T(\nabla_i^S N_{jk})dV\right]_{\Omega^e}^{t+\Delta t}}_{\left[\boldsymbol{K}_{kr}^T\right]_{\Omega^e}} \cdot \underset{\Omega^e}{\mathbf{A}}^{i+1}\left[\Delta U_r\right]_{\Omega^e}^{t+\Delta t}$$

(5.36)

$$0 \cong {}^i\left[\Delta f_k\right]_\Omega^{t+\Delta t} + {}^i\left[\boldsymbol{K}_{kr}^T\right]_\Omega^{t+\Delta t} \cdot {}^{i+1}\left[\Delta U_r\right]_\Omega^{t+\Delta t}$$

As a result, the linearized equilibrium equation is obtained, where $\left[\boldsymbol{K}_{kr}^T\right]_{\Omega^e}$ represents the tangent stiffness matrix for the elemental domain, $\left[\boldsymbol{K}_{kr}^T\right]_\Omega$ is the tangent stiffness matrix for the whole solid and $\tilde{\mathbf{C}}^T$ represents the tangent constitutive tensor in each point of the composite solid corresponding to the homogenized constitutive law. The residual forces $\Delta f_k\big|_\Omega$ are eliminated by the Newton-Raphson procedure until this residual tends to zero, which is known as the convergence of the non-linear system of equilibrium equations. (5.36).

Thus, the non-linear behavior problem of a composite solid is solved by using the finite element technique through the homogenization method. The procedure's result could be described as "general", in which the constitutive equation at macroscopic scale depends exclusively on the material microscopic behavior (micro scale).

5.2.8 Determination of the elastic constitutive tensor for composite materials

The homogenized elastic constitutive equation for composite materials in which any relative movement among components is assumed can be written as,

$$\widetilde{\boldsymbol{\sigma}}(x) = \widetilde{\mathbb{C}}(x) : \widetilde{\boldsymbol{E}}(x) \tag{5.37}$$

From the homogenized strain $\widetilde{\boldsymbol{E}}$ the stress can be obtained by assuming that the material behavior is elastic through the homogenized constitutive tensor $\widetilde{\mathbb{C}}(x)$. This is a fourth-order tensor with 81 components. Suquet assumes that the homogenized elastic constitutive tensor has the classic symmetric form provided that the components have a periodic distribution.

$$\widetilde{\mathbb{C}}_{ijkl} = \widetilde{\mathbb{C}}_{jilk} = \widetilde{\mathbb{C}}_{klij} \tag{5.38}$$

The properties of these materials reduce the complexity of the solution to the problem. Particularly, for orthotropic materials it is only necessary to obtain the nine components independently from the elastic constitutive tensor. On the other hand, the elastic homogenized stress tensor $\widetilde{\boldsymbol{\sigma}}$ results from the homogenized strain $\widetilde{\boldsymbol{E}}$ by using the constitutive tensor $\widetilde{\mathbb{C}}(x)$, that must be defined based on the information provided at micro scale (at periodic cells level).

The components of the homogenized elastic constitutive tensor are obtained by the perturbation procedure (activation of small displacements) in the cell boundaries to activate the different elastic constants of composites and to observe the result obtained.

Once the homogenized stress $\widetilde{\boldsymbol{\sigma}}$ and strain $\widetilde{\boldsymbol{E}}$ tensors are obtained, it is possible to calculate the elastic constants of the constitutive tensor $\widetilde{\mathbb{C}}(x)$ by applying the above mentioned perturbation procedure through the following expression,

$$\widetilde{\mathbb{C}}(x) = \widetilde{\boldsymbol{\sigma}}(x) : \left[\widetilde{\boldsymbol{E}}(x)\right]^{-1} \tag{5.39}$$

Obviously, this expression gives infinite solutions, as $\widetilde{\mathbb{C}}(x)$ is a fourth-order tensor whereas $\widetilde{\boldsymbol{\sigma}}(x)$ and $\widetilde{\boldsymbol{E}}(x)$ are second-order tensors. However, if the problem is solved in the main directions and perturbations are activated in each one of these directions one at the time, a constitutive tensor could be built up component by component and a single solution to the problem could be obtained.

For example, the symmetric part of the elastic constitutive tensor for two-dimensional problems in stress or plane strains with the following matrix form can be expressed as,

$$\widetilde{\mathbb{C}}(x) = \begin{bmatrix} \widetilde{\mathbb{C}}_{xxxx} & \widetilde{\mathbb{C}}_{xxyy} & 0 \\ \widetilde{\mathbb{C}}_{xxyy} & \widetilde{\mathbb{C}}_{yyyy} & 0 \\ 0 & 0 & \widetilde{\mathbb{C}}_{xyxy} \end{bmatrix} \qquad (5.40)$$

By applying the following perturbation field one by one,

$$\begin{aligned} \widetilde{\boldsymbol{E}}_1(x) &= \left\{ \widetilde{E}_{xx}, 0, 0 \right\} \\ \widetilde{\boldsymbol{E}}_2(x) &= \left\{ 0, \widetilde{E}_{yy}, 0 \right\} \\ \widetilde{\boldsymbol{E}}_3(x) &= \left\{ 0, 0, 2\widetilde{E}_{xy} \right\} \end{aligned} \qquad (5.41)$$

The stresses $\widetilde{\boldsymbol{\sigma}}\big(\widetilde{\boldsymbol{E}}_1(x)\big)$, $\widetilde{\boldsymbol{\sigma}}\big(\widetilde{\boldsymbol{E}}_2(x)\big)$, $\widetilde{\boldsymbol{\sigma}}\big(\widetilde{\boldsymbol{E}}_3(x)\big)$ are obtained for each one of these assumed strain fields. Then, the constitutive tensor coefficients are obtained as expressed by the following equation,

$$\begin{aligned} \widetilde{\mathbb{C}}_{xxxx} &= \widetilde{\sigma}_{xx}\big(\widetilde{\boldsymbol{E}}_1(x)\big)\big/ \widetilde{E}_{xx} \\ \widetilde{\mathbb{C}}_{xxyy} &= \widetilde{\sigma}_{xx}\big(\widetilde{\boldsymbol{E}}_2(x)\big)\big/ \widetilde{E}_{yy} \\ \widetilde{\mathbb{C}}_{yyyy} &= \widetilde{\sigma}_{yy}\big(\widetilde{\boldsymbol{E}}_2(x)\big)\big/ \widetilde{E}_{yy} \\ \widetilde{\mathbb{C}}_{yyxx} &= \widetilde{\sigma}_{yy}\big(\widetilde{\boldsymbol{E}}_1(x)\big)\big/ \widetilde{E}_{xx} \\ \widetilde{\mathbb{C}}_{xyxy} &= \widetilde{\sigma}_{xy}\big(\widetilde{\boldsymbol{E}}_3(x)\big)\big/ 2\widetilde{E}_{xy} \end{aligned} \qquad (5.42)$$

The analysis of the symmetry is not easy. By using periodic functions the following symmetry hypothesis can be satisfied,

$$\widetilde{\mathbb{C}}_{yyxx} = \widetilde{\mathbb{C}}_{xxyy} \qquad (5.43)$$

Consequently, this technique is used regardless of the strain shape or type applied to the cell. Thus, if another strain field different from the one previously assumed is applied and the cell is kept elastic, then a homogenized tensor like the previous one is obtained.

5.2.9 Determination of the quasi-tangent inelastic constitutive tensor for composite materials. Analytical determination

In order to solve a non-linear constitutive problem by the Newton-Raphson linearization technique, it is convenient to have the definition of the *tangent algorithm constitutive tensor* to improve the convergence velocity for solution of the problem and to achieve it with as few iterations as possible. Nevertheless, obtaining the *tangent algorithm constitutive tensor* is simple and more particularly when dealing with a composite material. This is because its response depends on the simultaneous participation of the behavior of various materials which respond to different constitutive models. On the other hand, the solution to the non-linear system of equations by the iterative-incremental method such as the modified Newton-Raphson method only requires an *elastic constitutive tensor* $\widetilde{\mathbb{C}}(x)$, despite the high computational cost involved. However, an alternative approach to these two extreme methods —*tangent algorithm tensor* or *elastic tensor* – is proposed, which consists of obtaining a "quasi-tangent algorithm tensor" with a convergence velocity similar to the velocity obtained by the tangent tensor. Through this approximate tensor a "quasi-tangent" structural stiffness $\boldsymbol{K}^T$ can be obtained for the whole structure by the information obtained in previous iterations. The "quasi-tangent" tensor can be obtained by different procedures (see: Dennis and More (1977)[58], Crisfield (1980)[59]), and another possible way to obtain it is presented here. It is assumed that after the homogenization, the constitutive law of the composite material in each point can be written in the following form,

$$\dot{\tilde{\sigma}}(x) = \tilde{\mathbb{C}}^T(x) : \dot{\tilde{E}}(x) \tag{5.44}$$

This law can also be written as follows,

$$\dot{\tilde{\sigma}} = \left(\tilde{\mathbb{C}} + \dot{\tilde{\mathbb{C}}} \right) : \dot{\tilde{E}} \tag{5.45}$$

$\tilde{\mathbb{C}}$ being the elastic constitutive tensor of the composite material and $\dot{\tilde{\mathbb{C}}}$ the temporal change that undergoes this elastic constitutive tensor to match the tangent constitutive tensor $\tilde{\mathbb{C}}^T$. The temporal variation of the elastic constitutive tensor can be obtained by equation (5.45), as follows,

$$\dot{\tilde{\sigma}} - \tilde{\mathbb{C}} : \dot{\tilde{E}} = \dot{\tilde{\mathbb{C}}} : \dot{\tilde{E}} \quad \Rightarrow \quad \dot{\tilde{\mathbb{C}}} = \left(\dot{\tilde{\sigma}} - \tilde{\mathbb{C}} : \dot{\tilde{E}} \right) : \dot{\tilde{E}}^{-1} \tag{5.46}$$

Like equation (5.39), this equation has infinite solutions, which would require a perturbation solution as mentioned in the previous section. Nevertheless, this is only aimed at obtaining an approximate tangent constitutive tensor and for this purpose it is sufficient to impose some restrictions to the temporal change of the constitutive tensor $\dot{\tilde{\mathbb{C}}}$, to reduce the indetermination degree and then obtain a unique solution. Therefore, it is assumed that tensor $\dot{\tilde{\mathbb{C}}}$ keeps the same symmetries as elastic tensor $\tilde{\mathbb{C}}$ and it is also assumed that some terms are null. In this case the solution can be easily obtained.

5.2.10 Micro-macro structural coupling

In order to solve the problem of composite materials with the homogenization method here presented, a micro-macro structural coupling is proposed based on the application of the finite element method at two scales (see Ghosh *et al.* (1996)[43] y Fish *et al.* (1997)[49]). A problem is solved by the finite element method at both scales such that the structural problem is solved at macro scale while the constitutive problem is solved at micro scale. Thus, each numerical integration point of the macro structure discretized in finite elements presents a boundary value problem at microstructural level. This means that the constitutive law of the behavior of the composite material is not analytical but it is expressed numerically through the finite element problem solved at micro scale. The internal state variables obtained in the composite material are the result of their own homogenization at micro scale.

Based on this scheme, the macro structure is solved along with a great number of cells representing the microstructure, as many times as the number of numeric integration points –Gauss points– the macrostructure has. The final solution at macro structural level is obtained once all the cells are solved.

Figure 3 shows a schematic flow diagram of the solution of the problem at both scales. The solution by the finite element method is very expensive. A parallelization algorithm

[58] Dennis J. E., and More J. J. (1977). *Quasi-Newton methods, motivation and theory*. SIAM. Review
[59] Crisfield M. (1980). *Numerical Methods for Nonlinear Problem*. Pineridge Press.

(PVM[57]) has been used for the macro-micro coupling solution of all the cells to reduce the calculation time. These processes at microstructural level synchronize themselves and exchange information among them as required.

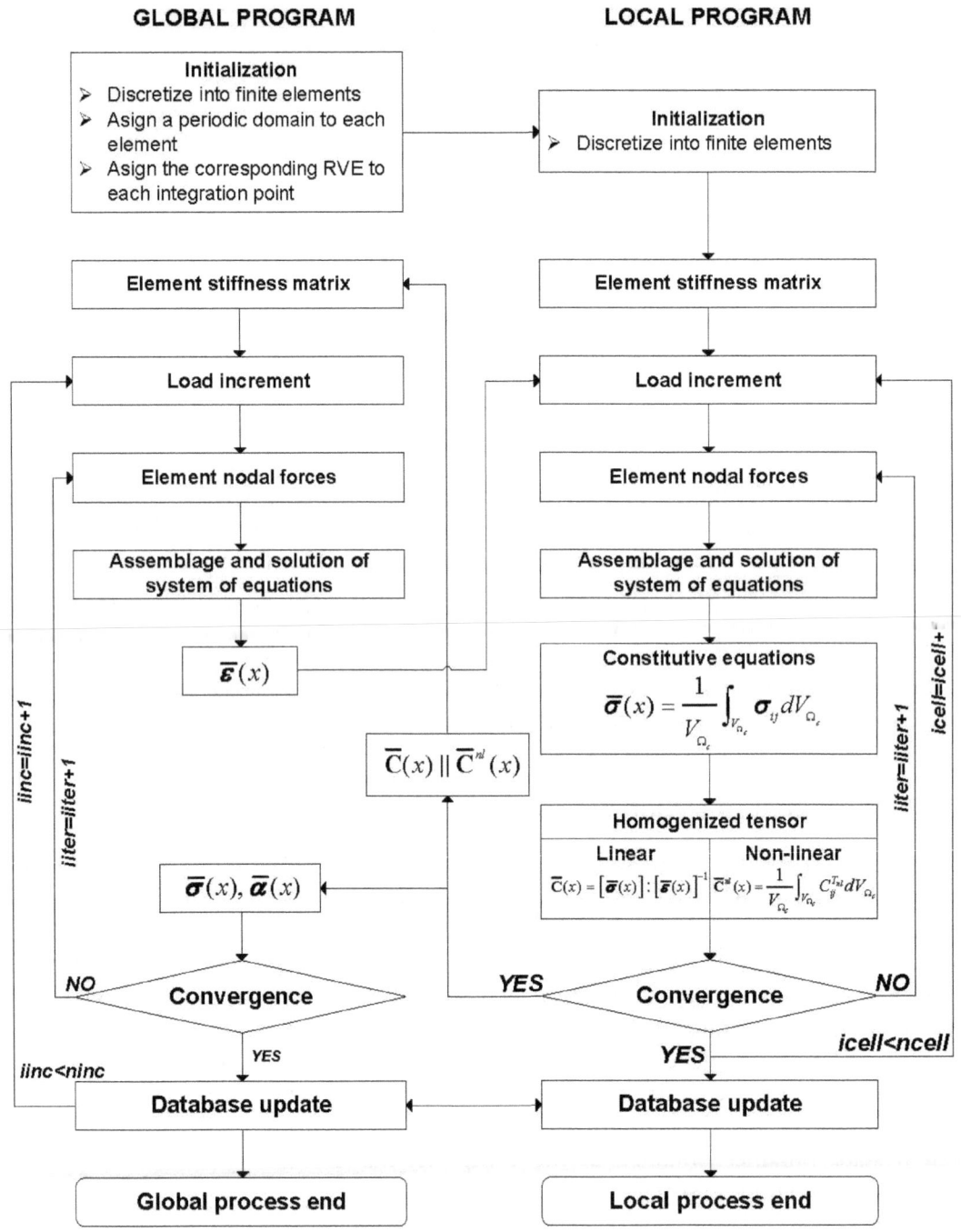

Figure 5.10 – Algorithm for the solution of composite material structures by the homogenization method at two scales through the finite element method.

A sequence of activities of the algorithm shown in Figure 5.10 is schematically presented as follows:

1. The process is started. The macro scale program (global program) executes initialization tasks and inputs the macrostructure information which also contains the microstructure file direction.

2. The homogenized elastic constitutive tensor is calculated $\mathbb{C}(x)$ through the micro scale program (local program or subprogram). The following activities are carried out:

    I. *The local program* is started; the databases are initialized by this program with a capacity of as many cells as integration points there are in the macrostructure. Then it reads the microstructure information.

    II. It computes the stiffness matrices of the finite elements of cell $\mathbf{K}$.

    III. The load increment is applied. The cells' sequential solution ("icelu") is carried out by imposing a pre-established macroscopic strain.

    IV. The nodal forces are determined.

    V. The microstructure system of equations is assembled and solved under periodic conditions.

    VI. The stresses are verified through the constitutive equation. In this case, the components are assumed to have elastic behavior (the stresses do not change). Next, the homogenized stress tensor is obtained. Then, if all the cells have been analyzed then return to step III.

    VII. The constants of the homogenized elastic constitutive tensor $\widetilde{\mathbb{C}}$ are determined. This information is sent to the *global program*.

3. The stiffness matrix $\left[\mathbf{K}_{kr}^{T}\right]_{\Omega^e}$ of each finite element in the macroscopic structure is obtained.

4. A new increment load is applied.

5. The nodal forces in each element are calculated.

6. The macro structure system of equations is generated and solved.

7. The constitutive equation of the homogenized composite material verifies that the stresses on each of the macro domain points are admissible. Such function is executed by the *local program*, therefore the macroscopic strain information of each integration point is sent. Then:

    I. The *local program* carries out the sequential solution of each cell by extracting the microscopic information from the database and applying the corresponding strain increment as a loading increment.

    II. The nodal forces are determined.

    III. The microstructure system of equations is assembled and solved under periodic conditions. The components of the respective constitutive equations verify that the microscopic stresses are admissible and in case they are not, they must be corrected according to the pre-established constitutive model. Then the stress and the homogenized tangent constitutive tensor are obtained.

    IV. The cell equilibrium condition is verified, in other words, for any iteration it is verified that $\Delta f_k \big|_{\Omega} = \underset{\Omega^e}{\mathbf{A}} \left[f_k^{\text{int}} - f_k^{\text{ext}} \right]_{\Omega^e}$. Then, the convergence to the correct solution is achieved when $\left\| \Delta f_k \big|_{\Omega} \right\| \to 0$.

V. When all the cells are solved the stresses and the homogenized constitutive tensor information is transmitted to *the global program*.

8. The *global program* assumes the magnitudes received, such as stress, constitutive tensor and internal variables as if they would come from a constitutive model. If the equilibrium is not reached, then return to step 5 and continue with the Newton-Raphson iterative technique.

9. Once convergence is achieved in the *global program* proceed with the loading increments and return to step 4 and continue until the last loading increment of the process is concluded.

10. As observed in the figure 5.10, the finite element code at both scales is formed by the problem solution coupling inside each one of the two levels The scheme on the left represents the FEM code that solves the problem at macroscopic level and the scheme on the right corresponds to the FEM code that solves the problem at microscopic level. It should be noted that the code at macro scale does not have the block representing the constitutive equations as the code at micro scale level is in charge of carrying out the homogenized constitutive equations functions of the composite material.

5.2.11 Local effects influence

The influence of the treatment of local effects such as point loads and particular boundary conditions on the homogenization theory is presented in this section. The main problem is the rigorous formulation of the homogenization theory which must satisfy the following ideal assumptions:

1. Composite materials have a periodic distribution of their component materials leading to their ideal division into equally sized domains Y called cells.

2. Each cell contains the internal structure of the composite material which is very small compared to the global structure of the composite material ($Y \ll \Omega$).

These two assumptions are contained in the local periodicity hypothesis and require that the stress and strain fields in the cell domain be the same for the neighboring cells. Based on this concept, the problem can be divided into two scales in which the microscopic variables have strong fluctuations while the macroscopic variables change smoothly. Consequently, this local periodicity hypothesis is the basis for most of the homogenization methods. Examples of this can be found in the average theory, in the formulation proposed in this paper and in both Zalamea's (Zalamea (2001)[11]) and Badillo's (Badillo (2012)[12]) PhD thesis. Some formulations do not accept the local periodicity hypothesis when strong gradients of the macroscopic variables occur or in the presence of local effects such as boundary conditions, fractures, etc. (Fish and Wagiman (1993)[33], Fish and Markolefas (1993)[34], Fish et alt. (1994)[36]). These other approaches try to obtain the solution to macro structural elastic problems by the homogenization theory by introducing perturbation terms in the displacement field when strong gradients are produced by local effects. In other approaches the overlapping of finite element meshes with high density, also known as multi-grid technique, is used in the domain where variables have a strong variation gradient. In these regions of the domain the periodic formulation introduces perturbation terms, too. These techniques are normally used in finite elements that are combined with error minimization algorithms. Moreover, the complexity of the problem is increased by this formulation. Obtaining better results by this technique rather than with the standard one cannot be guaranteed.

Strong gradients in the displacement field at a certain structural point involve a perturbation of these fields in the neighboring cells, which seems to contradict the periodicity hypothesis assumed. Nevertheless, it must be understood that the hypothesis only represents an idealization of the variation of the displacement field as it assumes that the macroscopic variables can only undergo very smooth changes or none at all. This would not admit a strong gradient in the stress ($\tilde{\sigma}_A(x)$ and $\tilde{\sigma}_B(x)$) (or strain) between the two points of the macrostructure –points A and B– (see Figure 5.11).

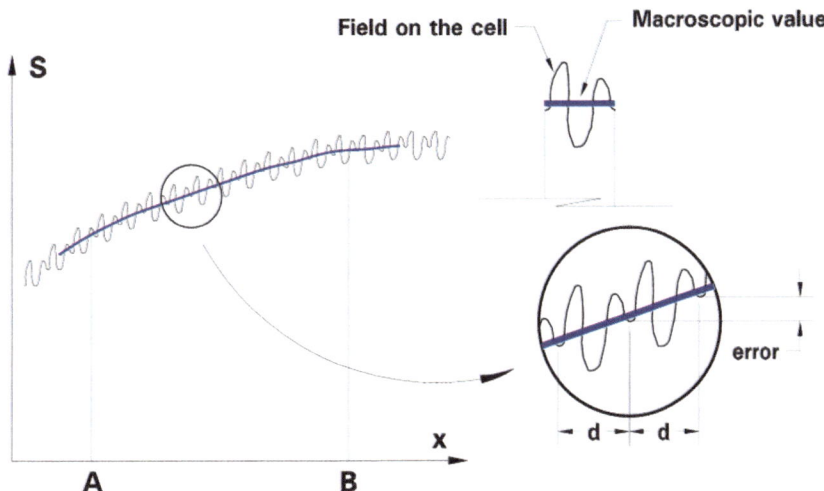

Figure 5.11 – Simplified representation of a quasi-periodic function.

On the other hand, the homogenization theory idealizes the problem based on the assumption that the cell dimensions tend to zero from the global scale point of view. Consequently, assume that between points A and B there are a great number of cells such that the stress variation between cell A and a neighboring cell is as small as wanted and continues the same way until point B is reached. The real problem differs from the ideal one in the finite number of cells between points A and B, as shown in Figure 5.11. For example, on the right hand-side of the same figure, a detail of the displacement field variation is shown indicating that the variation between the cells is very subtle. This variation can be understood as an error made along a long "period" d which also represents the cell length.

By admitting that the cell dimension decreases and the function field width –variation– does not change, then it is easy to see that the error decreases proportionally as dimension d decreases, represented as a periodic field in a very small periodic domain (in the limit when $d \to 0$). Thus, the local periodicity hypothesis is reached again.

For the exact stress field reproduction in the microstructure, a correction could be introduced by taking into consideration field variation by the cell finite dimension, which would be added to the periodic field so that it would reflect the change or variation inside the cell. This correction does not alter the cell internal variables of the global value; it only adjusts them so that the variation from one cell to the following one is continuous. It should be noted that this scale effect occurs because the homogenization theory assumptions are not completely satisfied. Obviously, if the cell is relatively very small such effect will be negligible and therefore disregarded.

On the other hand, the use of the periodicity hypothesis in the areas near the macrostructure boundaries has been questioned. This was tackled by Sanchez-Palencia

(Sanchez-Palencia (1987)[9]) based on the asymptotic developments. To analyze the boundary effect this author introduces an additional term $\boldsymbol{u}^{lc}$ of the microscopic strain order so that the displacement field is expressed as $\boldsymbol{u}^{\varepsilon} = \boldsymbol{u}^{0} + \varepsilon \boldsymbol{u}^{1} + \varepsilon \boldsymbol{u}^{lc}$. After analyzing it, it was found that at macroscopic level the effect of this additional term $\varepsilon \boldsymbol{u}^{lc}$ is negligible because its gradient vanishes very quickly. As a matter of fact, numerical experiments showed that the effect of this term is only significant in the cell located at the boundary (this is practically absorbed in one period (Dumontet (1986)[60]).

In conclusion, if it is admitted that the cell dimensions are very small with respect to the macrostructure, the error associated to the hypothesis of periodic approximation in the macroscopic domain boundary is negligible.

5.3 Test examples of the "homogenization theory of local periodicity"

5.3.1 Transversal behavior of a reinforced long fiber matrix – Simple tensile test

In this section an example presented by Jansson (1992)[61] and solved by the asymptotic expansion theory is used to validate the present formulation. Two different types of cells are presented (for other types of cell, check the reference Zalamea (2001)[11]). The first cell studied is a quadrilateral cell (see Figure 5.12). The second is a hexagonal cell of six periodic sides (see Figure 5.13).

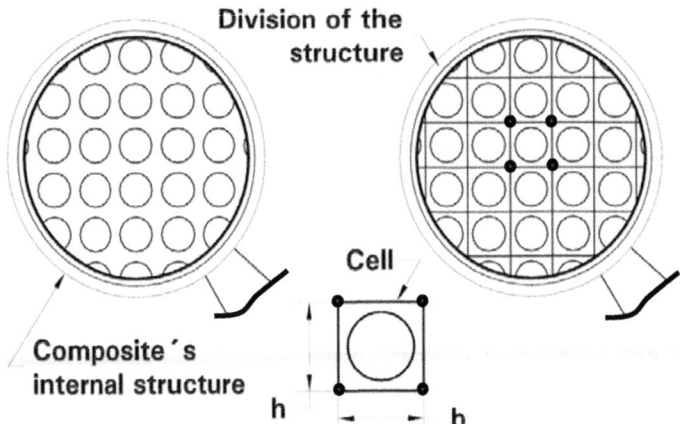

Figure 5.12 – Composite material's internal structure represented by square cells

---

[60] Dumontet H. (1986). *Local Effects in the Analysis of Structures. Chapter: Boundary layers stresses in elastic composites.* Elsevier.

[61] Jansson S. (1992). Homogenized nonlinear constitutive properties and local stress concentrations for composites with periodic internal structure. *Int. J. Solids Structures*, Vol. 29, No. 17, pp. 2181-2200.

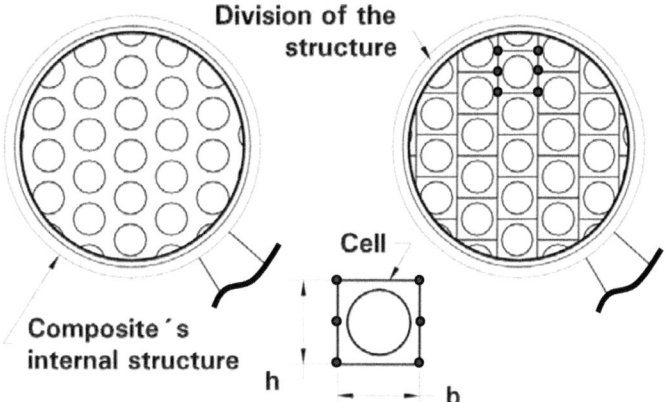

Figure 5.13 – Composite material's internal structure represented by hexagonal cells.

The composite material is made up by a ductile aluminum alloy matrix reinforced by long stiff alumina fibers. The bond between the fiber and the matrix is very strong (perfect adherence was assumed). The fiber and matrix volume proportions are 55% and 45%, respectively. The fiber diameter was 10.0 μmm. The cell dimensions are shown in table 5.1.

Material	Type of Cell	Fibers	Fibers Diam.	b	h
Composite	(Num. of Sides)	%	[μmm]	[μmm]	[μmm]
Quadrilateral	Quadrilateral	55	10.0	11.9499	11.9411
Hexagonal		55	10.0	11.1206	12.8410

Table 5.1 – Cell dimensions.

The fiber behavior is supposed to be elastic and isotropic while the aluminum matrix is represented by an isotropic elastoplastic material whose elastic discontinuity threshold matches the von Mises criterion. It is also admitted that the matrix has a non-linear kinematic hardening as well as an isotropic exponential hardening of the $\tilde{K}(\alpha) = \sigma^0 + H\alpha + (\sigma^{inf} - \sigma^0)(1 - \exp(-\delta\alpha))$ type, in which the kinematic hardening H=1000 MPa, and the difference between the initial yield stress (σ^0) and the saturation stress at infinite time (σ^{inf}) is 30Mpa at a saturation velocity of δ=300. The elastic properties of the material components are shown in Table 5.2.

	Young Modules	Poisson Rel.	Yield stress
Composite	E_{mod} [Mpa]	v	σ^0 [Mpa]
Matrix (Al_2O_3)	68900.0	55	94.0
Fiber	344500.0	55	-----

Table 5.2 – Elastic properties of the material components.

Special attention should be given to the numerical locking due to the incompressibility generated by the composite metallic matrix. Jansson used 9 node iso-parametric elements with selective reduced integration to avoid numerical locking. In the solution here presented the locking was avoided by implementing the "B-bar" method which is based on the mixed formulation of the finite element method in three fields: displacement, stress and strain.

A simple tension test is carried out by applying a macroscopic strain $\widetilde{E}_{xx}$ on the cell while a specimen is freed in a perpendicular direction of the cell strain such that the macroscopic stresses $\widetilde{\sigma}_{yy}$ obtained must be zero. This does not mean the stresses in transversal direction in each component are also zero. Moreover, as the square arrays of cells have a high degree of anisotropy, when they are turned 45^{0}, a different response is obtained. The behavior of the material is obtained by Ghosh *et al.* (1996)[43]. The stress-strain curves for the matrix and the fiber, for both a square array cell at 0° and 45^{0} and an hexagonal array cell, are shown in Figure 7. The results obtained from these cases coincide quite closely with the cited references (for more detail, see Zalamea's complete examples (2001)[11]).

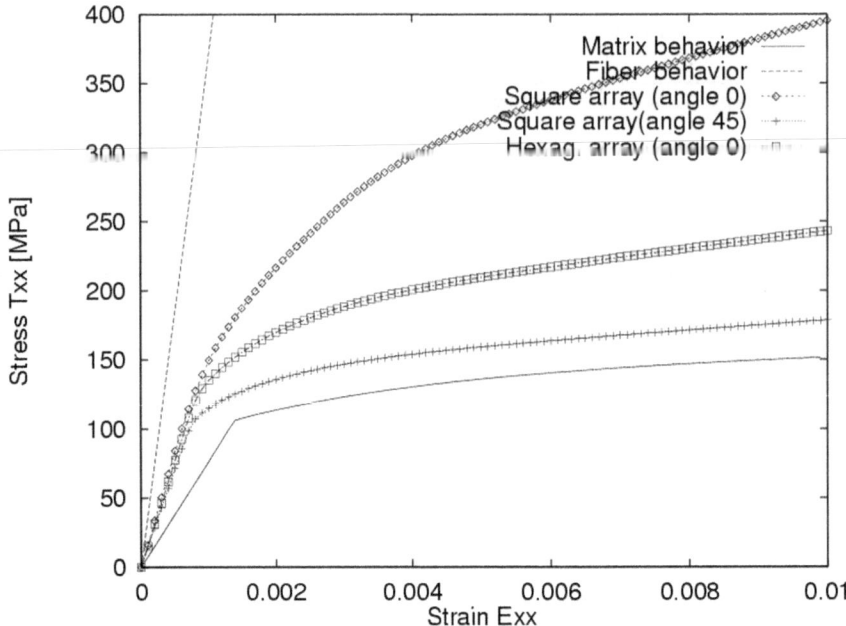

Figure 5.14 – Stress-strain curves in plane strain for: fiber, matrix. For quadrilateral cells at 0° and 45° and for hexagonal cells.

5.3.2 Thick cylinder subjected to internal pressure

This example consists of a cylindrical tube made of a composite material. Its mechanical characteristics are shown in Table 5.3. It is subjected to internal pressure which increases from 0 to a maximum of 100 MPa. (For more details about the original test, see the reference (Ghosh *et al.* (1996)[43]). The cylinder's symmetry enables us to study a quarter of the tube's cross section. Its domain is subdivided into 60 quadrilateral elements with a linear approximation. Its dimensions and boundary conditions are shown in Figure 5.15. This problem is solved at two scales by the local periodic homogenization technique mentioned in previous sections while Ghosh used the asymptotic expansion theory based on a two-scale representation for the micro structure using nonconventional finite elements called Voronoi elements.

Composite	Type of cell (num. Of sides)	Fibers %	Fibers diam. [μmm]	b [μmm]	H [μmm]
Quadrilateral	Quadrilateral	40	10.0	14.0125	14.0125
Hexagonal	Hexagonal	40	10.0	13.0401	15.0574

Table 5.3 – Cell dimensions.

Two cylindrical specimens made up by different composite materials have been studied:

- The first is solved by a quadrilateral cell as shown in Figure 5.12.

- The second tube is studied with a hexagonal cell as shown in Figure 5.13. In both cases the composite material is made up by 40% of fibers and 60% of matrix.

The mechanical properties of the component materials are identical to the ones used in the previous example and the cell dimensions are shown in Table 5.3.

The problem here presented is solved at two scales. Thus, for each iteration a macrostructure and an additional 240 cells must be solved (60 elements and 4 integration points each).

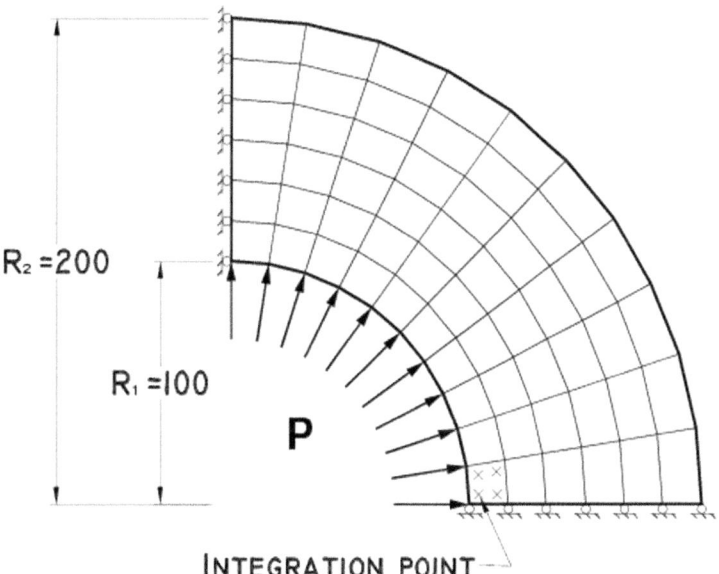

Figure 5.15 – Schematic representation of a cylindrical tube subjected to internal pressure (macroscopic structure).

The results obtained with the first composite are shown in Figure 5.16 In this figure four graphics of the macrostructure can be observed. Figure 5.17 shows four graphics of the microstructure of the first integration point (indicated in Figure 5.15). In each group of figures (macro and micro structures) a graphic 1 can be observed showing the structure subdivision into finite elements and the mesh displacement subjected to a pressure of 100

MPa. Graphic 2 shows the stress field represented as a function of the von Mises stress under an applied pressure of 10 MPa. In this case, the whole domain is subjected to an elastic behavior. It should be noted that the stress distribution on the tube is similar to the isotropic material distribution as the stress field is practically uniform and has a circumferential direction. Graphic 3 shows the von Mises stresses under an applied pressure of 50 MPa. The non-linear behavior starts in some points of the tube at this internal pressure. Finally, graphic 4 shows the von Mises stresses under an applied pressure of 100 MPa. In this case, part of the composite material has a non-linear behavior and the influence of the material anisotropy is strongly noticeable and produces stress concentration on parts of the domain. These results coincide closely with the results mentioned in the reference by Ghosh *et al.*, (1996) when the pressure reaches 100 PMa (see Figures 5.16 and 5.17). It should be highlighted that this coincidence is achieved although both formulations are very different from one another because Ghosh uses the Voronoi finite elements. The results obtained with hexagonal cells are very similar to the quadrilateral cell results. Therefore, composite materials made up by hexagonal cells can detect the non-linear anisotropy.

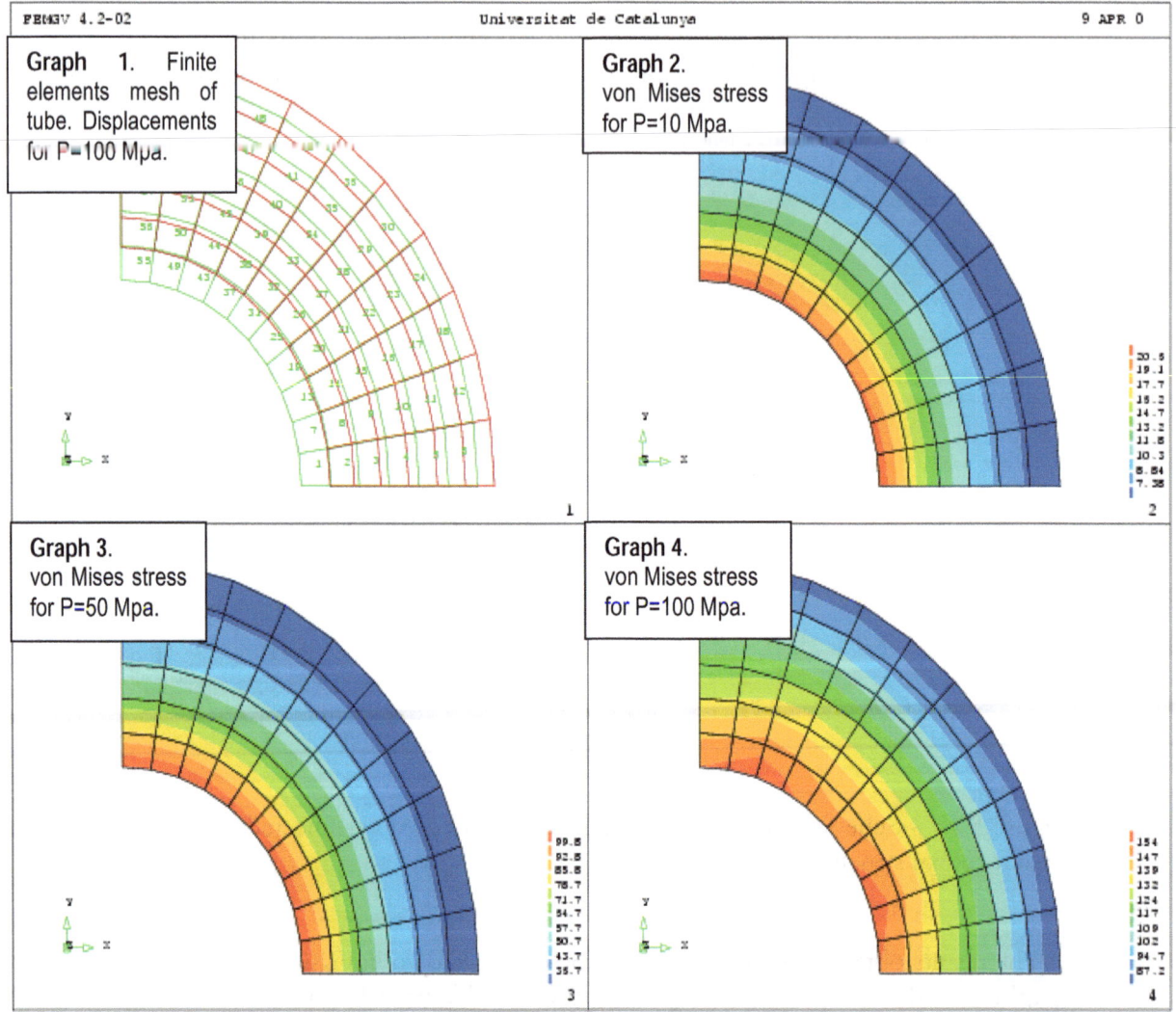

Figure 5.16 – Von Mises stresses in the composite material –macro scale– for three different stress levels.

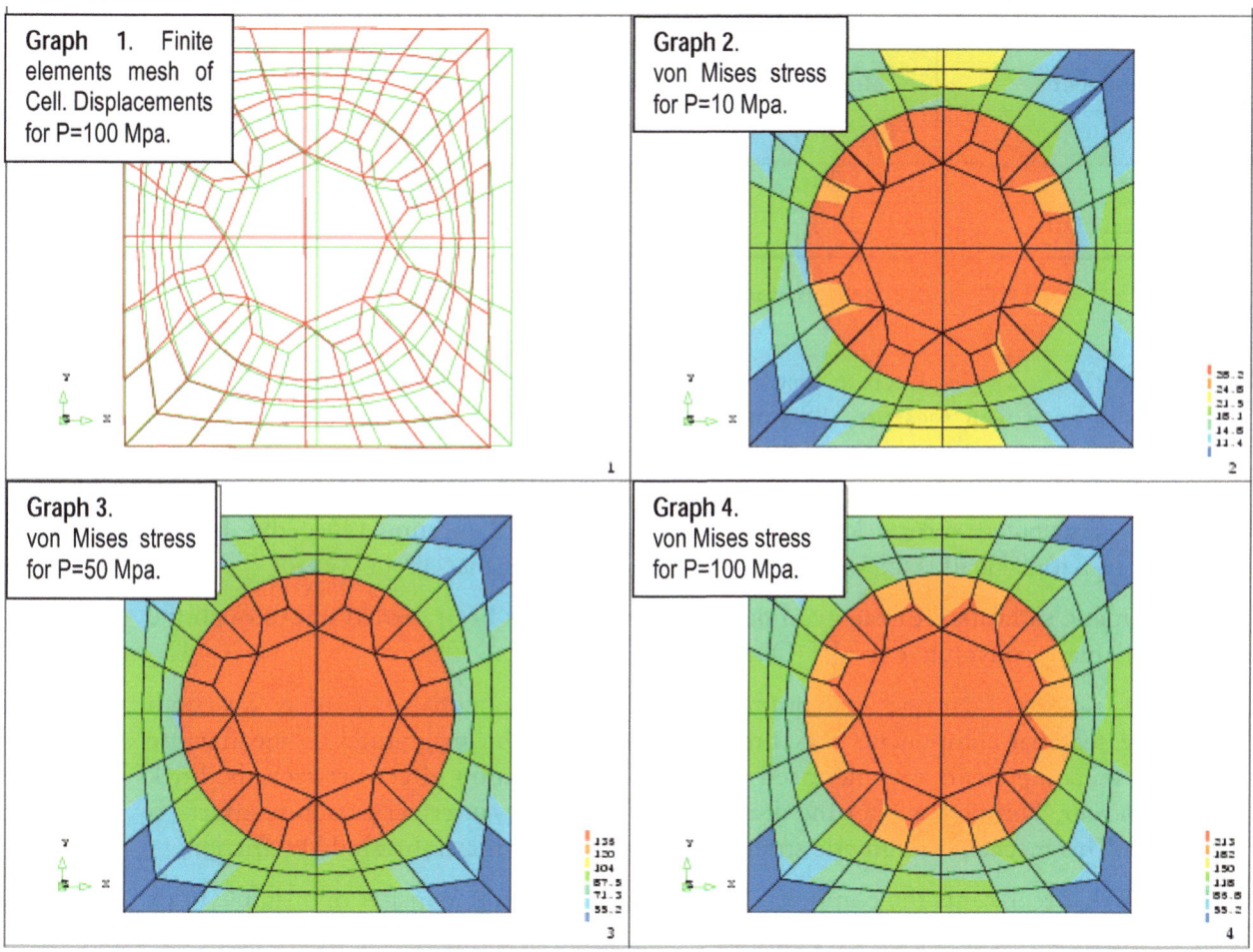

Figure 5.17 – Von Mises stresses in the microstructure corresponding to the first integration point of the composite material (see Figure 5.15) for three different stress levels.

5.3.3 Masonry homogenized, treated as a composite

As shown in the example above a traditional brick masonry is studied as a composite material. In fact, this masonry is a composite structure made up by different substructures –bricks, mortar joints–, rather than a composite material made up by different basic substances, which makes the simulation by constitutive models very difficult. Molins (1996)[62] studied some of the most important characterizations of this type of material and the simulation difficulties of its numeric-mechanical behavior. The example presented below consists of a masonry wall subjected to a shear load in a lab test (Lourenço (1996)[63] and Zijl et al. (1997)[64]), and for which the answer has been contrasted with numeric results. The properties of the component materials –brick, cement mortar– are detailed in Table

[62] Molins Borrell, C. (1986). *Structural analysis of historical constructions*. Capítulo de: *Characterization of the mechanical Behaviour of masonry*, pp. 86-122. Eds. P. Roca, J.L. González, A. Marí and E. Oñate. CIMNE, Barcelona.

[63] Lourenço, P. (1996). *Computational Strategies for Masonry Structures*. Delft University Press.

[64] Zijl, G.P.A.G. van, Lourenço, P.B. and Rots, J. G. (1997). Non-Associated Plasticity Formulation for Masonry Interface Behaviour. *Int. J. Comp. Plasticity*. pp. 1586-1593.

5.4. The perfect adherence between the mortar and the brick which simplifies the composite material constitutive modeling is assumed.

Material Component	Young's Modulus E_0 [N/mm²]	Poisson Relation [v]	Tensile Strength f_t [N/mm²]	Compression Strength f_c [N/mm²]
Brick	20000	0.15	5.0	15.0
Cement Mortar	2000	0.20	1.5	15.0

Table 5.4 - Elastic properties of the composite materials.

Figure 5.18 shows the fracture distribution on two masonry brick walls tested under shear loads. Each wall measures 990 mm along the base by 1000 mm high and has a window in its interior. The brick's dimensions are 210×52×100 mm. The bricks are bonded together by a 10 mm thick cement mortar. The lab test consists of fixing the wall base and applying a distributed vertical load on its upper part of 0.3 N/mm. Once this load is applied, a horizontal load is imposed on the top level under displacement control

The numeric simulation of each composite component has been carried out through the isotropic damage constitutive model (Oliver *et al.* (1990)[65]).

In the isotropic damage constitutive model, an exponential softening for both composite material components was first used, and then the magnitude of the fracture energy in its components was increased to represent the friction effect between the brick and the mortar. This occurs after the discontinuity threshold –interfacial fracture effect– is surpassed. This mechanical-numeric strategy introduces an artificial threshold by friction.

The cell that represents the composite material microstructure consists of a simple brick with two boundary mortar-layers 5 mm thick.

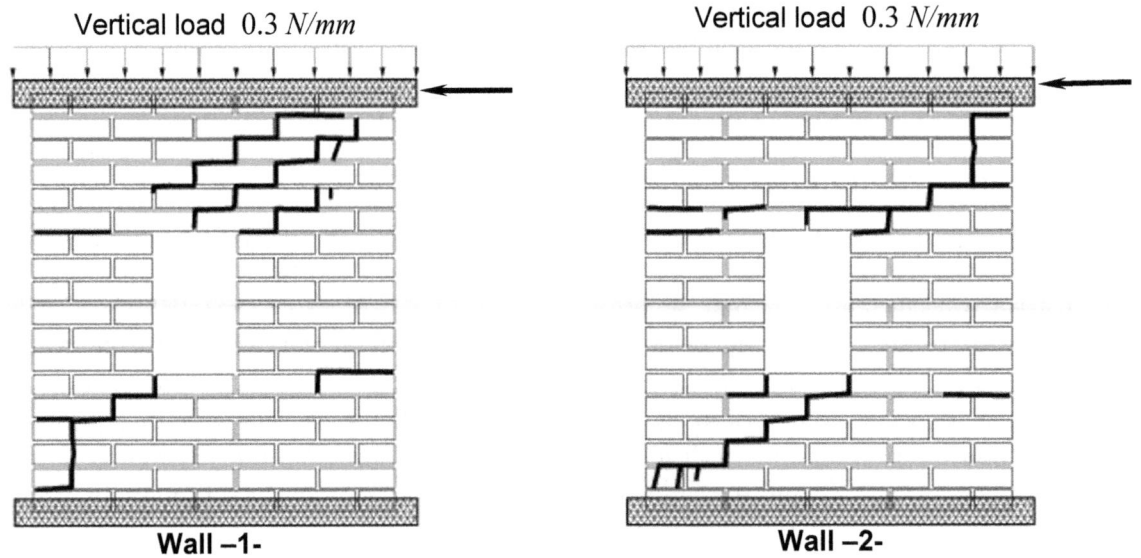

Figure 5.18 – Two fractured masonry walls under shear load.

---

[65] Oliver J., Cervera M., Oller S., and Lubliner J. (1990). Isotropic Damage Models and Smeared Crack Analysis of Concrete. *Second International Conference on Analysis and Design of Concrete Structures.* pp. 945-958.

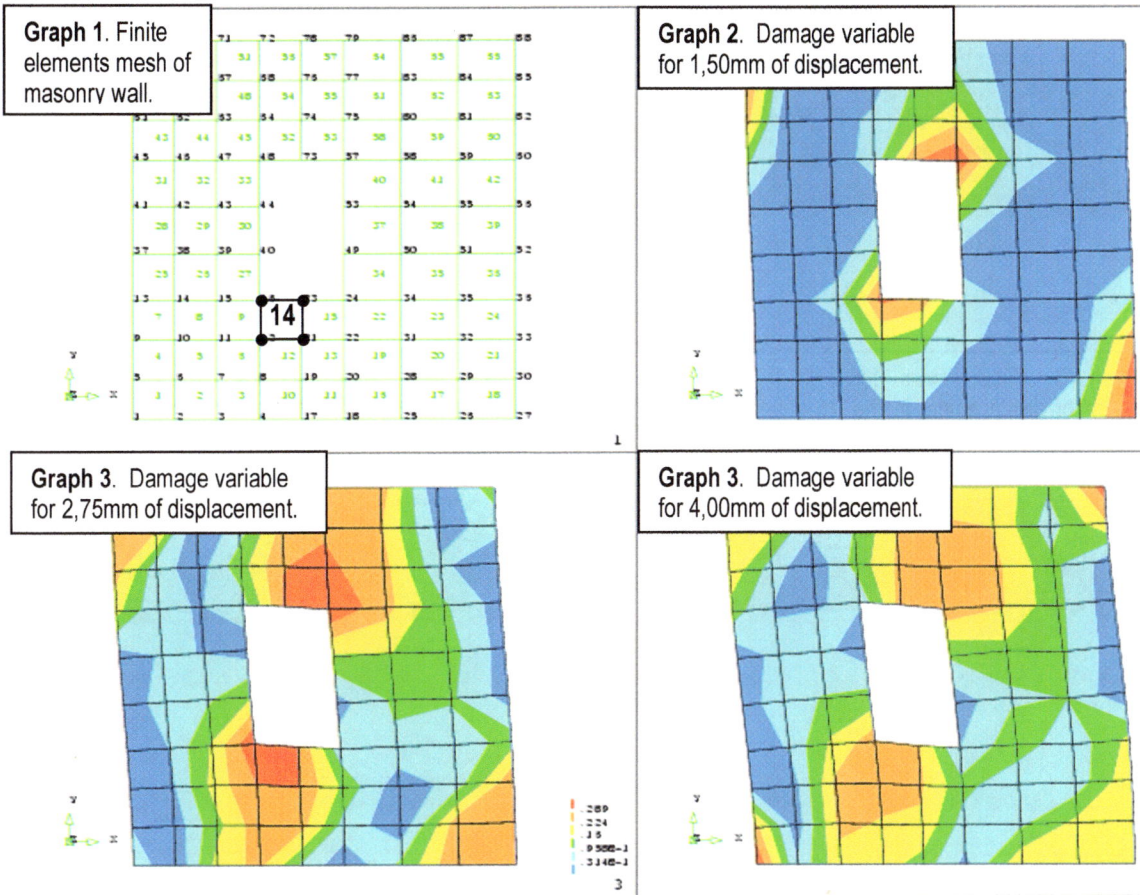

Figure 5.19 – Wall deformation for three different levels of load increments.

The wall is subdivided into 66 quadrilateral elements with four Gauss integration numeric points. Consequently, the macrostructure solution for each load increment is obtained in parallel through the 264 used cells.

The homogenization theory gives information about the two-scale results. Therefore, the masonry wall degradation or damage will be illustrated as well as the average damage in each of the cells. Graph 1 in Figure 5.11 shows the wall discretized in 66 quadrilateral finite elements. In the other graphs in this same figure (2, 3, 4) in the cell correspond to the microstructure. Observe that the damage variable has been represented in the deformed wall due to the displacement imposed. Graph 2 in Figure 5.11 shows the wall damage under horizontal displacement of 1.50 mm in the upper part of the wall. As shown, damage is observed first in the two opposite window corners produced by the shear stresses while tensile damage occurs in the lower right corner and in the upper left corner of the wall. Graph 3 in the same figure represents damage when horizontal displacement in the upper part of the wall reaches 2.75 mm. Graph 4, also in the same figure, shows the imposed displacement of 4.0 mm. The crushing damage in the lower left corner and in the upper right of the wall is also observed. The tension crack is observed in the mortar because it is the weakest component whereas the crushing by compression occurs in the two material components: brick and mortar.

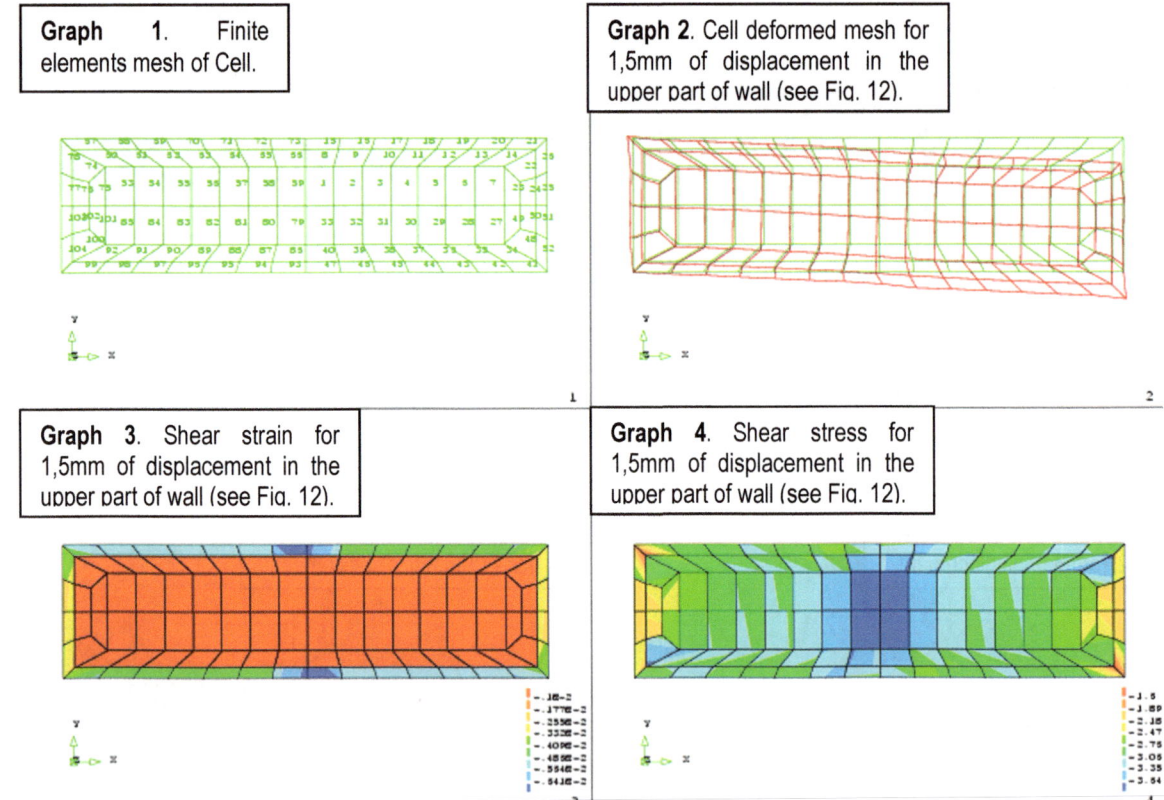

Figure 5.20 – Representation of a wall cell corresponding to the 3rd integration point of the finite element No. 14th (see Figure 5.19) for a displacement of 1.5 mm in the upper part of the wall. 1) Discretization by finite elements. 2) Deformed mesh. 3) Shear strains. 4) Shear stresses.

Thanks to the homogenization theory, the solution of the two-scale problem also gives information about the mechanical behavior at micro scale level.

As previously explained, the behavior of the discretized composite material is obtained by the mechanical solution at macro scale level for each integration point which in turn represents a cell in the micro domain. Consequently, the microscopic variable fields are obtained in each of these points. For example, Figure 5.20 shows four graphs of the cell of the composite material corresponding to the 3rd integration point of the 14th element of the macrostructure. Figure 5.19 shows when the applied displacement is 1.5 mm. Graph 1 in Figure 5.20 shows the cell domain subdivided into 104 quadrilateral finite elements of four (4) nodes. Graph 2 in the same figure shows the deformed mesh. Graph 3 represents the shear strains in the cell domain and Graph 4 the shear stresses in the basic cell.

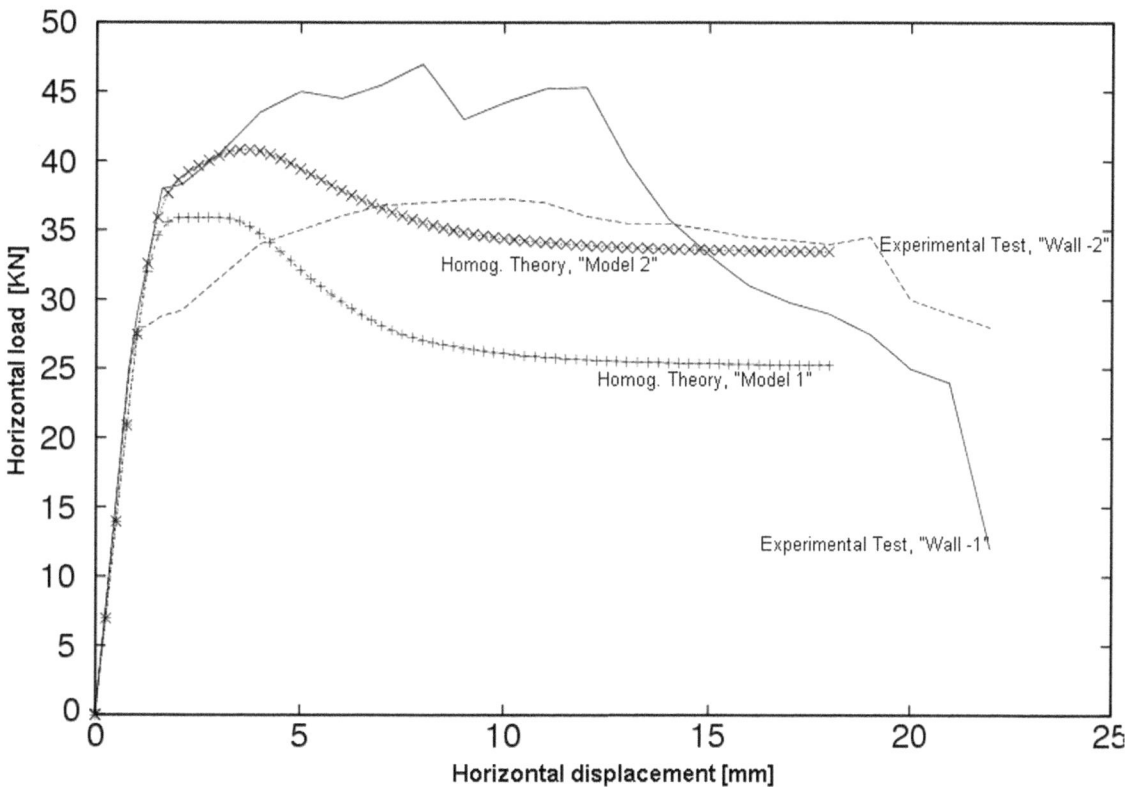

Figure 5.21 – Results compared to other numeric and experimental results.

Figure 5.21 shows the horizontal load curves measured in the wall base subjected to the imposed displacement on the upper part of the wall. The solution to the first damage of the numerical model using the homogenization theory coincides with the second numerical model and the experimental results in the elastic range. However, in the non-linear range, the strength deteriorates rapidly due to the mortar failure. In the second numeric model, the solution matched better the real results in the non-linear range. In this case the fracture energy in the mortar is increased to achieve the artificial friction effect between the mortar and the brick.

6 MASONRY-HOMOGENIZED COMPOSITE

6.1 Introduction and background

This chapter summarizes a singular approach for the homogenization of composite materials at micro scale level and more in particular for masonry (see López et al. (1999)[1], Oller et. al. (2002)[2], and its application to damage behavior (Quinteros et al.)[3]). The procedure is different from the procedures of the classic method mentioned in the previous chapter. However, after several manipulations, it could be classified as an "average homogenization method".

Masonry is the most widely used building material with the widest range of uses today as in the past. It can be found in many modern buildings nowadays. Masonry materials have been many and diverse throughout history: starting from simple rocks bonded with lime mortars, continuing later with huge marble blocks for the construction of great monuments during the height of the Renaissance architecture, and more recently refractory ceramic elements such as the materials used for the construction of furnaces, nuclear plants and thermal insulation spaceships.

The most widely used evaluation methods have strong simplifications over this "structural-material" behavior and some of these are included in the regulations of different countries. These norms are frequently obtained by empirical studies that tend to simplify the masonry behavior mechanisms and cause the structure over dimension. Another problem of using these structural methods is the unexpected crack formation and other mechanisms.

The development of more objective elements for the masonry structural analysis should be based on more advanced calculation methods such as the constitutive models of the mechanics of continuous media. The finite element methods are a powerful tool for masonry calculation but since this latter has a small size with respect to the global dimension of the structure they become unfeasible from the computational point of view.

---

[1] López J., Oller S., Oñate E., Lubliner J. (1999). A Homogeneous Constitutive Model for Masonry. *International Journal of Numerical Methods in Engineering.* Vol. 46, No.10, pp. 1651-1671.

[2] Oller S., Lubliner J., López J. (2002). *The masonry-An homogenized composite.* Chapter of book: *Structural Analysis of Composite Materials* (In Spanish), pp. 379, 410 - Ed. S. Oller. CIMNE, Barcelona.

[3] Quinteros R., Oller S., Nallim L. (2011). Nonlinear homogenization techniques to solve masonry structures problems. Composite Structures. Vol. 94, pp.724–730.

The need for simple, objective and fast calculation methods lead to the development of new formulations for the macro level treatment of masonry. Obtaining a balanced method pushes the development of the "homogenized constitutive model" presented in this work a step forward.

6.2 Masonry properties

Masonry can be seen from the composite material's perspective, among others. As a matter of fact, masonry is "a mixed structure" rather than a "composite material" but the frontier between these is established only by the component's scale and therefore, it could be said that it satisfies both qualifications. Attempts have been made to apply the classic homogenization theory (Anthoine (1997)[4]) as well as other homogenization theories such as the one presented in this work, originally based on the work by López et al. (1998)[5]. A description of the masonry formulation follows, which will be presented further and which could be described as a "simplified homogenization model".

The uniaxial behavior of the composite material will be described with respect to the material axes parallel and normal to the joint direction.

6.2.1 Masonry behavior under uniaxial compression

The masonry compressive strength normal to the joint has traditionally been seen as a relevant property of the material, at least until recently with the introduction of the numeric methods for masonry structures. A common test to obtain this compressive uniaxial strength is the prism assay with stack elements (see Figure 6.1). However, these strength parameters obtained by this type of tests are still uncertain. The most widely accepted uniaxial compression test to determine the uniaxial compressive strength of the masonry is made in the normal direction to the mortar joints (see Figure 6.1). This test is relatively long and expensive compared to standard tests carried out through concrete cylindrical specimens. Masonry uniaxial compression leads to triaxial compression in the mortar and to tension-compression in the brick. This test shows that initially vertical cracks are observed in the bricks along the structure's mid-lines and generally coincide with the continuation of the vertical mortar joints. As the strain grows, additional vertical cracks appear in the structure's small areas leading to a sliding crack failure in the structure.

[4] Anthoine, A. (1997). Homogeneization of periodic masonry: plane stress, generalized plane strain or 3D modelling. *Numerical Methods in Engineering*, Vol. 13, 319-326.
[5] López J., Oller S., Oñate E. (1998). *Masonry behavior evaluation using finite elements*. Monograph No. 46 CIMNE.

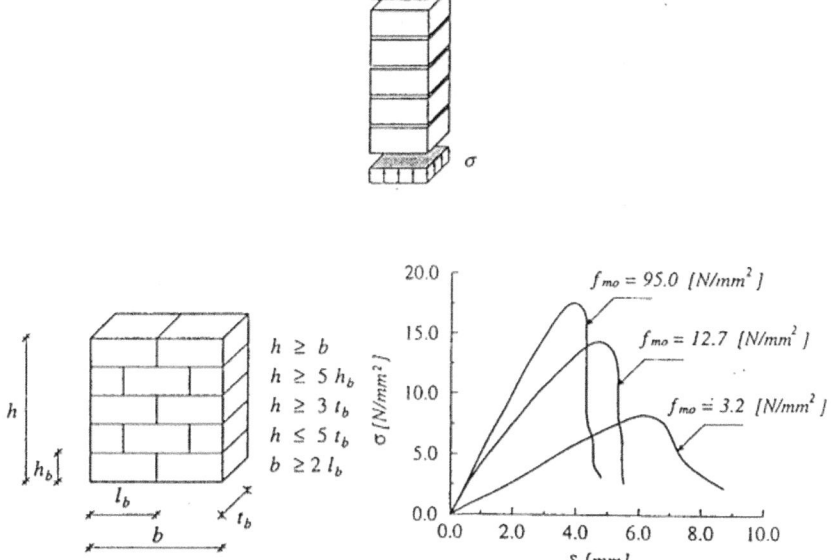

Figure 6.1 – Masonry uniaxial behavior subjected to normal loading to the horizontal joints plane

Masonry behavior subjected to uniaxial compression parallel to the mortar joints gets less attention. However, masonry is an anisotropic material-structure and the compressive strength under parallel loading to the mortar joints have decisive effects when determining the buckling load in the walls. The ratio between the uniaxial compressive strength parallel and normal to the joints ranges from 0.2 to 0.8. These ratios have been obtained for hollow bricks, mortar bricks and light-concrete blocks. For solid bricks the aforementioned ratio is 1.0.

6.2.2 Masonry behavior under uniaxial tension

For tension loading perpendicular to the mortar joints, the collapse is caused by a tension-strength loss in the mortar-brick interface. By not a very rigorous approximation, the masonry tension strength can be related directly to the tension strength of the bond between the joint and the brick.

In masonry built with low strength bricks and high tension strength joint bonds such as high strength mortars and bricks with a great number of hollows, failure occurs as the tension strength stresses in the brick are exceeded. The brick tension strength is assumed as the approximation of the masonry tension strength.

From the tests carried out in panels subjected to parallel tension to the joint, two different crack mechanisms have been observed. The first fails due to crack development in the vertical and horizontal joints and in "zigzag" form. In the second mechanism a perpendicular crack to the tension force is observed which goes on the mortar vertical joints and through the bricks.

In the first fracture mechanism, the masonry response is governed by the fracture energy of the vertical joints whereas in the horizontal joints the shear mechanism governs the cracking (see Figure 6.2). In the second mechanism, since all the cracks are vertical and involve both the joints and the brick, the fracture energy G_f of each material participates in the total cracking of the structure.

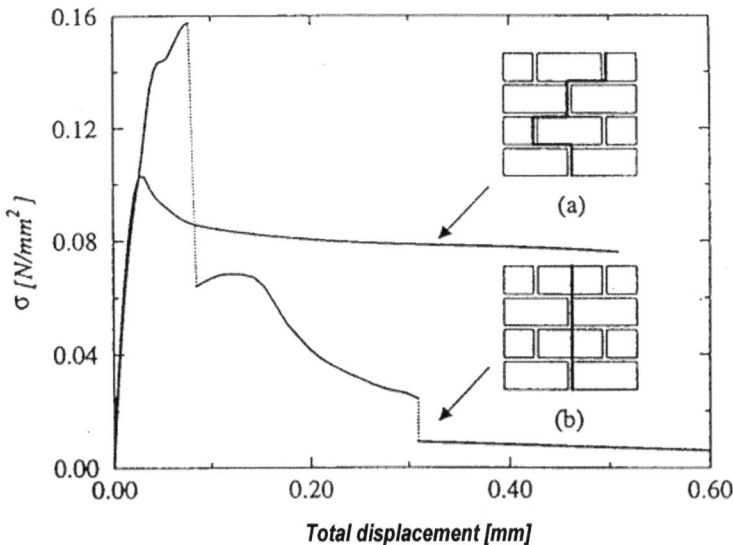

Figure 6.2 – Stress-displacement experimental curves for tension parallel to the mortar horizontal joints: (a) failure occurs due to staggering cracking parallel to the joints; (b) failure is vertical along the joints and brick.

6.2.3 Biaxial behavior

The masonry biaxial behavior under biaxial stress states cannot be completely described by a constitutive law under uniaxial loading conditions. The influence of biaxial stress state is important to find the resistant behavior which cannot be described only in terms of principal stresses because masonry is an anisotropic material. Therefore, the masonry biaxial-stress envelope can be described in terms of the orientation of the axes with respect to the material and the principal stresses, being θ the angle forming the principal stresses and the material axes. The most complete information gathered about masonry biaxial load proportionality can be observed in Figure 6.3, obtained by Page (1973)[6], (1981)[7], based on the tests carried out on a wall, presented at the end of this chapter.

---

[6] Page, A. W. (1973). *Structural brickwork-A literature review.* Engineering Bulletin No.CE4, Departament of Civil Engineering, Universidad de Newcastle, Australia.
[7] Page, A. W. (1981). The Biaxial Compressive Strength of Brick Masonry. *Proc. Instn. Civ.Engrs,* 71, (2), 893-906.

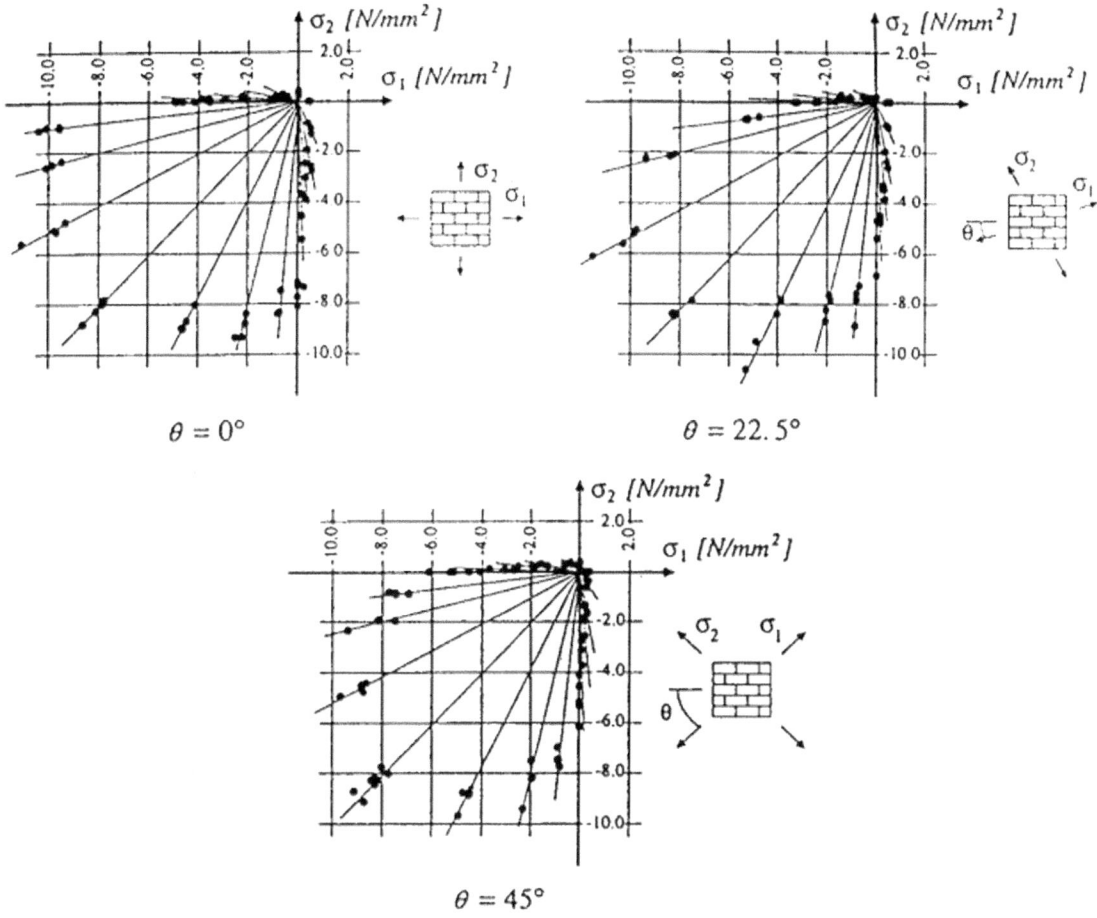

$\theta = 0°$ $\theta = 22.5°$

$\theta = 45°$

Figure 6.3 – Biaxial test on a masonry wall.

One of the most important results obtained by Page (1973)[6], (1978)[8], (1981)[7], was determining the different fracture or cracking mechanisms as a function of the load orientation with respect to the axes defined for the material (see Figure 6.4).

In uniaxial tension, failure occurs by the cracking and sliding of some bricks over others in the line formed by such couplings of cracks. The influence of lateral tension in the global tension strength is not known due to the lack of known experimental tests. Lateral compression makes the tension strength decrease due to the damage induced in the composite material by the micro cracking formation in the joints and bricks. In the case of combined tension-compression loads, failure occurs by both a cracking and sliding formation in the joints as well as by a combined mechanism that involve joints and bricks. Similar types of failure occur in uniaxial compression, but in this case smooth transitions towards other fracture mechanisms are observed. The most common failure mechanism in biaxial compression occurs as a result of the panel cracking in the structure mid-line area parallel to the loading plane. The increment of the compressive strength under biaxial compression can be explained by the friction development in the joints and in the mortar.

---

[8] Page, A. W. (1978). Finite element model for masonry. *Journal of the Structural Division*, ASCE, Vol. 104, No. ST8, Proc. Paper 13957, 1267-1285.

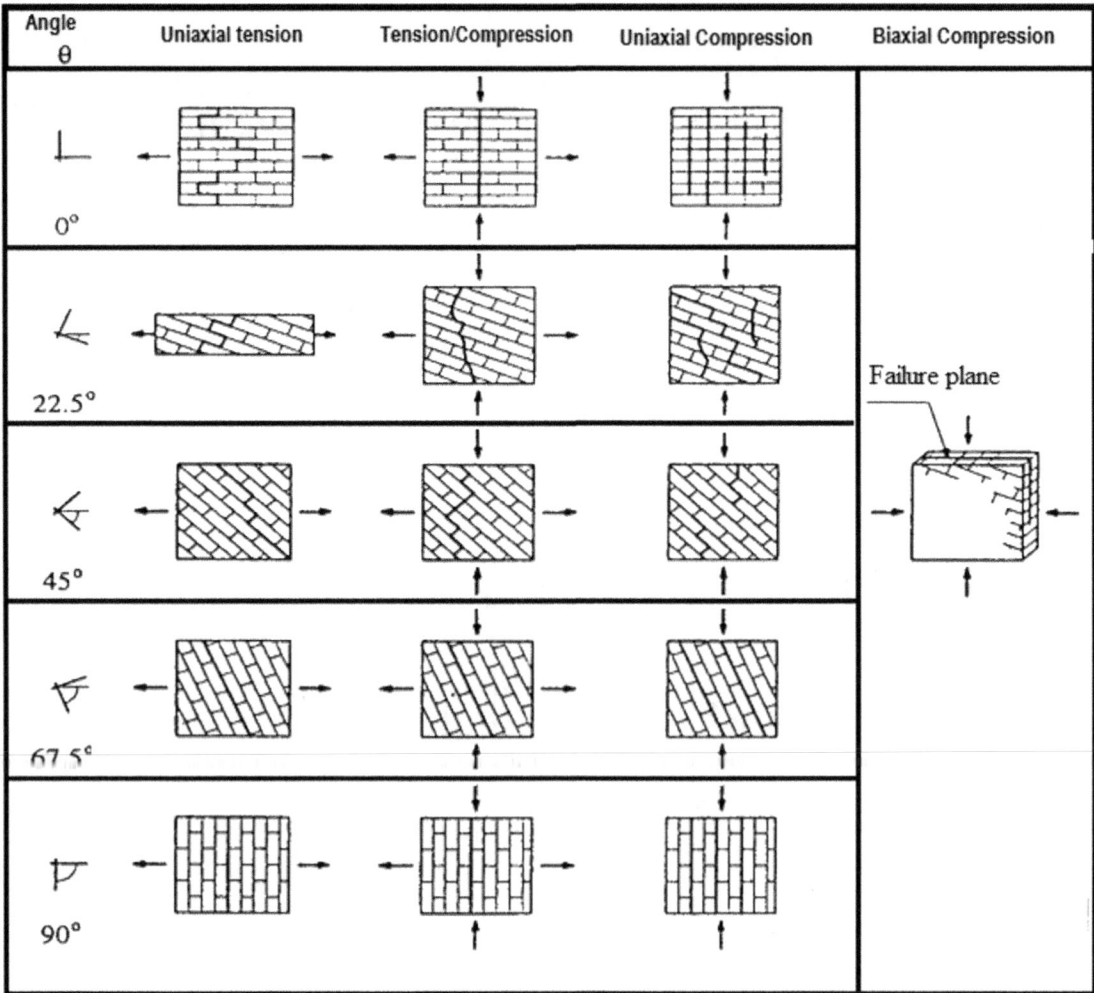

Figure 6.4 – Masonry cracking mechanisms.

6.2.4 Post-peak masonry behavior. Softening

Softening is a gradual decrease of the material strength properties in one point of a solid subjected to a monotonous increment of the imposed strains (Oller (2001)[9]). This is a characteristic feature of quasi brittle materials such as clay bricks, mortars, ceramic materials, rocks and concrete which fail due to a gradual internal decay caused by micro cracks in the component material interfaces. The mortar's micro cracks are due to shrinkage in the curing process and the presence of additives. Regarding clay bricks, defects and micro cracks occur as a result of the shrinkages during the sintering process of pieces. The initial stresses and cracks as well as the internal variations of stiffness and strengths cause a gradual increase of the cracks when the material is subjected to progressive strains. At the beginning the micro cracks are stable, which means they will only grow when the load is incremented. The cracks' interconnection and the excessive formation of some of them lead to the formation of "macro cracks". They are unstable, which means the load must decrease in order to avoid their uncontrolled increment. In controlled strain tests, the

---

[9] Oller, S. (2001). *Fracture Mechanics – A global approach*. CIMNE-Ed UPC.

macro cracking increment leads to softening and strains (cracks) in small areas while an unloading process is developed in the rest of the structure (Oller (1991)[10], (2001)[9]).

In view of the phenomenon described above, the following basic rules can be established for the numeric simulation of the constitutive behavior of a masonry point:

1. **In tension and shear mechanisms**, the softening process can be carried out as a cohesion loss using an elastoplastic model limited by the Mohr-Coulomb discontinuity surface.

2. **In compression cases,** the softening phenomenon strongly depends on the boundary conditions of the masonry structure as well as on the component dimensions.

Figure 5 shows the stress-displacement characteristic diagram for quasi brittle materials in pure shear case. By analyzing the graphics $\sigma - \delta$, the fracture energy G_f and crushing energy G_c are obtained, and by their suitable combination the corresponding dissipation shear G^{II} (see Oller (1991)[10]) is provided. Then the post-peak behavior of the material is obtained.

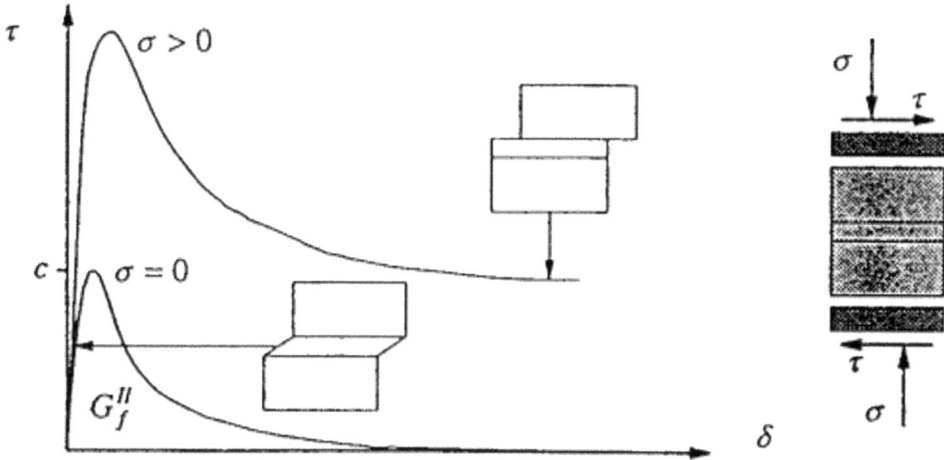

Figure 6.5 – Masonry behavior under pure shear states.

6.3 Different methods for masonry calculation

This section gives a quick look at the different methods based on finite elements for the calculation of masonry mechanical behavior. The bar-based approximate models is another calculation approach where the brick is represented by a bar while the mortar joints are represented by elastic supports at the bar's end. This type of model is not normally used as it moves off from the constitutive models' philosophy generated by solid mechanics.

Masonry is a material presenting directionality –anisotropy– among its properties because the mortar joints are like weakness planes in the material. In general, the approximation for numerical modeling can be focused on "micro models" to discretize the geometry in detail separating the bricks from the joints. On the other hand, a "macro model" approach can

---

[10] Oller, S. (1991). *Numerical Modelling of Frictional Materials*. Monograph No. 3, Ed. CIMNE. Barcelona.

also be carried out by assuming masonry as a composite material. Depending on the degree of detail required, the following modeling approaches (see Figure 6.6) can be used:

1. **Detailed micro modeling**. Bricks and mortars are represented by finite elements of continuum behavior, while the behavior of the mortar-brick interface is represented by the discontinuous finite elements.

2. **Simplified micro modeling**. In this case, the materials are represented by elements of continuum behavior while the behavior in the mortar joints and the mortar-brick interface are represented by discontinuities.

3. **Macro modeling**. The bricks, mortars and brick-mortar interface are represented by the same finite element. The homogenization technique, which will be discussed in this chapter, can be included within this type of discretization.

In the *first approximation*, the Young's modulus, the Poisson coefficient and, optionally, the inelastic properties of both materials must be taken into consideration regardless of the type of model used. The joint-brick interface represents a potential cracking surface with a fictitious initial stiffness to represent the contact and to avoid the penetration of one material into another. This produces the combined action of the brick, the mortar and the interface. In the *second approximation*, each joint has a mortar and two mortar-brick interfaces and they get joined in an interface that gathers these three elements to simplify the problem without changing the geometry. Thus, masonry is assumed as a set of elastic blocks separated by potential cracking surfaces (joints). Accuracy is lost when the mortar Poisson's effect is not taken into account. In the *third approximation* no geometrical distinction is made between bricks and joints because masonry is assumed as a continuous homogeneous and anisotropic material. None of these modeling strategies can "always" be considered to be better than the others as the application field for micro models and macro models is different.

The *micro models* are probably the best tool for understanding the behavior of masonry. The advantage of using them as an approximation is that they can consider different failure mechanisms (cracking, displacement). They have their application field in the study of the local behavior of structural details of masonry as well as in the study of the real behavior of the interface (discontinuities in the structure). These types of discontinuities are generally important in the global behavior of masonry structures. This type of modeling requires a large discretization in elements that involves a high computational cost regarding the addressed scale.

The *macro models* are suitable when the structure is made up by walls with dimensions sufficiently large to make the stresses through and along the elements essentially uniform. It is only necessary to know the behavior of the whole set. Examples of this type of approximation are found in the formulations based on the mixing theory and the homogenization theory. Obviously, from the computational point of view, the macro modeling is much more practical due to the short time and memory required, making the mesh generation easier, too. Moreover, this type of modeling is better when a compromise between accuracy and efficiency is required.

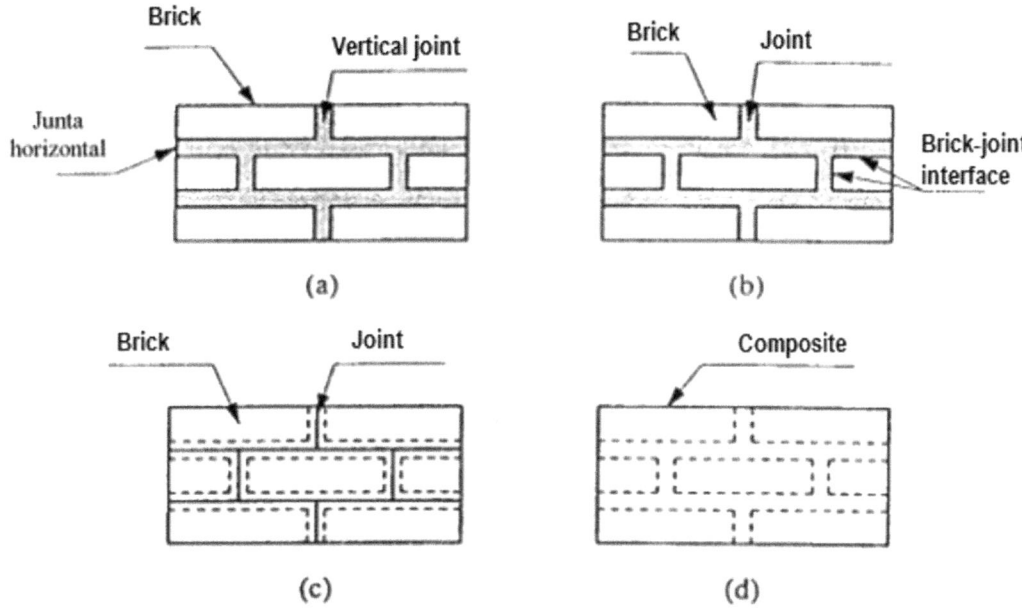

Figure 6.6 – Different masonry modeling strategies: (a) Simple masonry; (b) Micro model; (c) Simplified micro model; (d) Macro model.

Both masonry structures, *"micro models"* as well as *"macro models"*, require the materials' description through experimental work. However, masonry properties are influenced by a great number of factors such as the brick and mortar intrinsic properties, manpower quality, post production treatment, development, age, etc.

6.4 Constitutive model based on a particular case of the homogenization concept

In this section a non-linear constitutive model is presented based on an "average" technique to treat the masonry behavior following a particular formulation of the homogenization theory. The objective is to present a homogenized material containing the intrinsic geometrical and mechanical properties of masonry based on a simple definition of the constituent materials (brick and mortar). The resultant homogenized model will considerably reduce the discretization of the finite element mesh as the classic (brick, joint) detailed discretization that is required by micro modeling won't be necessary and therefore computer costs will be saved.

6.4.1 Constitutive model

The model is based on the description of the average behavior of a "basic cell" which is representative of a point of the masonry (Figure 6.7), from the detailed analysis of the forms and modes of behavior of such a "cell" under different load states. Figure 6.7 depicts the notation used to identify the dimensions to characterize the element.

Among the basic hypothesis imposed to support the formulation, two can be highlighted, mainly:

1. The geometry of a masonry wall allows the use of the plane stress hypothesis, provided the load is in the structure's plane.

2. Due to the brick and mortar joints arrangement, the orthotropic behavior of the whole set is accepted.

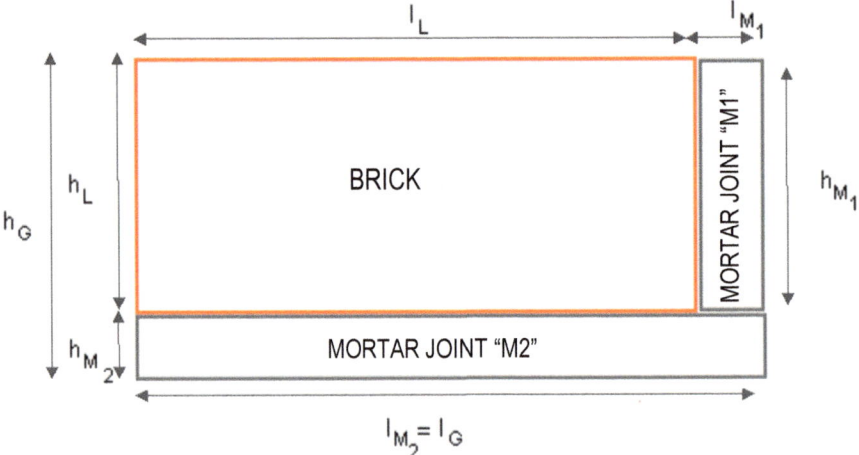

Figure 6.7 – Notation used for the dimensions of the components of the basic "cell".

The equations governing each of the different deformation modes of a basic masonry "cell" are formulated in the next section. It is necessary to keep in mind that hereinafter "*L*" stands for the brick, "*M1*" for the vertical mortar, "*M2*" for the horizontal mortar and "*G*" for the global dimensions of the –homogenized– set or the basic cell.

6.4.1.1 Equations of "*Mode 1*"

"Mode 1" of the cell behavior is established as the corresponding to the tension-compression state along "*x-axis*" (see Figure 6.8)

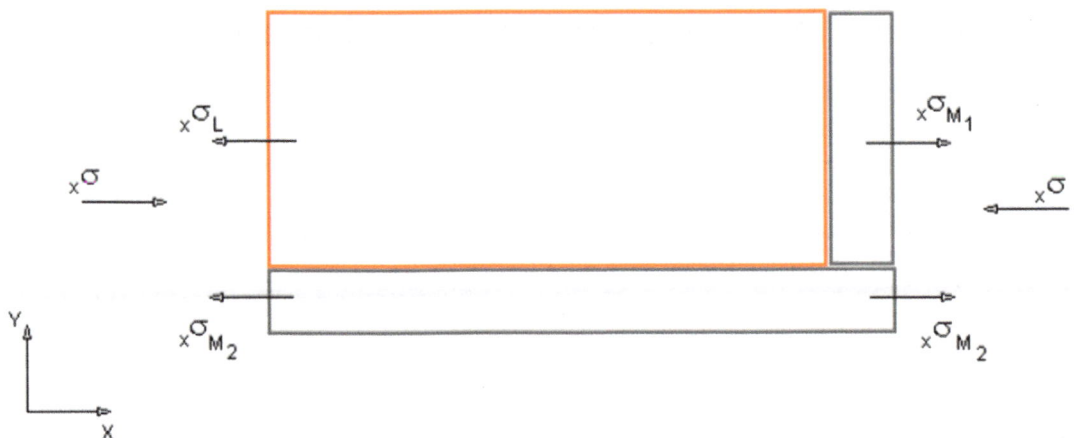

Figure 6.8 – Representation of *Mode 1* of the cell behavior.

- **Equilibrium condition of "*Mode 1*"**

Considering a uniform thickness *t=const.* for the whole cell, the following equilibrium equations are formulated,

$$_x\sigma \cdot h_G = {}_x\sigma_L \cdot h_L + {}_x\sigma_{M_2} \cdot h_{M_2} = {}_x\sigma_{M_1} \cdot h_{M_1} + {}_x\sigma_{M_2} \cdot h_{M_2}$$

$$h_{M_1} = h_L \Rightarrow {}_x\sigma_L \equiv {}_x\sigma_{M_1}$$

$$\Rightarrow \quad \begin{cases} {}_x\sigma = {}_x\sigma_L \cdot \dfrac{h_L}{h_G} + {}_x\sigma_{M_1} \cdot \dfrac{h_{M_1}}{h_G} \\[3mm] {}_x\sigma = {}_x\sigma_{M_1} \cdot \dfrac{h_{M_1}}{h_G} + {}_x\sigma_{M_2} \cdot \dfrac{h_{M_2}}{h_G} \end{cases} \tag{6.1}$$

where $_i\sigma_j$ represents the homogeneous stress state in the direction "i" of the component material "j" and h_i represents the height of the component material "i".

- **Compatibility condition of "*Mode 1*"**

By observing Figure 6.8, the following compatibility equation can be formulated,

$$\begin{cases} {}_x\dot{\varepsilon}_G \cdot l_G = {}_x\dot{\varepsilon}_{M_2} \cdot l_{M_2} \\[3mm] {}_x\dot{\varepsilon}_G \cdot l_G = {}_x\dot{\varepsilon}_L \cdot l_L + {}_x\dot{\varepsilon}_{M_1} \cdot l_{M_1} \end{cases} \tag{6.2}$$

where $_i\varepsilon_j$ represents the homogeneous deformation in the direction "i" of the component material "j" and l_i represents the length of the component material "i".

- **Constitutive equation of "*Mode 1*"**

For each component material of the basic "cell", the following constitutive relationship can be written,

$$_x\dot{\sigma}_i = {}_xE_i(\omega_i) \cdot ({}_x\dot{\varepsilon}_i - {}_x\dot{\varepsilon}_i^p)$$

$$_x\dot{\varepsilon}_i = \frac{{}_x\dot{\sigma}_i}{{}_xE_i(\omega_i)} + {}_x\dot{\varepsilon}_i^p \tag{6.3}$$

As observed above, in this last equation the Young's modulus depends on the damage parameter ω_i, as the elasticity modulus is degraded beyond the elastic branch.

- **Stress determination in each component**

By substituting equation (6.1) into (6.3) the total deformation of mortar M_2 is obtained. Thus,

$$_x\sigma_{M_2} = {}_x\sigma \cdot \frac{h_G}{h_{M_2}} - {}_x\sigma_L \cdot \frac{h_L}{h_{M_2}} \quad \Rightarrow \quad {}_x\dot{\sigma}_{M_2} = {}_x\dot{\sigma} \cdot \frac{h_G}{h_{M_2}} - {}_x\dot{\sigma}_L \cdot \frac{h_L}{h_{M_2}}$$

$$_x\dot{\varepsilon}_{M_2} = \left[\frac{{}_x\dot{\sigma}}{{}_xE_{M_2}} \cdot \frac{h_G}{h_{M_2}} - \frac{{}_x\dot{\sigma}_L}{{}_xE_{M_2}} \cdot \frac{h_L}{h_{M_2}} \right] + {}_x\dot{\varepsilon}_{M_2}^p \tag{6.4}$$

And substituting both the latter and equation (6.3) into expression (6.2), the global deformation is obtained and expressed in two different forms,

$$_x\dot{\varepsilon}_G \cdot l_G = \frac{_x\dot{\sigma}}{_x E_{M_2}} \cdot \frac{h_G}{h_{M_2}} \cdot l_{M_2} - \frac{_x\dot{\sigma}_L}{_x E_{M_2}} \cdot \frac{h_L}{h_{M_2}} \cdot l_{M_2} + {}_x\dot{\varepsilon}^P_{M_2} \cdot l_{M_2}$$

$$_x\dot{\varepsilon}_G \cdot l_G = \underbrace{\frac{_x\dot{\sigma}_L}{_x E_L} \cdot l_L + {}_x\dot{\varepsilon}^P_L \cdot l_L}_{_x\dot{\varepsilon}_L \cdot l_L} + \underbrace{\frac{_x\dot{\sigma}_L}{_x E_{M_1}} \cdot l_{M_1} + {}_x\dot{\varepsilon}^P_{M_1} \cdot l_{M_1}}_{_x\dot{\varepsilon}_{M_1} \cdot l_{M_1}} \qquad (6.5)$$

Matching the two equations above, the time variation of the brick stress is obtained,

$$\frac{_x\dot{\sigma}}{_x E_{M_2}} \cdot \frac{h_G}{h_{M_2}} \cdot l_{M_2} + {}_x\dot{\varepsilon}^P_{M_2} \cdot l_{M_2} - {}_x\dot{\varepsilon}^P_L \cdot l_L - {}_x\dot{\varepsilon}^P_{M_1} \cdot l_{M_1} = \frac{_x\dot{\sigma}_L}{_x E_L} \cdot l_L + \frac{_x\dot{\sigma}_L}{_x E_{M_1}} \cdot l_{M_1} + \frac{_x\dot{\sigma}}{_x E_{M_2}} \cdot \frac{h_L}{h_{M_2}} \cdot l_{M_2}$$

$$_x\dot{\sigma}_L = \frac{_x E_L \cdot {}_x E_{M_1} \cdot {}_x E_{M_2} \cdot h_{M_2}}{\underbrace{_x E_{M_1} \cdot {}_x E_{M_2} \cdot h_{M_2} \cdot l_L + {}_x E_L \cdot {}_x E_{M_2} \cdot h_{M_2} \cdot l_{M_2} + {}_x E_L \cdot {}_x E_{M_1} \cdot h_{M_2} \cdot l_{M_1}}_{^xA_L}} \cdot$$

$$\cdot \left[\frac{_x\dot{\sigma}}{_x E_{M_2}} \cdot \frac{h_G}{h_{M_2}} \cdot l_{M_2} + {}_x\dot{\varepsilon}^P_{M_2} \cdot l_{M_2} - {}_x\dot{\varepsilon}^P_L \cdot l_L - {}_x\dot{\varepsilon}^P_{M_1} \cdot l_{M_1} \right]$$

$$_x\dot{\sigma}_L = {}_x\dot{\sigma} \cdot \underbrace{\left({}^xA_L \cdot \frac{l_{M_2}}{_x E_{M_1}} \cdot \frac{h_G}{h_{M_2}} \right)}_{^xB_L} + \underbrace{{}^xA_L \cdot \left({}_x\dot{\varepsilon}^P_{M_2} \cdot l_{M_2} - {}_x\dot{\varepsilon}^P_L \cdot l_L - {}_x\dot{\varepsilon}^P_{M_1} \cdot l_{M_1} \right)}_{^x\dot{C}_L}$$

$$\boxed{{}_x\dot{\sigma}_L = {}_x\dot{\sigma} \cdot {}^xB_L + {}^x\dot{C}_L \equiv {}_x\dot{\sigma}_{M_1}} \qquad (6.6)$$

From equations (6.4) and (6.6) the time variation of the total stress in mortar M_2 is obtained,

$$_x\dot{\sigma}_{M_2} = {}_x\dot{\sigma} \cdot \frac{h_G}{h_{M_2}} - {}_x\dot{\sigma}_L \cdot \frac{h_L}{h_{M_2}} = {}_x\dot{\sigma} \cdot \frac{h_G}{h_{M_2}} - {}_x\dot{\sigma}_L \cdot {}^xB_L \cdot \frac{h_L}{h_{M_2}} - {}^x\dot{C}_L \cdot \frac{h_L}{h_{M_2}}$$

$$_x\dot{\sigma}_{M_2} = {}_x\dot{\sigma} \cdot \underbrace{\left(\frac{h_G - {}^xB_L \cdot h_L}{h_{M_2}} \right)}_{^xB_{M_2}} - \underbrace{{}^x\dot{C}_L \cdot \frac{h_L}{h_{M_2}}}_{^x\dot{C}_{M_2}}$$

$$\boxed{{}_x\dot{\sigma}_{M_2} = {}_x\dot{\sigma} \cdot {}^xB_{M_2} + {}^x\dot{C}_{M2}} \qquad (6.7)$$

- **Global constitutive equation of "*Mode 1*"**

From the second equation of expression (6.5) the global deformation for the masonry in "Mode 1" is written as

$$_x\dot{\varepsilon}_G = \frac{_x\dot{\sigma}_L}{_xE_L} \cdot \frac{l_L}{l_G} + {_x}\dot{\varepsilon}_L^P \cdot \frac{l_L}{l_G} + \frac{_x\dot{\sigma}_L}{_xE_{M_1}} \cdot \frac{l_{M_1}}{l_G} + {_x}\dot{\varepsilon}_{M_1}^P \cdot \frac{l_{M_1}}{l_G}$$

$$_x\dot{\varepsilon}_G = {_x}\dot{\sigma}_L \cdot \underbrace{\left(\frac{1}{_xE_L} \cdot \frac{l_L}{l_G} + \frac{1}{_xE_{M_1}} \cdot \frac{l_{M_1}}{l_G} \right)}_{^xD} + \underbrace{{_x}\dot{\varepsilon}_L^P \cdot \frac{l_L}{l_G} + {_x}\dot{\varepsilon}_{M_1}^P \cdot \frac{l_{M_1}}{l_G}}_{^x\dot{E}}$$

By substituting $_x\dot{\sigma}_L$ for its expression (equation (6.6)), the expression of the plastic deformation and the elasticity modulus corresponding to "*Mode 1*" of the masonry deformation is obtained as a function of the geometric and mechanical components of the "cell":

$$_x\dot{\varepsilon}_G = \left({_x}\dot{\sigma} \cdot {^xB_L} + {^x}\dot{C}_L \right) \cdot {^xD} + {^x}\dot{E}$$

$$_x\dot{\varepsilon}_G = \left({_x}\dot{\sigma} \cdot {^xB_L} \cdot {^xD} \right) + \left({^x}\dot{C}_L \cdot {^xD} + {^x}\dot{E} \right)$$

If we call $\dfrac{1}{_xE_G} = {^xB_L} \cdot {^xD}$ and $_x\dot{\varepsilon}_G^P = {^x}\dot{C}_L \cdot {^xD} + {^x}\dot{E}$

Then:

$$\boxed{\,_x\dot{\varepsilon}_G = \underbrace{\frac{_x\dot{\sigma}}{_xE_G} + {_x}\dot{\varepsilon}_G^P}_{_x\dot{\varepsilon}_G^e}\,}$$

(6.8)

6.4.1.2 Equations of "*Mode 2*"

"*Mode 2*" of the cell behavior is established as the corresponding to the tension-compression state along "*y-axis*" (see Figure 9).

By following an identical procedure to obtain the equations of "*Mode 1*" and keeping the same basic hypothesis, the equations in the "*y*" direction of the cell are obtained.

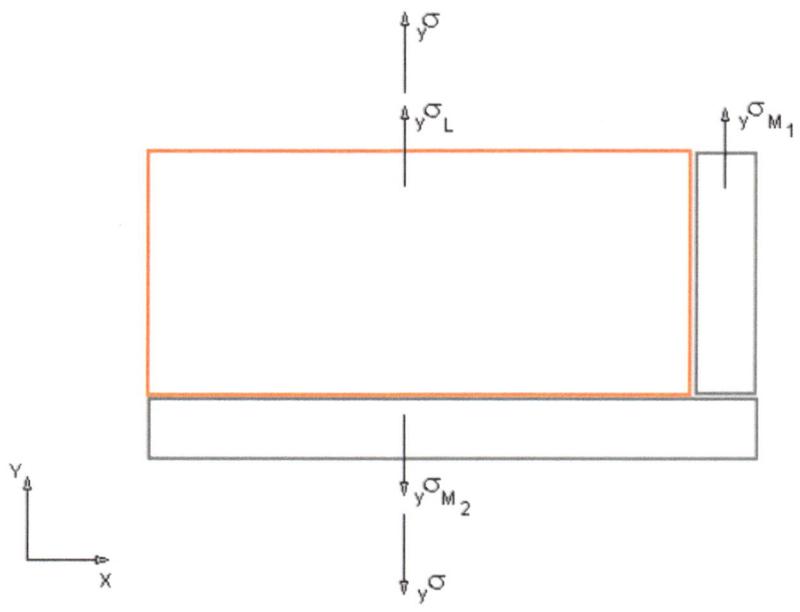

Figure 6.9 – Representation of *Mode 2* of the cell behavior.

- **Equilibrium equation of "*Mode 2*"**

Keeping in mind that the cell has a uniform thickness $t=const.$, the following equilibrium equations are obtained in the "y" direction,

$$_y\sigma \cdot l_G = {_y\sigma_L} \cdot l_L + {_y\sigma_{M_1}} \cdot l_{M_1} = {_y\sigma_{M_2}} \cdot l_{M_2}$$

$$l_{M_2} = l_G \quad \Rightarrow \quad _y\sigma \equiv {_y\sigma_{M_2}}$$

$$\Rightarrow \quad \begin{cases} _y\sigma = {_y\sigma_L} \cdot \dfrac{l_L}{l_G} + {_y\sigma_{M_1}} \cdot \dfrac{l_{M_1}}{l_G} \\[2mm] \\ _y\sigma = {_y\sigma_{M_2}} \end{cases} \tag{6.9}$$

where $_i\sigma_j$ represents the homogeneous stress state in the "i" direction of the component material "j", and l_i represents the length of the component material "i".

- **Compatibility condition of "*Mode 2*"**

From the cell in Figure 6.9 the following compatibility equations are formulated,

$$\begin{cases} _y\dot\varepsilon_G \cdot h_G = {_y\dot\varepsilon_L} \cdot h_L + {_y\dot\varepsilon_{M2}} \cdot h_{M_2} \\[2mm] _y\dot\varepsilon_G \cdot h_G = {_y\dot\varepsilon_{M_1}} \cdot h_{M_1} + {_y\dot\varepsilon_{M2}} \cdot h_{M_2} \end{cases} \tag{6.10}$$

where $_i\varepsilon_j$ represents the homogeneous strain in the direction "i" of the component material "j" and h_i represents the height of the component material "i".

- **Constitutive equation of "*Mode 2*"**

The general constitutive equation for each component of the basic "cell" is written as follows,

$$_y\dot\sigma_i = {_yE_i(\omega_i)} \cdot ({_y\dot\varepsilon_i} - {_y\dot\varepsilon_i^p})$$

$$_y\dot\varepsilon_i = \frac{_y\dot\sigma_i}{_yE_i(\omega_i)} + {_y\dot\varepsilon_i^p} \tag{6.11}$$

As in the previous mode, the elastic modulus is assumed in the direction of Y, susceptible of degradation. This is obtained through the damage variable ω_i.

- **Determining the stresses in each component**

By substituting equation (6.9) for (6.11), the total deformation in mortar's M_1 and M_2 are obtained. Then,

$$_y\sigma_{M_1} = {}_y\sigma \cdot \frac{l_G}{l_{M_1}} - {}_y\sigma_L \cdot \frac{l_L}{l_{M_1}} \quad\Rightarrow\quad {}_y\dot\sigma_{M_1} = {}_y\dot\sigma \cdot \frac{l_G}{l_{M_1}} - {}_y\dot\sigma_L \cdot \frac{l_L}{l_{M_1}}$$

$$_y\dot\varepsilon_{M_1} = \left[\frac{{}_y\dot\sigma}{{}_yE_{M_1}} \cdot \frac{l_G}{l_{M_1}} - \frac{{}_y\dot\sigma_L}{{}_yE_{M_1}} \cdot \frac{l_L}{l_{M_1}} \right] + {}_y\dot\varepsilon^{P}_{M_1} \tag{6.12}$$

$$_y\dot\varepsilon_{M_2} = \left[\frac{{}_y\dot\sigma_{M_2}}{{}_yE_{M_2}} \right] + {}_y\dot\varepsilon^{P}_{M_2} = \left[\frac{{}_y\dot\sigma}{{}_yE_{M_2}} \right] + {}_y\dot\varepsilon^{P}_{M_2}$$

By substituting these latter for equations (6.10), the global deformation expression is obtained by the two following forms,

$$_y\dot\varepsilon_G \cdot h_G = {}_y\dot\varepsilon_L \cdot h_L + {}_y\dot\varepsilon_{M_2} \cdot h_{M_2} = \left(\frac{{}_y\dot\sigma_L}{{}_yE_L} + {}_y\dot\varepsilon^{P}_L \right) \cdot h_L + \left(\frac{{}_y\dot\sigma}{{}_yE_{M_2}} + {}_y\dot\varepsilon^{P}_{M_2} \right) \cdot h_{M_2}$$

$$_y\dot\varepsilon_G \cdot h_G = {}_y\dot\varepsilon_{M_1} \cdot h_{M_1} + {}_y\dot\varepsilon_{M_2} \cdot h_{M_2} = \left(\frac{{}_y\dot\sigma}{{}_yE_{M_1}} \cdot \frac{l_G}{l_{M_1}} - \frac{{}_y\dot\sigma_L}{{}_yE_{M_1}} \cdot \frac{l_L}{l_{M_1}} + {}_y\dot\varepsilon^{P}_{M_1} \right) \cdot h_{M_1} + \left(\frac{{}_y\dot\sigma}{{}_yE_{M_2}} + {}_y\dot\varepsilon^{P}_{M_2} \right) \cdot h_{M_2}$$

Matching these last two equations, the temporal variation of the brick stress is obtained.

$$_y\dot\sigma \cdot \frac{l_G}{{}_yE_{M_1} \cdot l_{M1}} \cdot h_{M1} - \frac{{}_y\dot\sigma}{{}_yE_{M_2}} \cdot h_{M2} + \frac{{}_y\dot\sigma}{{}_yE_{M_2}} \cdot h_{M2} - {}_y\dot\varepsilon^{P}_L \cdot h_L - {}_y\dot\varepsilon^{P}_{M_2} \cdot h_{M_2} + {}_y\dot\varepsilon^{P}_{M_1} \cdot h_{M_1} + {}_y\dot\varepsilon^{P}_{M_2} \cdot h_{M_2} =$$

$$= \frac{{}_y\dot\sigma_L}{{}_yE_L} \cdot h_L + {}_y\dot\sigma_L \cdot \frac{l_L}{{}_yE_{M_1} \cdot l_{M1}} \cdot h_{M1}$$

$$_y\dot\sigma \cdot \left(\frac{l_G}{{}_yE_{M_1} \cdot l_{M1}} \cdot h_{M_1} - \frac{h_{M2}}{{}_yE_{M_2}} + \frac{h_{M2}}{{}_yE_{M_2}} \right) - {}_y\dot\varepsilon^{P}_L \cdot h_L + {}_y\dot\varepsilon^{P}_{M_1} \cdot h_{M_1} = {}_y\dot\sigma_L \cdot \left(\frac{h_L}{{}_yE_L} + \frac{l_L}{{}_yE_{M_1} \cdot l_{M1}} \cdot h_{M1} \right)$$

$$_y\dot\sigma_L = \underbrace{\left(\frac{{}_yE_L \cdot {}_yE_{M_1} \cdot l_{M_1}}{{}_yE_L \cdot l_{M_1} \cdot h_L + {}_yE_L \cdot l_L \cdot h_{M_1}} \right)}_{{}^yA_L} \cdot \left[\left(\frac{l_G}{{}_yE_{M_1} \cdot l_{M_1}} \cdot h_{M_1} \right) \cdot {}_y\dot\sigma - {}_y\dot\varepsilon^{P}_L \cdot h_L + {}_y\dot\varepsilon^{P}_{M_1} \cdot h_{M_1} \right]$$

$$_y\dot\sigma_L = {}_y\dot\sigma \cdot \underbrace{\left({}^yA \cdot \frac{l_G \cdot h_{M_1}}{{}_yE_{M_1} \cdot l_{M_1}} \right)}_{{}^yB_L} + \underbrace{{}^yA \cdot \left(-{}_y\dot\varepsilon^{P}_L \cdot h_L + {}_y\dot\varepsilon^{P}_{M_1} \cdot h_{M_1} \right)}_{{}^y\dot C_L}$$

$$\boxed{_y\dot\sigma_L = {}_y\dot\sigma \cdot {}^yB_L + {}^y\dot C_L}$$

From both the brick stress and equation (6.12), the stress is obtained in mortar M_1,

$$_y\dot\sigma_{M_1} = {}_y\dot\sigma \cdot \frac{l_G}{l_{M_1}} - {}_y\dot\sigma_L \cdot \frac{l_L}{l_{M_1}} = {}_y\dot\sigma \cdot \frac{l_G}{l_{M_1}} - \left({}_y\dot\sigma \cdot {}^yB_L + {}^y\dot C_L\right) \cdot \frac{l_L}{l_{M_1}}$$

$$_y\dot\sigma_{M_1} = {}_y\dot\sigma \cdot \underbrace{\left(\frac{l_G - {}^yB_L \cdot l_L}{l_{M_1}}\right)}_{{}^yB_{M_1}} - \underbrace{{}^y\dot C_L \cdot \frac{l_L}{l_{M_1}}}_{{}^y\dot C_{M_1}}$$

$$\boxed{_y\dot\sigma_{M_1} = {}_x\dot\sigma \cdot {}^yB_{M_1} - {}^y\dot C_{M_1}}$$

The stress in mortar M_2, is obtained directly from the condition imposed in equation (6.9), thus,

$$\boxed{_y\dot\sigma_{M2} = {}_y\dot\sigma}$$

- **Global constitutive equation of "*Mode 2*"**

From the global deformation equation previously described and from the stresses in the brick and mortar M_2, we obtain

$$_y\dot\varepsilon_G = {}_y\dot\varepsilon_L \cdot \frac{h_L}{h_G} + {}_y\dot\varepsilon_{M_2} \cdot \frac{h_{M_2}}{h_G}$$

$$_y\dot\varepsilon_G = \left(\frac{_y\dot\sigma_L}{_yE_L} + {}_y\dot\varepsilon_L^P\right) \cdot \frac{h_L}{h_G} + \left(\frac{_y\dot\sigma_{M_2}}{_yE_{M_2}} + {}_y\dot\varepsilon_{M_2}^P\right) \cdot \frac{h_{M_2}}{h_G} = \frac{_y\dot\sigma_L}{_yE_L} \cdot \frac{h_L}{h_G} + {}_y\dot\varepsilon_L^P \cdot \frac{h_L}{h_G} + \frac{_y\dot\sigma_{M_2}}{_yE_{M_2}} \cdot \frac{h_{M_2}}{h_G} + {}_y\dot\varepsilon_{M_2}^P \cdot \frac{h_{M_2}}{h_G}$$

Substituting $_y\dot\sigma_L$ and $_y\dot\sigma_{M2}$ into the above expressions, the time variation of the global strain —of the composite— is obtained as a function of the time variation of the applied stress $_y\dot\sigma$,

$$_y\dot\varepsilon_G = \left(_y\dot\sigma \cdot {}^yB_L + {}^y\dot C_L\right) \cdot \frac{h_L}{_yE_L \cdot h_G} + \frac{_y\dot\sigma \cdot h_{M_2}}{_yE_{M_2} \cdot h_G} + {}_y\dot\varepsilon_L^P \cdot \frac{h_L}{h_G} + {}_y\dot\varepsilon_{M_2}^P \cdot \frac{h_{M_2}}{h_G}$$

$$_y\dot\varepsilon_G = {}_y\dot\sigma \cdot \underbrace{\left(\frac{^yB_L \cdot h_L}{_yE_L \cdot h_G} + \frac{h_{M_2}}{_yE_{M_2} \cdot h_G}\right)}_{\dfrac{1}{_yE_G}} + \underbrace{\frac{^y\dot C_L \cdot h_L}{_yE_L \cdot h_G} + {}_y\dot\varepsilon_L^P \cdot \frac{h_L}{h_G} + {}_y\dot\varepsilon_{M_2}^P \cdot \frac{h_{M_2}}{h_G}}_{_y\dot\varepsilon_G^P}$$

By ordering this result, the homogenized constitutive law of the cell in the direction of *y*" is obtained,

$$\boxed{_y\dot\varepsilon_G = \underbrace{\frac{_y\dot\sigma}{_yE_G}}_{_y\dot\varepsilon_G^e} + {}_y\dot\varepsilon_G^p}$$

(6.13)

6.4.1.3 Equations of "*Mode 3*"

The third mode is the deformation mode caused by the action of a tangential stress on a plane "*x-y*" of the basic "cell". Figure 6.10 shows the discretized element as well as the deformation configuration of the "cell". It should be mentioned that the brick and the mortar undergo different distortions as they have different mechanical and geometrical properties. A global distortion is introduced on the homogenized cell and the behavior of the whole set and of each component is studied as well as the relation among them.

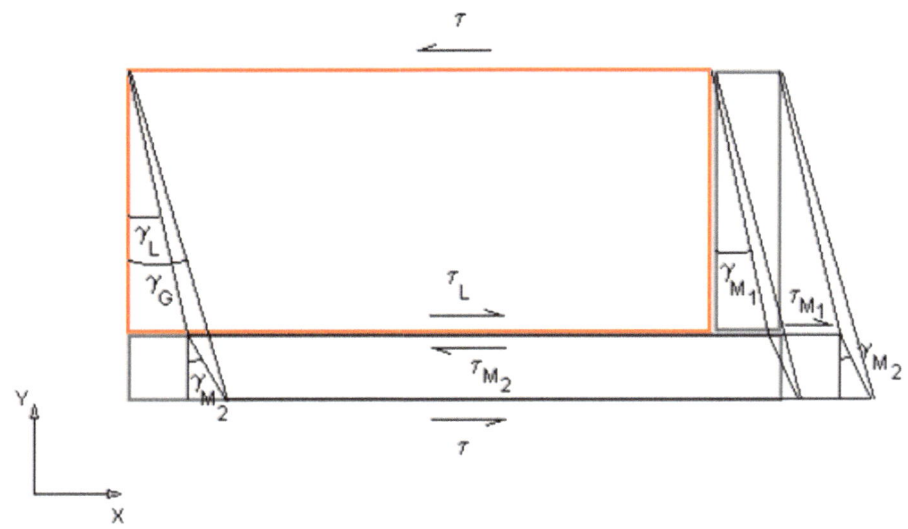

Figure 6.10 – Representation of *Mode 3* of the cell behavior.

Following the same procedure as with the first two deformation modes, the following relationships are written below:

- **Equilibrium condition in "*Mode 3*"**

By observing the stress state in the "cell", the following equilibrium conditions can be written as,

$$\dot{\tau} \cdot l_G = \dot{\tau}_L \cdot l_L + \dot{\tau}_{M_1} \cdot l_{M_1} = \dot{\tau}_{M_2} \cdot l_{M_2} \quad \Rightarrow \quad \dot{\tau} = \dot{\tau}_L \cdot \frac{l_L}{l_G} + \dot{\tau}_{M_1} \cdot \frac{l_{M_1}}{l_G} = \dot{\tau}_{M_2}$$

$$\Rightarrow \left. \begin{array}{l} \dot{\tau} = \dot{\tau}_L \cdot \dfrac{l_L}{l_G} + \dot{\tau}_{M_1} \cdot \dfrac{l_{M_1}}{l_G} \quad (a) \\[2ex] \dot{\tau} = \dot{\tau}_{M_2} \quad (b) \end{array} \right\} \qquad (6.14)$$

where τ_i represents the tangent stress of the component material "*i*" in the plane "*x-y*".

- **Compatibility condition in "*Mode 2*"**

By assuming the deformation mode imposed on the "cell" in Figure 6.10, the following kinematic compatibility equations are obtained,

$$_{xy}\dot{\gamma}_{M_1} \cdot h_{M_1} = {_{xy}\dot{\gamma}_L} \cdot h_L \quad \Rightarrow \quad \left\{ \begin{array}{l} _{xy}\gamma_{M_1} = {_{xy}\gamma_L} \\[1ex] h_{M_1} = h_L \end{array} \right.$$

where $_{xy}\gamma_j$ represents the distortion of the plane "x-y" of the component material "j". The compatibility of these movements involves the same distortion for the brick and mortar M_l. From this expression, the relation between the brick stress and the M_l is obtained as follows,

$$_{xy}\dot{\gamma}^e_{M_1} = \left[_{xy}\dot{\gamma}_{M_1} - _{xy}\dot{\gamma}^P_{M_1} \right] = \left[_{xy}\dot{\gamma}_L - _{xy}\dot{\gamma}^P_{M_1} \right] = \left[\left(_{xy}\dot{\gamma}^e_L + _{xy}\dot{\gamma}^P_L \right) - _{xy}\dot{\gamma}^P_{M_1} \right]$$

$$\dot{\tau}_{M_1} = _{xy}G_{M_1} \cdot _{xy}\dot{\gamma}^e_{M_1} = _{xy}G_{M_1} \cdot \left(\frac{\dot{\tau}_L}{_{xy}G_L} + _{xy}\dot{\gamma}^P_L - _{xy}\dot{\gamma}^P_{M_1} \right) \tag{6.15}$$

$$\dot{\tau}_{M_1} = \dot{\tau}_L \cdot \frac{_{xy}G_{M_1}}{_{xy}G_L} + _{xy}G_{M_1} \cdot \left(_{xy}\dot{\gamma}^P_L - _{xy}\dot{\gamma}^P_{M_1} \right)$$

- **Determining the stresses in each component**

By substituting expression (6.15) into expression (6.14)a, the time variation of the stress in the brick is obtained,

$$\dot{\tau} = \dot{\tau}_L \cdot \frac{l_L}{l_G} + \dot{\tau}_L \cdot \frac{_{xy}G_{M_1}}{_{xy}G_L} \cdot \frac{l_{M_1}}{l_G} + _{xy}G_{M_1} \cdot \frac{l_{M_1}}{l_G} \left(_{xy}\dot{\gamma}^P_L - _{xy}\dot{\gamma}^P_{M_1} \right) \equiv \dot{\tau}_{M_2}$$

$$\dot{\tau} = \dot{\tau}_L \cdot \left(\frac{l_L}{l_G} + \frac{_{xy}G_{M_1}}{_{xy}G_L} \cdot \frac{l_{M_1}}{l_G} \right) + _{xy}G_{M_1} \cdot \frac{l_{M_1}}{l_G} \cdot \left(_{xy}\dot{\gamma}^P_L - _{xy}\dot{\gamma}^P_{M_1} \right)$$

$$\dot{\tau}_L = \left[\dot{\tau} - \underbrace{_{xy}G_{M_1} \cdot \frac{l_{M_1}}{l_G} \cdot \left(_{xy}\dot{\gamma}^P_L - _{xy}\dot{\gamma}^P_{M_1} \right)}_{_{xy}\dot{B}_L} \right] \cdot \underbrace{\left(\frac{_{xy}G_L \cdot l_G}{_{xy}G_L \cdot l_L + _{xy}G_{M_1} \cdot l_{M_1}} \right)}_{_{xy}A_L}$$

and its compact expression is as follows,

$$\boxed{ \dot{\tau}_L = \dot{\tau} \cdot {}^{xy}A_L - {}^{xy}\dot{B}_L \cdot {}^{xy}A_L } \tag{6.16}$$

From the stress in the brick (6.16) and the equilibrium equation (6.14)a, the stress in the mortar M_l is obtained

$$\dot{\tau}_{M_1} = \dot{\tau} \cdot \frac{l_G}{l_{M_1}} - \dot{\tau}_L \cdot \frac{l_G}{l_{M_1}} = \dot{\tau} \cdot \frac{l_G}{l_{M_1}} - \left(\dot{\tau} \cdot {}^{xy}A_L - {}^{xy}\dot{B}_L \cdot {}^{xy}A_L \right) \cdot \frac{l_L}{l_{M_1}}$$

$$\dot{\tau}_{M_1} = \dot{\tau} \cdot \underbrace{\left(\frac{l_G}{l_{M_1}} - \frac{{}^{xy}A_L \cdot l_L}{l_{M_1}} \right)}_{{}^{xy}A_{M_1}} + \underbrace{\frac{{}^{xy}\dot{B}_L \cdot {}^{xy}A_L \cdot l_L}{l_{M_1}}}_{- {}^{xy}\dot{B}_{M_1}}$$

and its compact expression is as follows,

$$\boxed{ \dot{\tau}_{M_1} = \dot{\tau} \cdot {}^{xy}A_{M_1} - {}^{xy}\dot{B}_{M_1} } \tag{6.17}$$

- **Constitutive equation for each component**

From the stresses in each component obtained for Mode 3 of the strains in the previous section, its elastoplastic constitutive law can be presented as,

$$\dot{\tau}_i = {}_{xy}G_i(\omega_i) \cdot \left({}_{xy}\dot{\gamma}_i - {}_{xy}\dot{\gamma}_i^p \right) \Rightarrow {}_{xy}\dot{\gamma}_i = \frac{\dot{\tau}_i}{{}_{xy}G_i} + {}_{xy}\dot{\gamma}_i^p$$

- **Global constitutive equation for "*Mode 3*"**

Based on the compatibility condition for the whole cell, and on the constitutive equation of each component, the homogenized constitutive law in "*Mode 3*" for the whole "cell" is obtained. Thus,

$$_{xy}\dot{\gamma}_G \cdot h_G = \left({}_{xy}\dot{\gamma}_L \cdot h_L \right) + \left({}_{xy}\dot{\gamma}_{M_2} \cdot h_{M_2} \right) = \left({}_{xy}\dot{\gamma}_{M_1} \cdot h_{M_1} \right) + \left({}_{xy}\dot{\gamma}_{M_2} \cdot h_{M_2} \right)$$

$$_{xy}\dot{\gamma}_G = \left({}_{xy}\dot{\gamma}_L \cdot \frac{h_L}{h_G} \right) + \left({}_{xy}\dot{\gamma}_{M_2} \cdot \frac{h_{M_2}}{h_G} \right)$$

By introducing into this expression the constitutive expressions in each component and stresses (6.16) and (6.17), the constitutive equation in "*Mode 3*" for the whole "cell" is obtained.

$$_{xy}\dot{\gamma}_G = \left(\frac{\dot{\tau}_L}{{}_{xy}G_L} \cdot \frac{h_L}{h_G} + {}_{xy}\dot{\gamma}_L^p \cdot \frac{h_L}{h_G} \right) + \left(\frac{\dot{\tau}_{M_2}}{{}_{xy}G_{M_2}} \cdot \frac{h_{M_2}}{h_G} + {}_{xy}\dot{\gamma}_{M_2}^p \cdot \frac{h_{M_2}}{h_G} \right)$$

$$_{xy}\dot{\gamma}_G = \left(\left(\dot{\tau} \cdot {}^{xy}A_L - {}^{xy}\dot{B}_L \cdot {}^{xy}A_L \right) \cdot \frac{h_L}{{}_{xy}G_L \cdot h_G} + {}_{xy}\dot{\gamma}_L^p \cdot \frac{h_L}{h_G} \right) + \left(\dot{\tau} \cdot \frac{h_{M_2}}{{}_{xy}G_{M_2} \cdot h_G} + {}_{xy}\dot{\gamma}_{M_2}^p \cdot \frac{h_{M_2}}{h_G} \right)$$

$$_{xy}\dot{\gamma}_G = \dot{\tau} \cdot \frac{{}^{xy}A_L \cdot h_L}{{}_{xy}G_L \cdot h_G} + \dot{\tau} \cdot \frac{h_{M_2}}{{}_{xy}G_{M_2} \cdot h_G} - {}^{xy}\dot{B}_L \cdot {}^{xy}A_L \cdot \frac{h_L}{{}_{xy}G_L \cdot h_G} + {}_{xy}\dot{\gamma}_L^p \cdot \frac{h_L}{h_G} + {}_{xy}\dot{\gamma}_{M_2}^p \cdot \frac{h_{M_2}}{h_G}$$

Reordering the terms, the global distortion in the composite material can be expressed as,

$$_{xy}\dot{\gamma}_G = \dot{\tau} \cdot \underbrace{\left(\frac{{}^{xy}A_L \cdot h_L}{{}_{xy}G_L \cdot h_G} + \frac{h_{M_2}}{{}_{xy}G_{M_2} \cdot h_G} \right)}_{\dfrac{1}{{}_{xy}G_G}} + \underbrace{\left(- {}^{xy}\dot{B}_L \cdot {}^{xy}A_L \cdot \frac{h_L}{{}_{xy}G_L \cdot h_G} + {}_{xy}\dot{\gamma}_L^p \cdot \frac{h_L}{h_G} + {}_{xy}\dot{\gamma}_{M_2}^p \cdot \frac{h_{M_2}}{h_G} \right)}_{{}_{xy}\dot{\gamma}_G^p}$$

and its compact expression is as follows,

$$\boxed{{}_{xy}\dot{\gamma}_G = \dot{\tau} \cdot \frac{1}{{}_{xy}G_G} + {}_{xy}\dot{\gamma}_G^p} \tag{6.18}$$

6.4.1.4 Equations of "*Mode 4*"

The fourth behavior mode corresponds to the strain field outside the loading plane (plane "*x-y*"). Since in the model the plane-stress hypothesis is assumed, the strains in the plane "*x-z*" are due exclusively to the strains produced by the Poisson's effect. Moreover, since there are not any external tangent actions associated with planes ""*x-z* and "*y-z*" (plane stress hypothesis), the resultant tangent stresses τ_{xz} and τ_{yz} integrated on the whole "cell" are null and therefore it will be assumed that the associated distortions are also null.

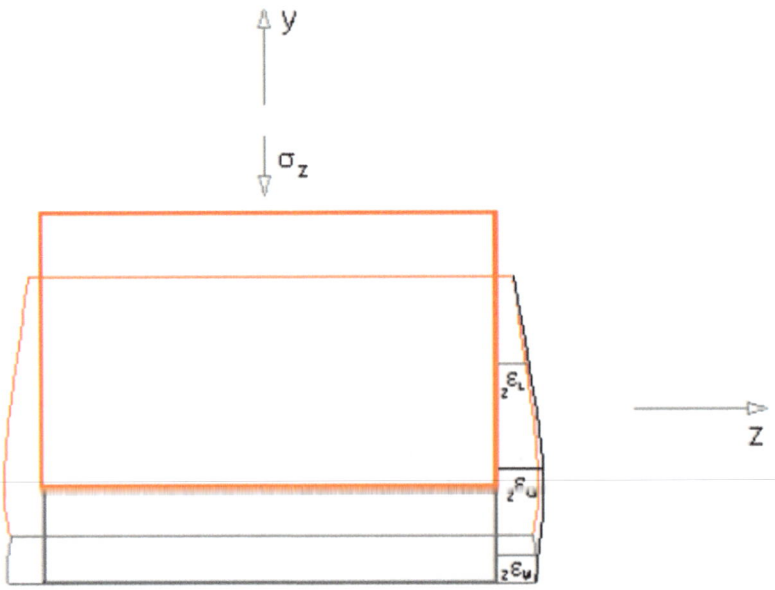

Figure 6.11 – Representation of *Mode 3* of the cell behavior.

Figure 6.11 shows that the brick and the mortar have deformations without continuity on their curves due to the different mechanical and geometrical properties of each component. In the homogenized model presented above, it has been adopted as a deformed shape that envelops the brick and mortar.

Unlike the methodology followed in the previous modes, in this case the expression is based on the secant stiffness tensor in orthotropic symmetry and the remaining terms are analyzed by this fourth deformation mode.

$$
\mathbf{C}^s = \begin{bmatrix}
\dfrac{_xE_G\cdot(-1+_{yz}\nu_G\cdot_{zy}\nu_G)}{D} & -\dfrac{_xE_G\cdot(_{yx}\nu_G+_{yz}\nu_G\cdot_{zx}\nu_G)}{D} & -\dfrac{_xE_G\cdot(_{zx}\nu_G+_{yx}\nu_G\cdot_{zy}\nu_G)}{D} & 0 & 0 & 0 \\
-\dfrac{_yE_G\cdot(_{xy}\nu_G+_{xzz}\nu_G\cdot_{zy}\nu_G)}{D} & \dfrac{_yE_G\cdot(-1+_{xz}\nu_G\cdot_{zx}\nu_G)}{D} & -\dfrac{_yE_G\cdot(_{zy}\nu_G+_{xy}\nu_G\cdot_{zx}\nu_G)}{D} & 0 & 0 & 0 \\
-\dfrac{_zE_G\cdot(_{xz}\nu_G+_{xy}\nu_G\cdot_{yz}\nu_G)}{D} & -\dfrac{_zE_G\cdot(_{yz}\nu_G+_{xz}\nu_G\cdot_{yx}\nu_G)}{D} & \dfrac{_zE_G\cdot(-1+_{xy}\nu_G\cdot_{yx}\nu_G)}{D} & 0 & 0 & 0 \\
0 & 0 & 0 & _{xy}G_G & 0 & 0 \\
0 & 0 & 0 & 0 & _{yz}G_G & 0 \\
0 & 0 & 0 & 0 & 0 & _{zx}G_G
\end{bmatrix}
$$

where the denominator of the axial stiffness terms:

$$D = -1 + {}_{xy}\nu_G \cdot {}_{yx}\nu_G + {}_{xz}\nu_G \cdot {}_{zx}\nu_G + {}_{yz}\nu_G \cdot {}_{zy}\nu_G + {}_{xy}\nu_G \cdot {}_{yz}\nu_G \cdot {}_{zx}\nu_G + {}_{xz}\nu_G \cdot {}_{yx}\nu_G \cdot {}_{zy}\nu_G$$

and the modules ${}_xE_G$, ${}_yE_G$, ${}_{xy}G_G$, are the elasticity terms obtained for each one of the three previously analyzed modes, equations (6.8), (6.13) and (6.18), respectively. The remaining modules ${}_zE_G$, ${}_{yz}G_G$, ${}_{zx}G_G$, will be obtained from a conceptual extension of the whole formulation developed here.

With this tensor defined in matrix form, the global constitutive relation is written as follows,

$$\boxed{\vec{\sigma}_G = \mathbf{C}^s \cdot \vec{\varepsilon}_G}$$

By isolating the component ${}_z\vec{\sigma}_G$ of the vector $\vec{\sigma}_G$ and assuming the plane stress hypothesis,

$$\sigma_z = -\frac{E_z}{D}\left({}_{xz}\nu_G + {}_{xy}\nu_G \cdot {}_{yz}\nu_G\right)_x\varepsilon_G - \frac{E_z}{D}\left({}_{yz}\nu_G + {}_{xz}\nu_G \cdot {}_{yx}\nu_G\right)_y\varepsilon_G + \frac{E_z}{D}\left(-1 + {}_{xy}\nu_G \cdot {}_{yx}\nu_G\right)_z\varepsilon_G = 0$$

$$\sigma_z = \frac{E_z}{D}\left[-\left({}_{xz}\nu_G + {}_{xy}\nu_G \cdot {}_{yz}\nu_G\right)_x\varepsilon_G - \left({}_{yz}\nu_G + {}_{xz}\nu_G \cdot {}_{yx}\nu_G\right)_y\varepsilon_G + \left(-1 + {}_{xy}\nu_G \cdot {}_{yx}\nu_G\right)_z\varepsilon_G\right] = 0$$

$${}_z\varepsilon_G = {}_z\varepsilon_G^e + {}_z\varepsilon_G^p = \frac{\left({}_{xz}\nu_G + {}_{xy}\nu_G \cdot {}_{yz}\nu_G\right)_x\varepsilon_G + \left({}_{yz}\nu_G + {}_{xz}\nu_G \cdot {}_{yx}\nu_G\right)_y\varepsilon_G}{\left(-1 + {}_{xy}\nu_G \cdot {}_{yx}\nu_G\right)}$$

$${}_z\varepsilon_G = {}_z\varepsilon_G^e + {}_z\varepsilon_G^p = \frac{\left({}_{xz}\nu_G + {}_{xy}\nu_G \cdot {}_{yz}\nu_G\right) \cdot \left({}_x\varepsilon_G^e + {}_x\varepsilon_G^p\right) + \left({}_{yz}\nu_G + {}_{xz}\nu_G \cdot {}_{yx}\nu_G\right) \cdot \left({}_y\varepsilon_G^e + {}_y\varepsilon_G^p\right)}{\left(-1 + {}_{xy}\nu_G \cdot {}_{yx}\nu_G\right)}$$

From this expression, and separating the elastic and plastic components, the global deformations are obtained along the "z" direction,

$${}_z\varepsilon_G^e = \frac{\left({}_{xz}\nu_G + {}_{xy}\nu_G \cdot {}_{yz}\nu_G\right)_x\varepsilon_G^e + \left({}_{yz}\nu_G + {}_{xz}\nu_G \cdot {}_{yx}\nu_G\right)_y\varepsilon_G^e}{\left(-1 + {}_{xy}\nu_G \cdot {}_{yx}\nu_G\right)} \qquad (6.19)$$

$${}_z\varepsilon_G^p = \frac{\left({}_{xz}\nu_G + {}_{xy}\nu_G \cdot {}_{yz}\nu_G\right)_x\varepsilon_G^p + \left({}_{yz}\nu_G + {}_{xz}\nu_G \cdot {}_{yx}\nu_G\right)_y\varepsilon_G^p}{\left(-1 + {}_{xy}\nu_G \cdot {}_{yx}\nu_G\right)} \qquad (6.20)$$

6.4.1.5 Summary of the homogenized mechanical parameters for a 3D "cell"

Based on the previous developments, the mechanical parameter values have been obtained for the homogenized masonry. Observe, in the previously presented equations, the sensitivity to the change of the properties of the component materials and dimensions of the "cell" (bricks and mortar joints, Figure 6.12).

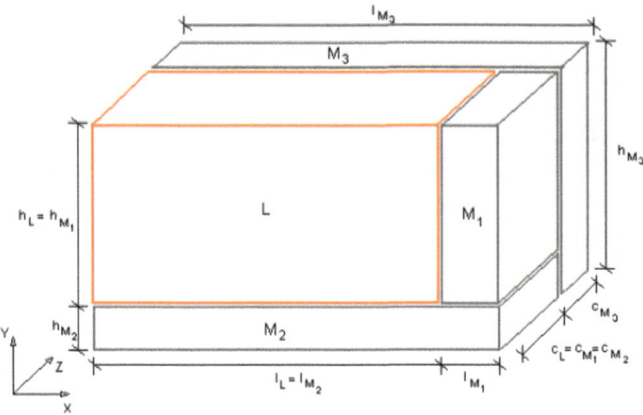

Figure 6.12 – Three-dimensional geometry representation of a masonry cell.

The expressions of the mechanical parameters using the developments previously carried out for each mode of the deformation are detailed below and for the three-dimensional case, a conceptual extension for the third dimension is made.

From expression (6.8) the longitudinal elasticity module is obtained in the global x-direction,

$$\boxed{_x E_G = \frac{1}{_x B_L \cdot {}^x D}} \quad \left| \begin{array}{l} {}^x D = \dfrac{l_L}{_x E_L \cdot l_G} + \dfrac{l_{M_1}}{_x E_{M_1} \cdot l_G} \quad ; \quad {}^x B_L = \dfrac{{}^x A_L \cdot l_{M_2} \cdot h_G}{_x E_{M2} \cdot h_{M_2}} \\[4mm] {}^x A_L = \dfrac{_x E_L \cdot_x E_{M_1} \cdot_x E_{M2} \cdot h_{M_1}}{_x E_{M_1} \cdot_x E_{M2} \cdot h_{M2} \cdot l_L +_x E_L \cdot_x E_{M2} \cdot h_{M2} \cdot l_{M_1} +_x E_L \cdot_x E_{M1} \cdot h_L \cdot l_{M2}} \end{array} \right. \tag{6.21}$$

From expression (6.13) the longitudinal elasticity module is obtained in the global y-direction,

$$\boxed{_y E_G = \frac{1}{\dfrac{_y B_L \cdot h_L}{_y E_L \cdot h_G} + \dfrac{h_{M_2}}{_y E_{M2} \cdot h_G}}} \quad \left\{ {}^y B_L = \dfrac{{}^y A_L \cdot l_G \cdot h_{M_1}}{_y E_{M_1} \cdot h_{M_1}} \quad ; \quad {}^x A_L = \dfrac{_y E_L \cdot_y E_{M_1} \cdot l_{M_1}}{_x E_{M_1} \cdot l_{M_1} \cdot h_L +_x E_L \cdot l_L \cdot h_{M_1}} \right. \tag{6.22}$$

To obtain $_z E_G$ a conceptual extrapolation of the development is carried out in "Mode 1". Thus, the longitudinal elastic module in the global z-direction is expressed as,

$$\boxed{_z E_G = \frac{1}{_z B_L \cdot {}^z D}} \quad \left| \begin{array}{l} {}^z D = \dfrac{c_L}{_z E_L \cdot c_G} + \dfrac{c_{M_1}}{_z E_{M_1} \cdot c_G} \quad ; \quad {}^z B_L = \dfrac{{}^z A_L \cdot c_{M_2} \cdot h_G}{_z E_{M2} \cdot h_{M_2}} \\[4mm] {}^z A_L = \dfrac{_x E_L \cdot_x E_{M_1} \cdot_x E_{M2} \cdot h_{M_1}}{_x E_{M_1} \cdot_x E_{M_2} \cdot h_{M2} \cdot c_L +_x E_L \cdot_x E_{M2} \cdot h_{M2} \cdot c_{M3} +_x E_L \cdot_x E_{M1} \cdot h_L \cdot c_{M2}} \end{array} \right. \tag{6.23}$$

From expression (6.18) the transverse elastic module in the global "xy" plane results:

$$\boxed{_{xy}G_G = \cfrac{1}{\cfrac{_{xy}A_L \cdot h_L}{_{xy}G_L \cdot h_G} + \cfrac{h_{M_2}}{_{xy}G_{M_2} \cdot h_G}}} \quad \left\{ _{xy}A_L = \cfrac{_{xy}G_L \cdot l_G}{_{xy}G_L \cdot l_L + _{xy}G_{M_1} \cdot l_{M_1}} \right. \tag{6.24}$$

By proceeding the same way for the remaining directions, we obtain:

Transverse elasticity module in the global "yx",

$$\boxed{_{yx}G_G = \cfrac{1}{\cfrac{_{yx}A_L \cdot l_L}{_{yx}G_L \cdot l_G} + \cfrac{l_{M_2}}{_{yx}G_{M_2} \cdot l_G}}} \quad \left\{ _{yx}A_L = \cfrac{_{xy}G_L \cdot h_G}{_{xy}G_L \cdot h_L + _{xy}G_{M_1} \cdot h_{M_2}} \right. \tag{6.25}$$

Transverse elasticity module in the global "xz" plane,

$$\boxed{_{xz}G_G = \cfrac{1}{\cfrac{_{xz}A_L \cdot c_L}{_{xz}G_L \cdot c_G} + \cfrac{c_{M_3}}{_{xz}G_{M_3} \cdot c_G}}} \quad \left\{ _{xz}A_L = \cfrac{_{xz}G_L \cdot l_G}{_{xz}G_L \cdot l_L + _{xz}G_{M_1} \cdot l_{M_1}} \right. \tag{6.26}$$

Transverse elasticity module in the global "zx" plane,

$$\boxed{_{zx}G_G = \cfrac{1}{\cfrac{_{zx}A_L \cdot l_L}{_{zx}G_L \cdot l_G} + \cfrac{l_{M_1}}{_{zx}G_{M_1} \cdot l_G}}} \quad \left\{ _{zx}A_L = \cfrac{_{zx}G_L \cdot c_G}{_{zx}G_L \cdot c_L + _{xz}G_{M_3} \cdot c_{M_3}} \right. \tag{6.27}$$

Transverse elasticity module in the global "zy" plane,

$$\boxed{_{zy}G_G = \cfrac{1}{\cfrac{_{zy}A_L \cdot h_L}{_{zy}G_L \cdot h_G} + \cfrac{h_{M_2}}{_{zy}G_{M_2} \cdot h_G}}} \quad \left\{ _{zy}A_L = \cfrac{_{zy}G_L \cdot c_G}{_{zy}G_L \cdot c_L + _{xy}G_{M_3} \cdot c_{M_3}} \right. \tag{6.28}$$

Transverse elasticity module in the global "yz" plane,

$$\boxed{_{yz}G_G = \cfrac{1}{\cfrac{_{yz}A_L \cdot c_L}{_{yz}G_L \cdot c_G} + \cfrac{c_{M_3}}{_{zx}G_{M_3} \cdot c_G}}} \quad \left\{ _{yz}A_L = \cfrac{_{yz}G_L \cdot h_G}{_{yz}G_L \cdot h_L + _{yz}G_{M_2} \cdot c_{M_2}} \right. \tag{6.29}$$

As the basic cell components are oriented according to the main directions, the following elasticity relations must be met,

$$_{xy}G_G = \alpha_{xy} \cdot \sqrt{_xE_G \cdot _yE_G} \qquad\qquad _{yx}G_G = \alpha_{yx} \cdot \sqrt{_yE_G \cdot _xE_G}$$

$$_{yz}G_G = \alpha_{yz} \cdot \sqrt{_yE_G \cdot _zE_G} \qquad\qquad _{zy}G_G = \alpha_{zy} \cdot \sqrt{_zE_G \cdot _yE_G}$$

$$_{xz}G_G = \alpha_{xz} \cdot \sqrt{_xE_G \cdot _zE_G} \qquad\qquad _{zx}G_G = \alpha_{zx} \cdot \sqrt{_zE_G \cdot _xE_G}$$

For geomaterials within the elastic field, it is verified that:

$$\boxed{\nu_{ij} = \frac{2 \cdot_{ij} G_G}{\sqrt{_i E_G \cdot_j E_G}} - 1} \left\{ \alpha_{ij} = \frac{1}{2}\left(1 + \nu_{ij}\right) \quad ; \quad _{ij}G_G = \frac{1}{2}\left(1 + \nu_{ij}\right) \cdot \sqrt{_i E_G \cdot_j E_G} \right. \tag{6.30}$$

As a result, the following Poisson coefficients are obtained for the different direction:

$$
\begin{array}{ll}
_{xy}\nu_G = \dfrac{2 \cdot_{xy} G_G}{\sqrt{_x E_G \cdot_y E_G}} - 1 & \qquad _{yx}\nu_G = \dfrac{2 \cdot_{yx} G_G}{\sqrt{_y E_G \cdot_x E_G}} - 1 \\[3ex]
_{xz}\nu_G = \dfrac{2 \cdot_{xz} G_G}{\sqrt{_x E_G \cdot_z E_G}} - 1 & \qquad _{zx}\nu_G = \dfrac{2 \cdot_{zx} G_G}{\sqrt{_z E_G \cdot_x E_G}} - 1 \\[3ex]
_{yz}\nu_G = \dfrac{2 \cdot_{yz} G_G}{\sqrt{_y E_G \cdot_z E_G}} - 1 & \qquad _{zy}\nu_G = \dfrac{2 \cdot_{zy} G_G}{\sqrt{_z E_G \cdot_y E_G}} - 1
\end{array}
$$

where $_{ij}\nu_G$ represents the homogenized Poisson coefficient in the "i-j" plane.

6.4.1.6 Homogenized plastic flow

Once the elastic limit defined by the yield criterion for geomaterials such as the modified Mohr-Coulomb's (Oller (1991)[10]) is surpassed, the material shows plastic deformations influenced by the anisotropy directions. Therefore, it is necessary to define the inelastic model in the anisotropic field, which is a complex situation especially dealing with a composite material made up of geomaterials. To do this and to simplify the problem, a yield and implicit plastic potential criterion[*] are defined in the anisotropic space through an appropriately chosen isotropic image of it. This space transformation is perfectly defined by the *Space Mapping Theory* (Oller *et. al.* (1995)[11], (2003)[12]), which establishes a general definition of anisotropic threshold criteria through isotropic thresholds.

First, the Mohr-Coulomb's associated isotropic plasticity hypothesis will be used, then the yield and potential surfaces $\mathbb{F}^\tau(\overline{\tau}, g, \alpha, f^\tau) \equiv \mathbb{G}^\tau(\overline{\tau}, g)$ coincide in the fictitious isotropic space, where $\mathbb{F}^\tau(\overline{\tau}, g, \alpha, f^\tau) = 0$ and $\mathbb{G}^\tau(\overline{\tau}, g) = \text{cte}$ represent the yield and plastic potential functions in the updated configuration, $\overline{\tau}$ is the Kirchhoff 's stress tensor, g is the metric tensor in the space configuration, α is the number of internal variables and f^τ is the strength tensor in the threshold limit of the elastic material in the updated configuration.

The plastic yield condition in the isotropic stress space $\overline{\tau}$ requires that the following inequality be met, where

---

[*] **Note**: The "implicit" formulation here refers to a formulation that is not expressly formulated but through another "explicitly" defined expression among which there is a biunivocal correspondence.

[11] Oller, S., Botello, S., Miquel, J., Oñate, E. (1995).An isotropic elastoplastic model based on an isotropic formulation. *Engineering Computations*, Vol. 12, 245-262.

[12] Oller s., Car E. and Lubliner J. (2003). Definition of a general implicit orthotropic yield criterion. *Computer Methods in Applied Mechanics and Engineering*. Vol. 192, No. 7-8, pp. 895-912.

$$\mathbb{F}^{\tau}(\overline{\tau}, g, \alpha, f^{\tau}) = f(\overline{\tau}) - f_G^{\tau}(\kappa^P) \leq 0 \tag{6.31}$$

where $f_G^{\tau}(\kappa^P) \equiv c_G(\kappa^P)$ is the homogenized cohesion of the material function of the damage parameter κ^P and $\overline{\tau}$ is the stress tensor in the fictitious isotropic space.

The value of the homogenized cohesion is obtained from a generalization of the equilibrium equation in compression ("*Mode 2*", equation (6.3)). Thus,

$$c_G(\kappa^P) = c_L(\kappa^P) \cdot \frac{l_L}{l_G} + c_{M_1}(\kappa^P) \cdot \frac{l_{M_1}}{l_G} \tag{6.32}$$

The damage parameter κ^P represents the dissipated energy normalized to the unit (Oller, (1991)[10]),

$$\kappa^P = \int_t \frac{\Xi^P}{\left[g_L + g_{M_1} + g_{M_2}\right]_{PR}} \, dt$$

in which Ξ^P is the defined dissipation based on the second law of thermodynamics (Lubliner (1990)[13]), $[g_i]_{PR}$ is the maximum energy to dissipate by the component "i" of the composite material, according to the PR mechanical process to which it is subjected (tension, compression, pure shear or intermediate states).

Once the expression is satisfied (6.31), the homogenized total deformation in each direction can be expressed as a sum of an elastic and a plastic component.

$$_i\varepsilon_G = {}_i\varepsilon_G^e + {}_i\varepsilon_G^P \tag{6.33}$$

where "i" is the component of the global deformation of the "cell" according to the orthotropy direction. The different components of the plastic deformation are obtained from the development previously presented in this chapter and their terms will be explained here as follows. Based on "*Mode 1*" of the deformation, equation (6.8), the component of the plastic deformation is obtained in the direction "x"

$$_x\varepsilon_G^P = {}^xA_L \cdot \left({}_x\varepsilon_{M_2}^P \cdot l_{M_2} - {}_x\varepsilon_L^P \cdot l_L - {}_x\varepsilon_{M_1}^P \cdot l_{M_1} \right) \cdot \left(\frac{1}{{}_xE_L} \cdot \frac{l_L}{l_G} + \frac{1}{{}_xE_{M_1}} \cdot \frac{l_{M_1}}{l_G} \right) + \left({}_x\varepsilon_L^P \cdot \frac{l_L}{l_G} + {}_x\varepsilon_{M_1}^P \cdot \frac{l_{M_1}}{l_G} \right)$$

Since all the materials that made up the cell are geomaterials, the same yield and potential surfaces can be adopted for all of them. For this particular case the Mohr-Coulomb's has been chosen. Thus, the following simplification can be carried out in the previous equation,

$$_x\varepsilon_G^P = A_x^P \cdot \left(l_{M_2} - l_L - l_{M_1} \right) \cdot {}_x\varepsilon_G^P \Big|_{Mohr} + \left(\cdot \frac{l_L}{l_G} + \cdot \frac{l_{M_1}}{l_G} \right) \cdot {}_x\varepsilon_G^P \Big|_{Mohr}$$

$$\tag{6.34}$$

$$_x\varepsilon_G^P = \left[A_x^P \cdot l_{M_2} + \left(\frac{1}{l_G} - A_x^P \right) \cdot \left(l_L + l_{M_1} \right) \right] \cdot {}_x\varepsilon_G^P \Big|_{Mohr}$$

where,

---

[13] Lubliner J. (1990). *Plasticity Theory*. Macmillan Publishing, U.S.A.

$$A_x^p = \frac{{}_xE_L\cdot{}_xE_{M_1}\cdot{}_xE_{M_2}\cdot h_{M_2}}{{}_xE_{M_1}\cdot{}_xE_{M_2}\cdot h_{M_2}\cdot l_L + {}_xE_L\cdot{}_xE_{M_2}\cdot h_{M_2}\cdot l_{M_2} + {}_xE_L\cdot{}_xE_{M_1}\cdot h_{M_2}\cdot l_{M_1}}\left(\frac{l_L}{{}_xE_L\cdot l_G} + \frac{l_{M_1}}{{}_xE_{M_1}\cdot l_G}\right)$$

Following the same procedure, from "*Mode 2*", equation (6.13), the component "*y*" of the global plastic deformation is obtained as,

$$_y\varepsilon_G^p = A_y^p\cdot\left(-{}_y\varepsilon_L^p\cdot h_L + {}_y\varepsilon_{M_1}^p\cdot h_{M_1}\right) + \left({}_y\varepsilon_L^p\cdot\frac{h_L}{h_G} + {}_y\varepsilon_{M_1}^p\cdot\frac{h_{M_2}}{h_G}\right)$$

Taking the same potential surface for the cell materials, then

$$_y\varepsilon_G^p = A_y^p\cdot\left(-h_L + h_{M_1}\right)\cdot\left.{}_y\varepsilon_G^p\right|_{Mohr} + \left(\frac{h_L}{h_G} + \frac{h_{M_2}}{h_G}\right)\cdot\left.{}_y\varepsilon_G^p\right|_{Mohr}$$

$$_y\varepsilon_G^p = \left[A_y^p\cdot h_{M_1} + \left(\frac{1}{h_G} - A_y^p\right)\cdot h_L + \frac{h_{M_2}}{h_G}\right]\cdot\left.{}_y\varepsilon_G^p\right|_{Mohr} \qquad (6.35)$$

where:

$$A_y^p = \frac{{}_yE_L\cdot{}_yE_{M_1}\cdot l_{M_1}}{{}_yE_{M_1}\cdot l_{M_1}\cdot h_L + {}_yE_L\cdot l_L\cdot h_{M_1}}\cdot\frac{h_L}{{}_yE_L\cdot h_G}$$

As proceeded with the two previous components of the global plastic deformation, the global plastic distortion is obtained,

$$_{xy}\gamma_G^p = A_\gamma^p\cdot\left({}_{xy}\gamma_L^p - {}_{xy}\gamma_{M_1}^p\right) + {}_{xy}\gamma_L^p\cdot\frac{h_L}{h_G} + {}_{xy}\gamma_{M_2}^p\cdot\frac{h_{M_2}}{h_G}$$

And from here, after substituting the Mohr-Coulomb's flow, the following expression is obtained,

$$_{xy}\gamma_G^p = A_\gamma^p\cdot\left(\left.{}_{xy}\gamma_G^p\right|_{Mohr} - \left.{}_{xy}\gamma_G^p\right|_{Mohr}\right) + \left.{}_{xy}\gamma_G^p\right|_{Mohr}\cdot\frac{h_L}{h_G} + \left.{}_{xy}\gamma_G^p\right|_{Mohr}\cdot\frac{h_{M_2}}{h_G}$$

$$\boxed{_{xy}\gamma_G^p = \left(\frac{h_L + h_{M_2}}{h_G}\right)\cdot\left.{}_{xy}\gamma_G^p\right|_{Mohr}} \qquad (6.36)$$

where

$$A_\gamma^p = G_{M_1}\cdot\frac{l_{M_1}}{l_G}\cdot\left(\frac{G_L\cdot l_G}{G_L\cdot l_L + G_{M_1}\cdot l_{M_1}}\right)\cdot\frac{h_L}{G_L\cdot h_G}$$

From expression (6.20) the plastic flow along the z-direction is directly obtained

$$\boxed{_z\varepsilon_G^p = \frac{\left({}_{xz}\nu_G + {}_{xy}\nu_G\cdot{}_{yz}\nu_G\right)\cdot\left.{}_x\varepsilon_G^p\right|_{Mohr} + \left({}_{yz}\nu_G + {}_{xz}\nu_G\cdot{}_{yx}\nu_G\right)\cdot\left.{}_y\varepsilon_G^p\right|_{Mohr}}{\left(-1 + {}_{xy}\nu_G\cdot{}_{yx}\nu_G\right)}} \qquad (6.37)$$

From expressions (6.34), (6.35), (6.36) and (6.37) the plastic deformation flow is obtained for a composite represented by the homogenized "cell" as a function of the Mohr-Coulomb's flow. Thus,

$$\vec{\varepsilon}_G^p = \mathbf{M}^p \cdot \vec{\varepsilon}^p \Big|_{Mohr} = \left\{ {}_x\varepsilon_G^p, {}_y\varepsilon_G^p, {}_{xy}\gamma_G^p, {}_z\varepsilon_G^p \right\}^T \tag{6.38}$$

being:

$$\vec{\varepsilon}^p \Big|_{Mohr} = \left\{ {}_x\varepsilon_G^p \Big|_{Mohr}, {}_y\varepsilon_{G_{Mohr}}^p \Big|, {}_{xy}\gamma_{G\,Mohr}^p \Big|, {}_z\varepsilon_G^p \Big|_{Mohr} \right\}^T$$

$$\mathbf{M}^p = \begin{bmatrix} \left[A_x^p \cdot l_{M_2} + \left(\dfrac{1}{l_G} - A_x^p \right) \cdot \left(l_L + l_{M_1} \right) \right] & 0 & 0 & 0 \\[2em] 0 & \left[A_y^p \cdot h_{M_1} + \left(\dfrac{1}{h_G} - A_y^p \right) \cdot h_L + \dfrac{h_{M2}}{h_G} \right] & 0 & 0 \\[2em] 0 & 0 & \dfrac{h_L + h_{M2}}{h_G} & 0 \\[2em] \dfrac{{}_{xz}\nu + {}_{xy}\nu \cdot {}_{yz}\nu}{-1 + {}_{xy}\nu \cdot {}_{yx}\nu} & \dfrac{{}_{yz}\nu + {}_{xz}\nu \cdot {}_{yz}\nu}{-1 + {}_{xy}\nu \cdot {}_{yx}\nu} & 0 & 0 \end{bmatrix}$$

The plastic flow vector is obtained as:

$$\vec{g}_G = \frac{\dot{\varepsilon}_G^p}{\dot{\lambda}} \tag{6.39}$$

This latter $\dot{\lambda} = \dfrac{f(\tau) - c_G}{H + g_G : \mathbf{C} : f_G}$ being the classic plastic consistency factor. From all this it can be concluded that the homogenization leads to a plastic behavior with a potential function not associated with the yield.

6.5 Formulation checkout

Page's experimental results (1978)[8] are used to check the functioning of the model presented here. This test is one of the most widely used tests for the calibration of numerical methods in masonry. Lourenço (1992)[14] also used this test to calibrate his model. Therefore, although the results obtained in this work will be contrasted with the experimental model, they will also take Lourenço's numerical model as a reference.

The dimensions of the masonry panel are $75.7 \times 45.7 \times 5.4 \ cm^3$, using bricks of $12.2 \times 3.7 \times 5.4 \ cm^3$ and joints $0.5 \ cm$ thick. The mechanical characteristics of the materials used are summarized in Table 6.1. The test gets stresses and vertical deformation values in a gauge line arranged to $18.65 \ cm$ from the bottom. The **P** load is applied by a jack and distributed over a steel beam in the central area, having a length distribution of $38.1 \ cm$ (see Figure 6.13).

[14] Lourenço, P. B. (1996). *Computational Strategies for Masonry Structures*, Doctoral Dissertation, Technological University of Delft. Delf University Press.

Property	Value
Elasticity module for a parallel load to the brick plate base load E_x (in kp/cm²)	59200
Elasticity module for a perpendicular load to the brick plate base load E_y (in kp/cm²)	75500
E_y/E_x	1.35
Poisson's coefficient in the plane	0.167
Brick compressive strength (in kp/cm²)	362.5
Mortar compressive strength (in kp/cm²)	32
Internal friction angle	30°
Mortar elasticity module (in kp/cm²)	8041

Table 6.1 – Masonry mechanical parameters.

Since there were no records of the material's fracture and crushing energy values, the values 8.32 kp/cm and 207.36 kp/cm have been adopted as the brick and mortar fracture values, respectively, and the values 384 kp/cm and 592 kp/cm have been assumed for the crushing energy values for the mortar and brick, respectively. These values are assumed high in order to improve the convergence of the constitutive equation for high stress levels. The main objective is to obtain all the post-peak response and therefore these energy values are acceptable.

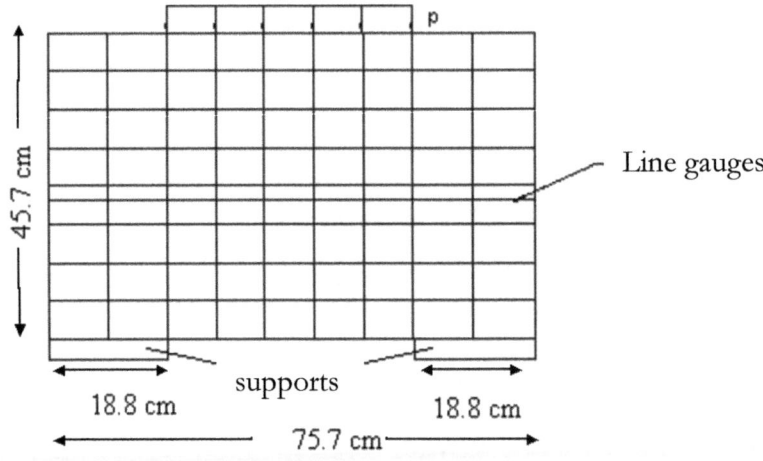

Figure 6.13 – Geometrical scheme of the Page's test.

The test has been carried out for loads of 20 KN, 40 KN, 60 KN and the resultant stress results are shown in Figure 6.14 and Figure 6.15. The strain gauges coincide with the position of the upper Gauss points in elements of the intermediate line. Therefore, the controlled results have been taken for such Gauss points.

The generated mesh has 72 quadrilateral elements of four nodes and four Gauss points per element (see Figure 6.13). The use of a low number of elements is aimed at observing the model's response to the small number of mesh elements because a very dense mesh would not highlight the advantages of using a simplified micro model in which joints and bricks are discretized by different elements. The advantage of this type of homogenization model is saving computational time and simplifying the mesh generation process.

Below is a comparison of the stress state in three tests —Page (1978)[8], Lourenço (1996)[14] and the model homogenization presented in this chapter— along the gauges line. The load is gradually increased and then the characteristic results are obtained.

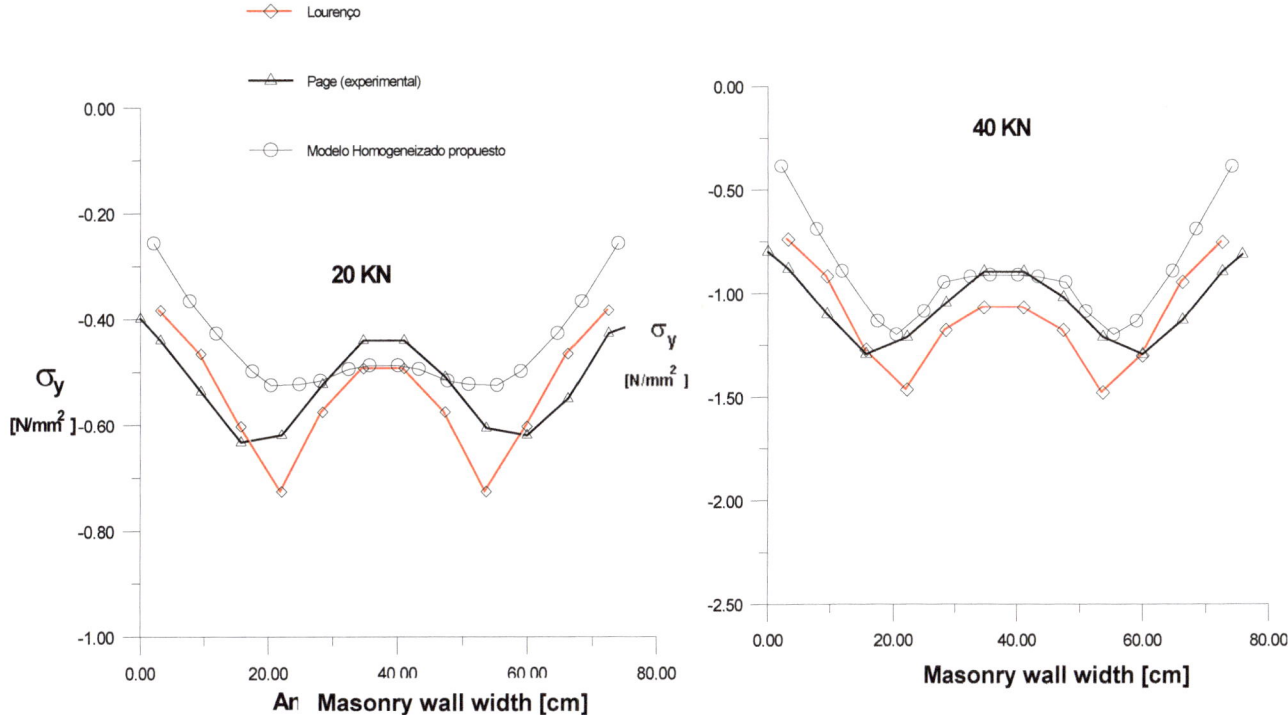

Figure 6.14 – Stress vs. masonry-wall width curve for *20 KN* and *40 KN*

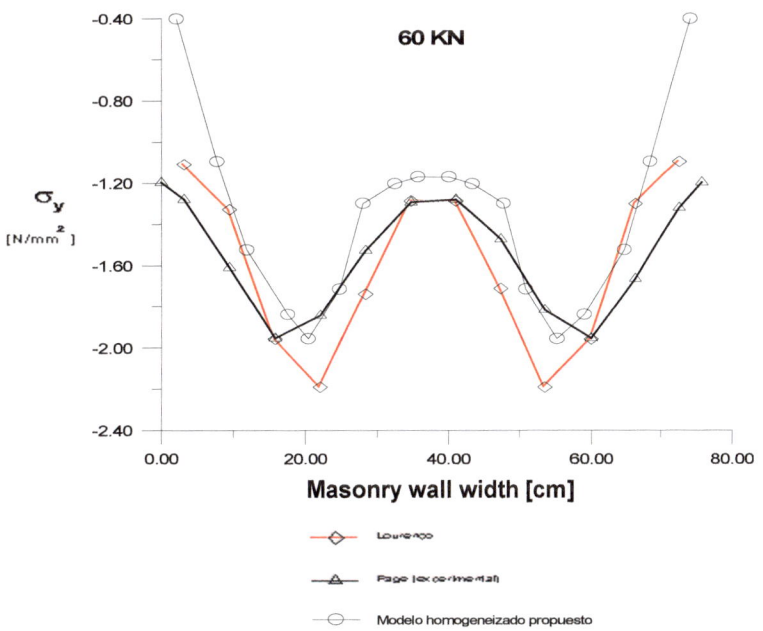

Figure 6.15 – Stress vs. masonry-wall width curve for *60 KN*

From the curves, it can be concluded that the approximation is good excepting at the ends. This is probably due to the local effects in the experimental area or because of the

Gauss's point, where the measure had been taken, does not coincide with the exact position of the gauge, as there is no record of its position. Moreover, the lack of discretization at the ends can produce this poor accuracy.

7 NON-LINEAR BUCKLING OF REINFORCED COMPOSITES

7.1 Introduction

This chapter summarizes part of the work published by J. Puig (2001)[1] and Puig et al. (2002)[2], whose motivation has been the solution of this complex phenomenon within the theoretical framework described in previous chapters.

Actually, the number of specialized works dealing with elastic instability due to compression in the fibers using finite elements is very low. In fact, this can be due to the lack of a suitable model to deal with the complex interaction phenomena between the fibers and the matrix, which are usually found in these composites. An approximate numerical solution to the local non-linear instability problem is provided in this chapter. Consequently, a constitutive model of stiffness-loss due to *micro-buckling* in long fibers in composites is presented. This model is established from a generalization of Kachanov's isotropic damage (see Oliver et al (1990)[3] and Oller (1991)[4], (2001)[5]).

The resultant formulation belongs to the context of anisotropic composite materials and it is incorporated into a finite element code. This formulation numerically simulates the complex phenomenon of local instability of composite's long fibers and, at the same time, treats this local instability within the global instability represented by a large-strain formulation. This formulation is based on the basic-substances mixing theory, general anisotropy, large strains and non-linear behavior of each basic substance of the composite. Given the amount of work involved in the treatment of the buckling problem, only the local instability and its global influence on the structure will be presented herein.

---

[1] Puig, J.M. (2001). *Resolución del Problema de Inestabilidad Elástica por Compresión en Materiales Compuestos con Fibras Largas*. Tesis de Especialidad, Universidad Politécnica de Cataluña. Barcelona, España.

[2] Puig J.M, Car E., Oller S. (2002). *Solución numérica para el pandeo inelástico de materiales compuestos reforzados con fibras largas*. Capítulo del libro: *Análisis y cálculo de estructuras de materiales compuestos*, – pp. 295, 320 - Ed. S. Oller. Centro Internacional de Métodos Numéricos en Ingeniería. Barcelona

[3] Oliver, J.; Cervera, M.; Oller, S. & Lubliner, J. (1990). Isotropic damage models and smeared crack analysis of concrete. In N. Bicanic & H. Mang (eds), *Computer Aided Analysis and Design of Concrete Structures; Proceedings 2nd International Conference* 2: 945-958.

[4] Oller, S. (1991). *Modelización Numérica de Materiales Friccionales*. Monografía No. 3, Ed. Centro Internacional de Métodos Numéricos en Ingeniería. Barcelona.

[5] Oller, S. (2001). *Fractura mecánica – Un enfoque global*. CIMNE-Ediciones UPC.

7.2 Problem description and state-of-the-art

The problem examined is found in those reinforced composite materials with fibers that are parallel to each other and oriented along the structure's longitudinal axis. These fibers behave coupled together and are confined because of the matrix effects. The reinforcement in this kind of materials is calculated so that the composite offers an optimal response against traction stresses, as these slender materials reinforced with fibers are not expected to be subjected to tractions. However, unexpected situations where the material could be subjected to traction must be considered.

As observed in the formulation, the model has not been designed for any fiber-matrix combination in particular. Thus, it is really general. It can be applied to advanced composite materials as well as to classic materials of civil engineering such as reinforced concrete (the scale can vary but the concept is the same).

7.2.1 Euler critical load

Undoubtedly, the starting point should be the instability solution due to compression of slender structures made of elastic, homogeneous and isotropic materials[6]. The non-homogeneous differential equation for ideal structures (see Figure 7.1) can be written as,

$$EIy''(x) + N \cdot y = F(Y_A, q, x) \tag{7.1}$$

where E is the elastic modulus; I is the cross section inertia when bending; N is the compression load applied on the structure's axis; y is the vertical distance between a point at the deformed to the same point at the not deformed position and F is a function depending on the particular loading state of the structure. In case there is no perpendicular loading or any other action causing bending, then equation (7.1) is reduced to its homogeneous expression

$$EIy''(x) + N \cdot y = 0 \tag{7.2}$$

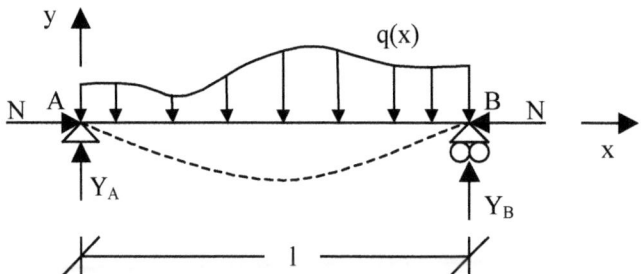

Figure 7.1 – Ideal structure: compression and bending.

and its solution is known as the Euler critical load

$$N_E = \frac{n^2 \pi^2 EI}{l^2} \tag{7.3}$$

l is the structure's length.

Generally in structures and in particular for those set up by anisotropic composites, good approximations can be obtained by using finite elements having implemented large-strain kinematics. However, instability due to compression of long-fibers embedded into a matrix is a highly-complex and difficult problem to study. The main difficulty is that both fiber and matrix properties affect the response (non-local problem). Both the fiber's geometry, which by nature is highly slender, and the confinement exerted by the matrix decisively affect the problem. The anisotropy of the majority of the fibers used in advanced composites also determines the response of these materials under compression.

The relevance of the fiber buckling problem is that it occurs at low stresses as compared to those predicted by the shell theory (dent) or by the elastic instability beam theory.

7.2.2 Rosen's model

The first attempt to propose an analytical model was made by Rosen (1965)[6], based on the idea that the failure in axial compression would be related to the fibers' curvature, as their behavior is similar to that of the buckling of slender columns. Indeed, in structures tested until failure due to the instability of fibers, the phenomenon is microscopically developed as a system of fiber micro-folds. Based on this evidence, he differentiated two possible buckling modes depending on the fiber volume of the composite: "extensional" mode (or out-of-phase) that is characteristic of low volumes of fibers and "shear" mode (or in-phase) for large volumes of fibers (Figure 7.2). On the basis of these modes and the mixing theory, the first theory to predict the material failure due to fiber instability was formulated. Specifically, for the model of misalignment in shear ("shear") or in-phase, Rosen obtained the following approximation

$$\sigma_u = \frac{G_m}{1-V_f} \tag{7.4}$$

where σ_u is the ultimate stress for compression failure of the composite; G_m is the matrix shear modulus and V_f is the fiber volume of the composite. Although Rosen's work was insufficient to reliably predict failure loads, the basic ideas of his formulation, based on the micro-fold modes, was the starting point for most subsequent work.

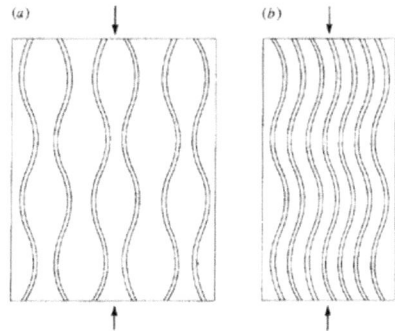

Figure 7.2 – Schematic representation of the misalignment in shear modes (a) out-of-phase and (b) in-phase in longitudinal-compression unidirectional sheets (Hull, (1987)[7])

[6] Rosen, B.W. (1965). *Mechanics of composite strengthening*. Fibre Composite Materials.
[7] Hull, D. (1987). *Materiales Compuestos*. Editorial Reverté, S.A., Barcelona, España.

7.2.3 Micro-mechanical models

7.2.3.1 Finite element formulation

In subsequent works after Rosen, several research lines have been discussed. One of them addresses a model to reproduce fiber micro-buckling and its kinematics. Thus, Balacó de Morais (1996)[8] proposed a 2-D and 3-D finite element formulation, based on the hypothesis that the fiber local instability drives the failure due to the longitudinal compression of unidirectional sheets. The original model was simplified in further works (Balacó de Morais (2000)[9]) until the linearization of the model. In any case, the starting hypothesis is that the in-phase fiber waving can be approximated with enough accuracy by using a sinusoidal curve.

$$y_0 = a_0 \left[1 - \cos\left(\frac{2\pi x}{\lambda} \right) \right] \tag{7.5}$$

The fiber curl is completely defined by means of only two parameters, namely, the wavelength λ and the maximum angle of deviation with respect to the hypothesis of perfectly aligned fibers:

$$\tan \theta_{max} = \left(\frac{dy}{dx} \right)_{max} = \frac{2\pi a_0}{\lambda} \tag{7.6}$$

In the original finite element model a perfect elastoplastic behavior with a Von Mises plastic criterion was considered for the matrix. In the same work it is recognized that this is a major simplification of the real behavior of polymeric matrices, which is much more complex. However, the hypothesis of assessing if the values obtained were realistic and determining the most important variables affecting the compression process was good enough.

Nevertheless, the multi-fiber model was too expensive (computationally). Thus, this model was used mainly to test a 2-D basic-cell simplified model. In the 3-D case, the procedure was the same but two cases for spatial distribution of the considered fibers were differentiated: hexagonal and square.

The results obtained show that the wavelength λ of the fiber initial wavy is not especially relevant. In contrast, it was found that the elastic modulus of the matrix and the volumetric participation of the fibers in the composite significantly affected the sheet compression strength, which was higher for high values of the wavelength. However, generally, the predicted values using realistic data for the model's definition are far superior to those obtained by experiments.

7.2.3.2 Simplified formulation

Due to the high computational costs of the developed model, further works presented simplified versions of the model, resulting in a linearized formulation (see Balacó de Morais (2000)[9]). The starting point of such formulation is based on the same hypotheses of the original model, i.e., an initial curl of the fiber is assumed as shown in equation (7.5). Its

[8] Balacó de Morais, A. (1996). Modelling lamina longitudinal compression strength of carbon fibre composite laminates. *Journal of Composite Materials* 30(10): 1115-1131.

[9] Balacó de Morais, A. (2000). Prediction of the layer longitudinal compression strength. Journal of Composite Materials 34(21): 1808-1820.

main parameter was the deviation angle given in equation (7.6). Similarly, it is assumed that the deformed shape of the fiber, at any instant of the loading process, can be written as:

$$y = a\left[1 - \cos\left(\frac{2\pi x}{\lambda}\right)\right] \tag{7.7}$$

The potential energy of the system is defined as,

$$\Pi = U_f + U_m - W \tag{7.8}$$

where U_f is the strain energy due to buckling; U_m is the strain energy of the matrix due to shear and W is the work done by the compression applied loads. In a previous work by Balacó de Morais & Marques (1997)[10], it was shown that the deformation energy of the fiber due to buckling was negligible and then equation (7.8) becomes

$$\Pi = U_m - W \tag{7.9}$$

The work done by the applied loads, assuming the hypothesis of small displacements, can be written as

$$W = \sigma_c (h + 2c)\frac{\pi^2}{\lambda}\left(a^2 - a_0^2\right) \tag{7.10}$$

where h and $2c$ are the transverse dimensions of the fiber and the matrix, respectively, and σ_c is the stress in the composite, obtained by the mixing theory

$$\sigma_c = V_f \sigma_f + \left(1 - V_f\right)\sigma_m \tag{7.11}$$

Furthermore, the shear-strain energy of the matrix can be expressed as,

$$U_m = 2c\frac{G_m}{\left(1 - V_f\right)^2}\frac{\pi^2}{\lambda}\left(a^2 - a_0^2\right) \tag{7.12}$$

By substituting equations (7.10) and (7.12) into equation (7.9), and applying the variational principle of minimum potential energy:

$$\frac{\partial \Pi}{\partial a} = 0 \tag{7.13}$$

yields:

$$a = \frac{a_0}{1 - \left[\dfrac{1 - V_f}{G_m}\right]\sigma_c} \tag{7.14}$$

It should be noticed that when instability arrives then $a \to \infty$ and Rosen's formula given in equation (7.4) is then recovered. It is also important to remark that the linear nature of the formulation is unsuitable for the micro-buckling failure criterion. Instead, it is assumed that failure occurs when the matrix starts to yield. Accordingly, an elastic-perfect- plastic

[10] Balacó de Morais, A. & Marques, A. T. (1997). A micromechanical model for the prediction of the lamina longitudinal compression strength of composite laminates. *Journal of Composite Material* 31(14): 1397-1412.

shear-stress-strain curve with identical strain energy of the current curve is used, as depicted in Figure 7.3.

The accepted failure criterion is Drucker-Prager's , which in this particular case can be expressed as

$$\sqrt{\sigma_m^2 + 3\tau_m^2} - \frac{\tan\beta}{3}\sigma_m - \sqrt{3}\tau_{um} = 0 \tag{7.15}$$

where τ_{um} is the shear strength of the matrix and the friction angle is

$$\tan\beta = 3\left[1 - \sqrt{3}\frac{\tau_{um}}{\sigma_{um}}\right] \tag{7.16}$$

σ_{um} is the compression strength of the matrix.

Making several assumptions and substituting the above expressions, a fourth-order polynomial is then obtained, as shown below

$$C_i\sigma_{uc}^i = 0 \tag{7.17}$$

with i=1-4, which can be solved by using numerical methods.

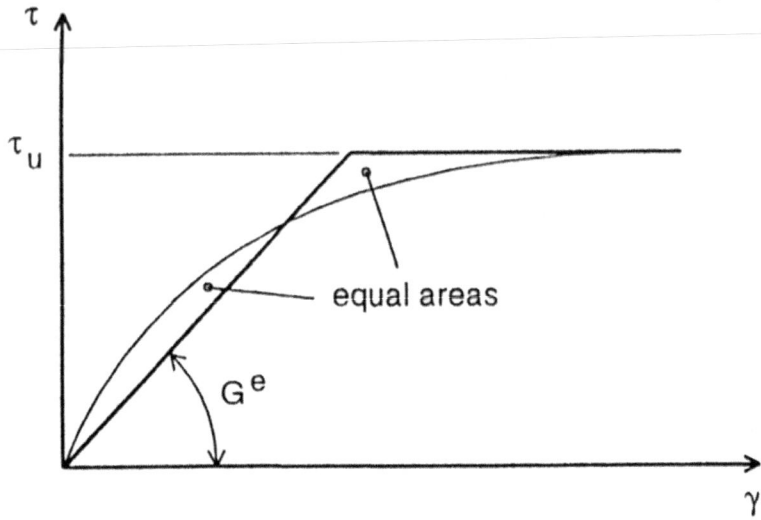

Figure 7.3 – Elastic-perfect plastic curve with the same strain energy (Balacó de Morais, 2000)

It can be shown that, in the case of the elastic-perfect plastic approximation, the failure will firstly occur when the maximum shear stress of the matrix is equal to its failure shear stress. That is,

$$\left(\tau_m\right)_{max} = \tau_{um} \tag{7.18}$$

Having all this in mind, and considering that curl angles are small ($\tan\theta \cong \theta$), an explicit expression for the failure compression is obtained

$$\sigma_{uc} = \frac{\tau_{um}}{\theta_0 + \frac{1-V_f}{1+V_f}\frac{\tau_{um}}{G_m^e}} \tag{7.19}$$

Usually, the values obtained with these formulations are good approximations to results obtained with experiments, except for glass fiber laminates. In this case, some of the hypotheses made would not be met.

In conclusion, micro-mechanical models aimed at obtaining the composite failure compression load through approximations to the real kinematics of fiber buckling are quite expensive and, therefore, simplified formulations to explicitly get these values must be derived.

7.2.4 Mechanical damage models

Other research lines found in references are based on the considerations of the imperfections as the key variable to formulate a valid model. In this sense, Barbero & Tomblin (1996)[11] combined a statistical distribution of the imperfections with the classic Kachanov's mechanical damage model. Thus, the fiber strength decreases due to the presence of imperfections in the matrix, and the larger the reduction of strength the larger the fiber volume.

The starting point of the model is finding that the micro-buckling of fibers perfectly aligned (Rosen's model) is a sensitive to imperfections problem. It means that low quantities of imperfections (misalignments) can lead to large reductions of the buckling load and then to a reduction of the compression strength predicted by Rosen's model. However, it has been observed that this sensitivity to imperfections only occurs in composites showing non-linear response to shear. Therefore, those models showing a linear response to shear do not exhibit this sensitivity.

Based on the experimental tests carried out by the authors, the strain-stress response to shear was approximated by the following equation,

$$\tau = \tau_u \tanh\left(\frac{G_{LT}}{\tau_u}\gamma\right) \tag{7.20}$$

where G_{LT} is the initial rigidity to shear and τ_u is the asymptotic value of the shear stress obtained in the experiments.

From the same tests, the adjustment of the normal distributions of the obtained misalignments was

$$f(x) = \frac{1}{\Sigma\sqrt{2\pi}}\exp\left(\frac{-x^2}{2\Sigma^2}\right); \qquad -\infty < x < +\infty \tag{7.21}$$

where Σ is the standard deviation and x is the random continuous variable, in this case the misalignment angle.

The micro-buckling phenomenon occurs at the same load for both positive and negative misalignments angles. Therefore, it is necessary to convert the normal symmetric distribution into a semi-normal distribution.

In this type of distribution, the random variable is given by $\alpha = |x|$ where x is the random variable of the normal distribution. Using the new variable α the density function of the semi-normal distribution is given by

---

[11] Barbero, E.J. & Tomblin, J.S. (1996). A damage mechanics model for compression strength of composites. *International Journal of Solids Structures* 33(29):4379-4393.

$$f(\alpha) = \frac{1}{\Sigma}\sqrt{\frac{2}{\pi}}\exp\left(\frac{-\alpha^2}{2\Sigma^2}\right); \qquad \alpha \geq 0 \tag{7.22}$$

This equation has two different meanings in the context of fiber reinforced materials. Firstly, it represents the probability that one randomly chosen fiber has a misalignment of value α. But, more importantly, assuming that the number of fibers of a section is very high, equation (7.22) shows the ratio of the number of fibers with a misalignment α on the overall number of fibers. Due to the statistical nature of the model, it is not possible to identify which fibers have a particular value of misalignment, but it is possible to ensure that a number of them, proportional to $f(x)$, are present in the cross section.

The relationship between the buckling stress (compression strength) and the magnitude of the imperfections (misalignments) is known in the theory of stability as the curve of sensitivity to imperfections. The work by Barbero & Tomblin (1996)[12] also checked that the predictions are very sensitive to the equation for modeling the strain-stress response to shear. Then, although equation (7.20) was derived to obtain a better representation of the materials analyzed, it was necessary to develop a model based on the stability theory.

Therefore, a new model based on a representative volume element (RVE) (Figure 7.4) was developed using the representation of the shear response given in equation (7.20). It incorporates a realistic representation of the sinusoidal shape used to model the fiber. The shear energy of the RVE is represented in the energy formulation used.

The model is based on the principle of the total potential energy and, for the sake of simplicity, the axial effects are neglected. Also, it is assumed that micro-buckling occurs in the extensional mode (or in-phase mode), which is a hypothesis widely accepted by models using the theory of stability. The assumption in this hypothesis is that this type of micro-buckling is observed in materials with a high volume of fibers and are the most widely used in engineering applications. By substituting and operating, an explicit relationship between the applied stresses in terms of the misalignment angle α and the shear strain γ is obtained.

$$\sigma = \frac{\tau_u}{2(\gamma+\alpha)} \cdot \frac{\left(\sqrt{2}-1\right)\cdot\left(e^{\sqrt{2}\gamma G_{LT}/\tau_u} - e^{2\gamma G_{LT}/\tau_u}\right) + \left(\sqrt{2}+1\right)\cdot\left(e^{(2+\sqrt{2})\gamma G_{LT}/\tau_u} - 1\right)}{1 + e^{2\gamma G_{LT}/\tau_u} + e^{\sqrt{2}\gamma G_{LT}/\tau_u} + e^{(2+\sqrt{2})\gamma G_{LT}/\tau_u}} \tag{7.23}$$

The results obtained by this formulation are shown in Figure 7.5 for two different responses against shear. Note that if the shear behavior is linear (constant shear stiffness), the model does not predict any critical stress value (maximum). On the contrary, if the representation of equation (7.20) is used, a maximum value is reached, which corresponds to the compression strength for the considered misalignment angle. Given the complexity of finding an explicit expression of this maximum, an implicit solution of the form $F(\sigma_{CR}, \alpha)=0$ is provided. If an explicit solution $\sigma_{CR} = f(\alpha)$ is wanted, the implicit solution can be approximated through:

$$\sigma_{CR} = (G_{LT} - k_1)\cdot\exp\left(k_2\sqrt{\alpha}\right) + k_1\exp\left(k_3\sqrt{\alpha}\right) \tag{7.24}$$

where k_1, k_2 and k_3 are constants calculated by the curve of the implicit solution. Equation (7.24) shows the following characteristics: when $\alpha \to 0$, $\sigma_{CR} \to G_{LT}$, that matches Rosen's predicted value, as well as when $\alpha \to \pi/2$, $\sigma_{CR} \to 0$.

---

[12] Barbero, E.J. & Tomblin, J.S. (1996). A damage mechanics model for compression strength of composites. *International Journal of Solids Structures* 33(29):4379-4393.

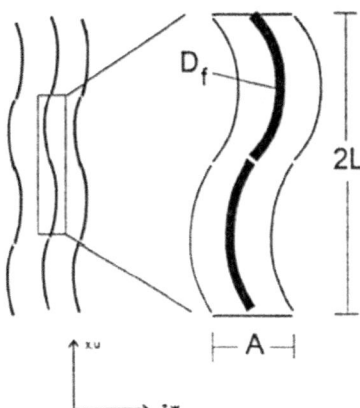

Figure 7.4 – Representative volume element (RVE) (Barbero & Tomblin, 1996)

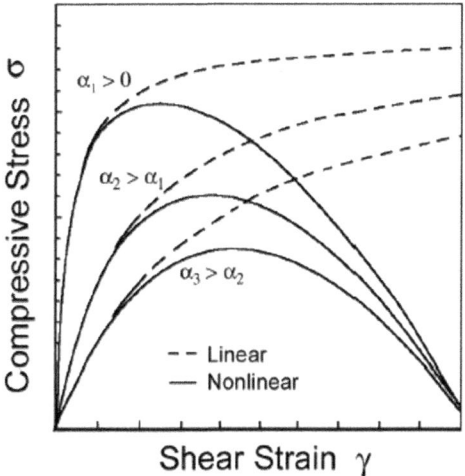

Figure 7.5 – Equilibrium states showing the compressive applied stress as a function of the shear strain for three values of the misalignment angle, considering linear behavior (dashed line) and non-linear behavior (continuous line) against shear. (Barbero & Tomblin, 1996).

The final definition of the damage model depends on some considerations. In a previous work by Tomblin (1994)[13], it was shown that a buckled fiber has no post-buckling (or post-critical) strength. However, this statement neither assumes that a buckled fiber is permanently damaged, nor that the process is irreversible. It just shows that the load capacity of a buckled fiber is lower than the applied load. The fiber can fail due to the curvature when lateral deformation is very large. Therefore, the one dimensional damage model does not require that the fiber be in a permanent damaged state. It is enough to assume that all the buckled fibers are unable to bear higher loads because they do not have post-buckling strength. While a fiber buckles, the applied load is redistributed amongst the remaining fibers that have not yet been buckled.

At any instant of the specimen loading, the applied load (at the initial instant the load is applied to the total fiber area) is equal to the effective stress, at the time considered, applied to the unbuckled fibers. That is,

[13] Tomblin, J. S. (1994). *Compressive Strength Models for Pultruded Glass Fiber Reinforced Composites.* PhD thesis, West Virginia University, Morgantown, WV, U.S.A.

$$\sigma_{app} = \sigma_{CR}(\alpha) \cdot [1 - \omega(\alpha)] \tag{7.25}$$

where $0 \le \omega(\alpha) \le 1$ is the area of the buckled fibers per unit area of initial fibers. The area of buckled fibers $\omega(\alpha)$ is proportional to the area under the curve of semi-normal distribution (Figure 7.6) located beyond the misalignment angle α that corresponds to the actual value of effective stress. Therefore, $\omega(\alpha)$ can be written as

$$\omega(\alpha) = 1 - F(\alpha) = 1 - \int_0^\alpha f(\alpha)d\alpha = \int_\alpha^\infty f(\alpha)d\alpha \tag{7.26}$$

Equation (7.25) has a maximum that corresponds to the maximum stress that can be applied to the composite. This is a unique value and it provides a value of the compression strength which considers the sensitivity curve to imperfections and the random distribution of fiber misalignments.

The comparison of the model predictions with the experiments have shown that the differences were very sensitive to the way in which the shear properties of the specimens were measured. Also, a clear dependence of the compression strength with the initial shear stiffness was found.

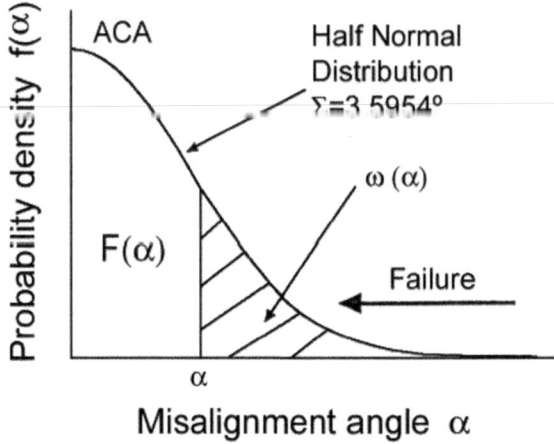

Figure 7.6 – Density distribution of semi-normal probability of the misalignments and interpretation of the fiber buckling evolution (Barbero & Tomblin, 1996).

Also, it was found that the compressive strength clearly decreases when the imperfections of fibers increase, and which do not rise at higher fiber-volumes as it also increases the probability of presenting imperfections.

In the next section, a model based on the mechanics of continuous media is presented. This model is able to generalize the numerical simulation of the instability phenomenon in composites.

7.3 Model of stiffness-loss due to buckling in reinforced long-fiber composites

7.3.1 Introduction

An approximation to the stiffness-loss due to buckling through the mechanics of continuous media is presented in this section. The drawback of this model is the analysis of the kinematic problem as a constitutive approach. However, the same problem can be seen from two points of view:

1. The kinematic that studies the global stiffness-loss.

2. The constitutive that studies the strength-loss at a point due to the global stiffness-loss.

Both points of view are assumed as valid in this model.

Also, it should be remarked that the model was not conceived for any particular material type. Thus, the formulation is as general as possible. It can be applied both to advanced composite materials and to classic materials in civil engineering such as reinforced concrete. The concept is the same and only the scale varies.

The theory does not assume any determined buckling mode. This allows the model to be used for any fiber volume.

7.3.2 General definition of the fiber model

The proposed model is a modification of the Kasyanov isotropic damage model (Oliver et al. (1990)[3]). This modification involves including a new internal scalar variable, which defines the buckling behavior, as shown in the equation below

$$\boldsymbol{\sigma} = \left(1 - d_f\right) \cdot \left(1 - d_p\right) \cdot \mathbb{C} : \boldsymbol{\varepsilon} \tag{7.27}$$

where $\boldsymbol{\sigma}$ is the stress tensor, $\boldsymbol{\varepsilon}$ is the strain tensor, $\mathbb{C}$ is the initial constitutive tensor, d_f is the mechanical damage variable, and d_p is the variable of stiffness-loss due to buckling. Variable d_p has lower and upper bounds, taking values between zero and one, $0 < d_p < 1$, at each end. The resultant model is therefore an isotropic damage model. The mechanical damage variable d_f is also bounded and $0 < d_f < 1$ is obtained the same way as in the original model (see Oliver et al. (1990)[3]). Its basic property is irreversibility and its function within the model is to evaluate the material degradation when the material strength is exceeded.

The stiffness-loss variable due to buckling d_p, in opposition to the mechanical damage variable, is characterized mainly by its reversibility. It is useful for the fiber stiffness-loss variable due to instability as a material-strength loss at a constitutive level, affected by the fiber matrix interaction. To avoid confusion, it should be remarked that from a phenomenological point of view buckling is not a constitutive phenomenon. But an equivalence relation is made in this model between the kinematic problem of buckling and the constitutive problem generated by this stiffness-loss reversible variable d_p. This "macroscopic simulation" procedure allows the study of the instability of a large number of fibers, regardless the behavior of isolated fibers. Thus, the influence of this phenomenon over the whole set of fibers is considered.

7.3.3 Definition of the stiffness-loss variable due to buckling

The definition of d_p is complemented by including the damage model into the generalized mixing theory and the mapping-space technique to deal with general anisotropy, proposed by Car (2000)[14]. The representation of the fiber as a set of slender columns, as proposed by Rosen (1965)[6], is also considered. Unlike this model, no "a priori" mode of micro folds of the material fiber is assumed, but the instability mode naturally yields from the formulation itself. The most general definition is shown below:

$$d_p = f_1(\text{fiber}) \cdot f_2(\text{matrix}) \tag{7.28}$$

where f_1(fiber) is a function based on the characteristic properties of the fiber and f_2(matrix) depends on the matrix properties. It is mandatory to consider the first function because a model of fiber instability must consider the mechanical properties of the fibers. The inclusion of the second function is intended to simulate the movement constraint exerted by the matrix over the fiber, thus establishing a boundary condition for the fibers that depends on the evolution of the mechanical process (internal variable).

The basic idea behind this definition is to exploit the situation generated by the generalized mixing theory for the numerical solution of the interaction among the components. This theory is based on the decomposition of the composite material into its components, the integration of the constitutive equation of each of them and the determination of the composite behavior as a combination of the response of each material component by considering the volumetric participation of each component. The interaction among the different materials occurs in the recomposition of the response. Consequently, the idea is to consider the fiber as a system of independent columns acting together but without the cohesive matrix (Figure 7.7). These columns are geometrically perfect and flawless. Under loading, the fibers will have a natural tendency to buckle due to their large slenderness. The effect of the movement restriction and, in general, of the confinement exerted by the surrounding matrix is considered in two levels: through the recomposition of the mixing theory and the variable of stiffness-loss due to buckling. This assumes a major factor of interaction among the material components.

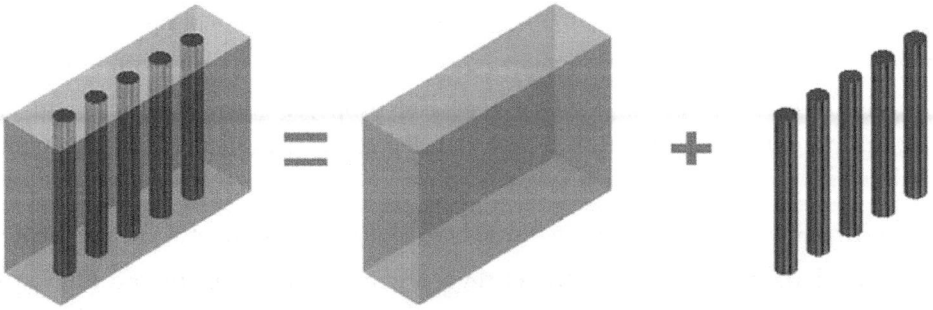

Figure 7.7 – Decomposition of the composite material.

[14] Car, E.J. (2000). *Modelo Constitutivo Continuo para el Estudio del Comportamiento Mecánico de los Materiales Compuestos*. PhD thesis, Universitat Politècnica de Catalunya. Barcelona, España.

7.3.3.1 Fiber participation

Fiber participation is considered by means of a relationship between the applied stress and the comparison stress (f_1(fiber)=$\sigma_{aplic}/\sigma_{comp}$). This formulation for the fiber is obtained by the classic expression of the Euler critical stress, modified to consider the reduction of the elastic modulus due to the fiber's degradation. These applied and comparison stresses are defined as

$$\sigma_{app} = \left(1-d_f\right)\cdot\left(1-d_p\right)E_0\varepsilon \tag{7.29}$$

where σ_{app} is the fiber longitudinal stress, E_0 is the fiber longitudinal modulus, ε is the fiber longitudinal elastic deformation, and

$$\sigma_{comp} = \frac{n_p^2\pi^2 I}{A\cdot l^2}\left(1-d_f\right)\cdot\left(1-d_p\right)E_0 \tag{7.30}$$

where σ_{comp} is the comparison stress, n_p is the buckling mode, $I=\pi d^4/64$ is the inertia of the fiber cross section with respect to the bending axis, where d is the fiber's nominal diameter, and l is the fiber's nominal length. The buckling length is generally defined as

$$\bar{l} = \frac{l}{n_p} \tag{7.31}$$

The comparison stress (see equation 7.30) has been obtained by substituting the material elastic modulus of the Euler critical stress (equation 7.3) by the damaged modulus corresponding to each load increment. This change leads to a difference between both stresses: the comparison stress depends on the loading process and therefore it cannot be defined in advance while the Euler classic expression does not depend on the loading process. The basic idea, however, is the same in both cases: to provide a reference stress for the definition of the applied stress on the material to create instability and the way it will be produced.

The buckling length of the model (equation 7.31) might be interpreted as a suitable modification of an original nominal length l. In each particular case, the analyst must decide the most suitable value for the best representation of the problem. In particular, for a typical composite material, length l will be equal to the fiber's nominal length. Similarly, in the analysis of a reinforced concrete beam or column, the suitable value would be the distance between stirrups. The subsequent evolution of the buckling length will be defined by the loading process, but further complementary adjustments will not be necessary.

The evolution of the buckling modes n_p depends on the loading process. For loading states where fiber length increases with displacements (growing branch of the stress-strain diagram), the buckling mode should be such that the comparison stress is equal to or greater than the stress reached in the previous step of the loading process. Thus

$$n_p^{t+1}/\sigma_{comp}^{t+1} \geq \sigma_f^t \tag{7.32}$$

where $t+1$ is the load step corresponding to the current time, t is the previous loading step and σ_f is the fiber stress at the convergence. If the fiber cannot take higher levels of stress related to new strain increments (decreasing branch of the stress-strain diagram) it is assumed that fiber buckles and, therefore, the ratio between the applied stresses σ_{app} and the comparison stresses σ_{comp} is equal to 1. In this case, the fiber buckling mode can be approximated as

$$n_{\lim} = \frac{4l}{\pi d} \sqrt{\varepsilon} \tag{7.33}$$

where $n_{\lim}$ is the limit buckling mode corresponding to the decreasing branch of the stress-strain diagram.

In summary, the fiber participation is defined as

$$f_1(\text{fibra}) = \frac{\sigma_{\text{aplic}}}{\sigma_{\text{comp}}} = \left(\frac{4l}{n_p \pi d}\right)^2 \varepsilon \tag{7.34}$$

7.3.3.2   Matrix participation

The second part of the definition is based on the idea that the variable of stiffness-loss due to buckling should consider the effect of the movement restriction and the confinement exerted by the matrix. This effect must be considered because the mixing theory is not a valid tool for its representation.

Therefore, it is necessary for the matrix to be represented by Kachanov's isotropic damage model. The mechanical damage variable corresponding to the matrix, d_m, as a measure of material degradation is also used in this model as a representative variable of the matrix confinement ability over the fiber. If the matrix is formed by only one material we can define

$$f_2(\text{matrix}) = d_m \tag{7.35}$$

If the matrix is formed by more than one component, the definition in equation (7.9) is generalized as follows

$$f_2(\text{matrix}) = \sum_{j=1}^{n_m} k_{m,j} d_{m,j} \tag{7.36}$$

where n_m is the number of components in the matrix, $d_{m,j}$ is the mechanical damage of the j-th matrix material and $k_{m,j}$ is the coefficient of local volumetric participation of the j-th matrix material, which is defined as

$$k_{m,j} = \frac{V_{m,j}}{V_m} = \frac{V_{m,j}}{V - V_f} \tag{7.37}$$

where V_m, V_f, $V_{m,j}$ and V are, respectively: matrix volume, fiber volume, j-th matrix material volume and total composite volume. That is, the mixture theory of substances is applied to the mechanical damage of matrix component materials, then establishing a composition of behaviors.

Finally, the convenience (or not) of modifying the definition given in equation (7.36) should be considered as follows:

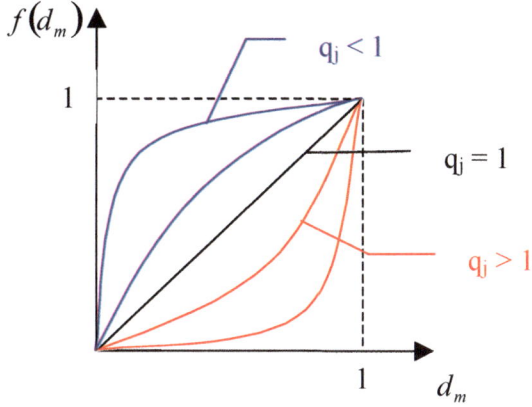

Figure 7.8 – Representation of the matrix confinement ability using an exponent q_j.

$$f_2(\text{matriz}) = \sum_{j=1}^{n_m} k_{m,j}\left(d_{m,j}\right)^{q_j} \tag{7.38}$$

where q_j is a regulating factor of the effect the matrix exerts on the fiber. That is, if $q_j > 1$, the material is able to effectively restrain the movement during the first steps of the material degradation process (Figure 7.8). The higher the damage increases, the higher the buckling risks. On the contrary, if $q_j < 1$ the material rapidly loses the ability to restrain the movement as soon as the degradation begins.

However, the subject is probably much more complex than it is explained here. The exact adjustment of such coefficients possibly depends on the material combination. That is, the same matrix with two different types of fibers could exhibit a different behavior caused by the fiber's instability, because the adherence mechanisms are different due to their own process of composite fabrication. However, extreme cases should be avoided and not considered. This means that the definition of a single value for all possible cases would be as impractical as trying to define a single value for each and every possible material combination. The definition of an average value for each material, regardless of the combination of materials and the fiber type, would be enough for the majority of the cases. Including this variable q_j into the model has a clear drawback. It would be an empirical (or semi empirical) parameter. The best option would be to define such coefficients as a function of the properties of the matrix components. However, value approximation from experimental tests would be easier. If one acts in this latter form, the model depends on an experimental variable, which would lose the generality of the formulation above presented. Therefore, before trying to define such a parameter, the advantages of including it (or not) into the model should be discussed.

7.3.3.3   Proposal of a variable for stiffness-loss due to buckling

Up to now, the stiffness-loss variable definition due to the buckling proposed herein can be written as

$$d_p = \left(\frac{4l}{n_p \pi d}\right)^2 \varepsilon \left[\sum_{j=1}^{n_m} k_{m,j} d_{m,j}\right] \tag{7.39}$$

Thus, the stiffness-loss variable due to buckling depends on the geometric characteristics of the fiber (length and diameter), on the evolution of the mechanical characteristics of the matrix (weighted sum of the mechanical damage of the matrix components) and on the loading step considered (acting strain and evolution of the buckling mode).

7.4 Main characteristics of the model

From the variable's own definition for stiffness-loss due to buckling d_p (equation 7.39) some conclusions regarding the behavior of the model defined in equation (7.27) can be drawn. Firstly, if there is no matrix degradation, fiber buckling is impossible. This is clearly stated in equation (7.39). If there is no matrix degradation, there is no matrix mechanical damage and thus the stiffness-loss variable d_p value is zero. The stiffness-loss will occur only when the first damage threshold in the matrix is reached.

As observed in equation (7.27), the total stiffness-loss of the fiber strength might occur as a consequence of the mechanical damage d_f in the fiber, the stiffness-loss due to buckling d_p or a combination of both effects. Research indicates that, in these cases, the failure is due to a combination of both phenomena. When the process starts, the fibers can be subjected to both large displacements and curvatures that will not be able to withstand, leading to the fiber's physical failure. This sentence could be demonstrated using a simple example.

The example is a single 4-noded rectangular finite element with 4 Gauss points (plane stress state) and subjected to imposed displacements, as displayed in Figure 7.9. The fibers are arranged parallel to the stress direction. The material properties are collected in Tables 7.1 and 7.2

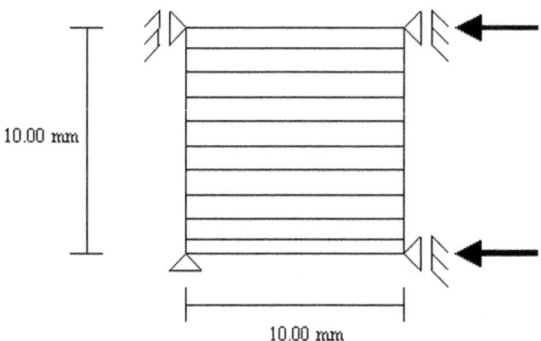

Figure 7.9 – Problem sketch and boundary conditions.

Property	Value
Model type	Kachanov's isotropic damage
Young modulus (MPa)	$3.5 \cdot 10^4$
Compressive strength (MPa)	10
Internal friction angle	30°
Crush compression energy (N/m)	$2.0 \cdot 10^3$
Matrix volume (k_m)	0.40

Table 7.1 – Matrix mechanical properties (concrete).

Property	Value
Model type	Several: Kachanov's isotropic damage Stiffness-loss due to buckling Coupled Kachanov's isotropic damage and buckling
Longitudinal Young's modulus (MPa)	$2.1 \cdot 10^5$
Transverse Young's modulus (MPa)	$2.33 \cdot 10^4$
Compressive strength (MPa)	200
Poisson's coefficient	0.00
Crush compression energy (N/m)	$5.0 \cdot 10^3$
Fiber volume (k_f)	0.60

Table 7.2 – Fiber's mechanical properties (steel).

The fiber has been modeled by three different models to compare different responses. The first one is Kachanov's isotropic damage model. The second one is a simplified model that only takes into consideration the stiffness-loss due to buckling but not the possible mechanical damage. Finally, the model of stiffness-loss due to buckling is used, taking into consideration the coupling of both the mechanical damage variable and the buckling variable.

It should be highlighted that the aim of the example is not the possible global buckling simulation of the specimen (talking about a finite-element buckling is useless) but the stiffness-loss due to fiber instability at microscopic level. As seen before, although both phenomena share similar theoretical basis, they represent very different structural behaviors. Thus, while the general buckling of the structure causes a stiffness-loss of all the material components, the local instability of fibers leads to a sudden stiffness-loss of the reinforcement material causing a stress transfer to the matrix. These stresses often exceed by far matrix strength, leading to the failure of the whole composite. In summary, the global buckling of the structure causes a stiffness-loss from composite to components, while local instability leads to a stiffness-loss from components to composite.

To complement the description above and although it is not treated herein, it should be mentioned that the mixing theory also considers matrix damage not only from the point of view of its influence on the fiber's buckling, but also as the deterioration of the matrix itself. Among the possible causes leading to this degradation is the damage produced by the global buckling.

The results are shown in Figure 7.10. Kachanov's isotropic damage model is unable to simulate the fiber degradation due to instability (Figure 7.1) that occurs as a consequence of the stresses and strains which are much lower than the first damage threshold. In other words, the mechanical damage variable is a measure of the stiffness-loss due to increments of the applied stresses on the fiber but it is unable to represent the instability phenomena caused by the loss of the matrix strength.

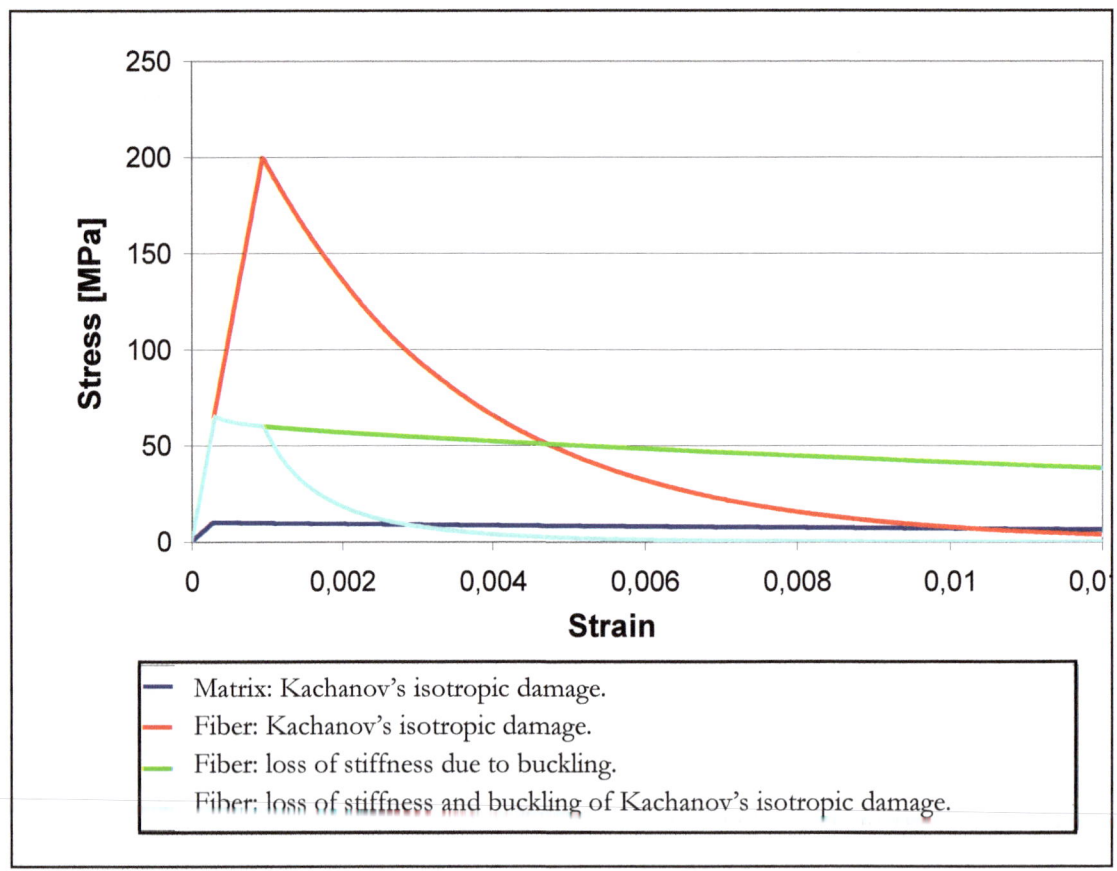

Figure 7.10 – Comparison of fiber responses in different models.

The stiffness-loss variable due to buckling d_p is not able to represent the total buckling phenomenon, since it cannot consider the cracks and failures occurring in the matrix and fibers caused by large deformations. The only way to reproduce this complex phenomenon is by combining the matrix and fiber damage together with the stiffness-loss due to buckling. In the case under study, where the matrix strength is much lower than the fiber's (the most common case in advanced composites), a stiffness-loss in the fiber begins when the first damage threshold of the matrix is reached. Then, the peak value of the fiber stress is much lower than the material yield stress. When the mechanical damage of the fiber begins, the multiplier effect of the stiffness-loss due to buckling is observed: the strength loss decreases significantly and quickly, and displays stress levels which are much lower than those obtained by Kachanov's damage model (see Puig (2001) for a more detailed description).

Reversibility is one of the main properties of the stiffness-loss variable due to the buckling defined in equation (7.39) that differs from the classical damage variable. During a monotonically increasing loading process, the variable of stiffness-loss due to buckling will also increase. Specifically, the values of the buckling modes n_p, strain ε and matrix damage d_m must be updated. If an unloading process begins, the matrix damage and the buckling mode remain constant: the first one due to its irreversibility and the second one because of the boundary conditions derived from matrix degradation. Indeed, the matrix confinement over the fiber can be represented, following a structural scheme, by micro boundary conditions induced from one material to another. The changes in the physical properties of the matrix due to its own degradation lead to variations in the micro boundary conditions and, therefore, the fiber buckling mode must also change. During the unloading process the matrix cannot recover the original material properties, keeping constant both the reached damage and the micro boundary conditions induced into the fiber. Consequently, the buckling mode must also be constant.

The total deformation in the composite, at each load step, is predicted before the mixing theory algorithm is applied. Then, the deformation at each component is obtained by a compatibility equation between fiber and matrix (Car (2000)). This procedure does not depend on the loading or unloading processes. Therefore, strain ε will vary from the last value reached in the loading process until null value if a complete unloading is performed. Then, the stiffness-loss variable due to buckling d_p will behave in the same way. That is, in a loading process until the $r = t+i$ step, we will have

$$d_p^{t+1} \in \left[0, d_p^r\right]; \ \forall (t+1) \in [1, r] \tag{7.40}$$

If a complete unloading process ranges between step $r+1$ until step $s = r+k$, we have

$$d_p^{r+j} \in \left[d_p^r, 0\right]; \ \forall (r+j) \in [r, s] \tag{7.41}$$

Thus, the value of the stiffness-loss variable due to buckling d_p is totally reversible. However, the loading process is not.

Indeed, Figure 7.11 shows a complete loading cycle, the fiber unloading and reloading. The loading process is described by the O-A-B-C curve, the unloading by the C-O parabola and the reloading by the same parabola O-C. At this point, if the loading is maintained, the material will continue the degradation just before the unloading.

The **first conclusion** from this loading-unloading-reloading process is that the unloading follows a non-linear path. It can be shown that the corresponding expression is (Puig, 2001),

$$\sigma_{\text{loading-unloading}} = \mathbb{C}(\varepsilon) : \varepsilon \tag{7.42}$$

The **second conclusion** is that the fiber's instability is partially not recoverable, because this instability is produced after matrix damage. In other words, the stiffness-loss due to instability is a phenomenon directly related to the compressive load. If it decreases, the material must recover the stiffness lost (equation 7.41 shows that the stiffness-loss variable is fully recoverable). However, it should be noted that fiber loading has taken place with inconstant matrix stiffness during the mechanical process, because it is subjected to the degradation considered into the mechanical damage variable. In the unloading process, the matrix is unable to recover its original properties, so that the fiber unloading process must follow a different path from the one of the loading process.

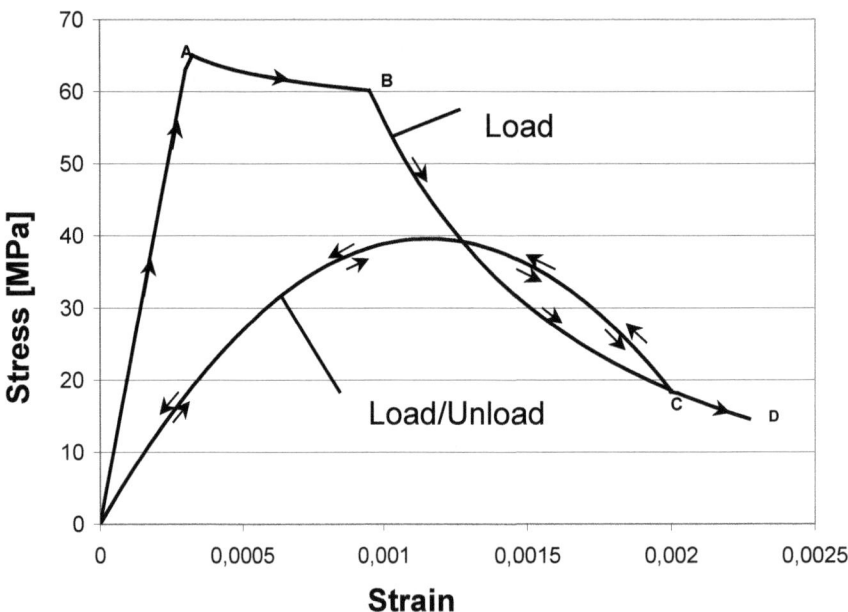

Figure 7.11 – Loading-unloading-reloading cycle of the fiber.

The **third conclusion** is that in an unloading-reloading process, once the stiffness-loss due to buckling has begun, the fiber cannot resist like the original material, even in the absence of mechanical damage of the fiber. As shown in Figure 7.10, if an unloading process begins before reaching the first mechanical-damage threshold of the fiber, Kachanov's isotropic damage model will predict that the fiber will stay within the elastic range. Therefore, the fiber would behave like the original material regardless of the matrix degradation process starting. Instead, if the stiffness-loss due to buckling is also considered, Figure 7.11 shows that the fiber does not behave the same way. And if the unloading occurs when the fiber's mechanical damage has reached the first threshold, the model predicts that the fiber's response will also be lower than the response of Kachanov's isotropic damage model, because the stiffness-loss variable is different from zero (despite the fact that it decreases) until the unloading is completed.

The **fourth and more important conclusion** that is drawn from the previous one is that the buckling phenomenon in long fibers runs under a dissipative process. Its details can be accessed in Puig (2001).

7.5 Energy dissipation

As observed in Figure 7.11 and more clearly in Figure 7.12 the buckling phenomenon in long fibers occurs under a dissipative mechanical process. Figure 7.12 displays a complete loading-unloading cycle at the fiber. The deformation energy of the loading process (area under the O-A-B curve) is bigger than the energy of the unloading process (area under the B-O curve). The difference is the dissipated energy. As seen before, the reloading runs by the same curve of the unloading (O-C curve in Figure 7.11). In other words, the unloading-reloading processes are not dissipative.

The dissipated energy is physically justified because the fiber's boundary conditions are different during the loading and unloading processes. Indeed, the micro boundary conditions induced by the matrix confinement on the fiber evolve along the loading process due to the degradation of the matrix. This means that this degradation prevents the matrix from exerting confinement with the same efficacy as the undamaged material. Since the boundary conditions are not the same, the deformation energies coming from both processes will also be different.

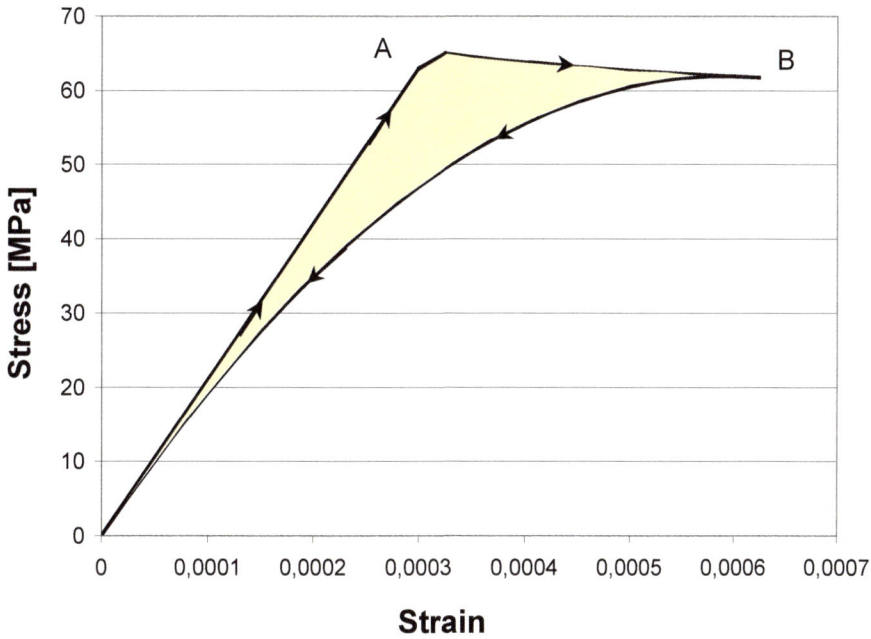

Figure 7.12 – Load and unload cycle of the fiber. Energy dissipated.

For the sake of clarity, consider the process described in Figure 7.13, where a slender column of length l has the lateral displacements restrained at its ends. The column is subjected to a loading process and then it buckles. In this situation, the critical load for each spans, and therefore for the structure, is expressed as follows

$$N_{E,a} = \frac{\pi^2 EI}{(l/2)^2} = \frac{4\pi^2 EI}{l^2} \tag{7.43}$$

In the buckled configuration, consider that the middle support fails and it is not possible to restrain the displacement. Then, the Euler critical load is written as

$$N_{E,b} = \frac{\pi^2 EI}{l^2} \tag{7.44}$$

Equations (7.43) and (7.44) can also be interpreted as the Euler critical loads for the first and second buckling modes of a slender column of length l (equation (7.3)). If we consider ideal structures, the column depicted in Figure 7.13.b will naturally take the deformed shape and the load corresponding to the second buckling mode. However, in a real situation with real structures, the column would take the first buckling mode because it represents a less-energetic equilibrium situation. The change in the structure will be observed at constitutive level as the deformations increase and stresses decrease, as shown in Figure 7.13.c. If the structure is unloaded, it is evident that the path in the stress-strain diagram will not be the same as in the loading process. Therefore, *the energy dissipation is caused by the changes in the structure's boundary-conditions.* This reasoning can be summarized as that the boundary conditions in both loading and unloading processes are different and that two elastic processes having different boundary conditions will produce different strain energies.

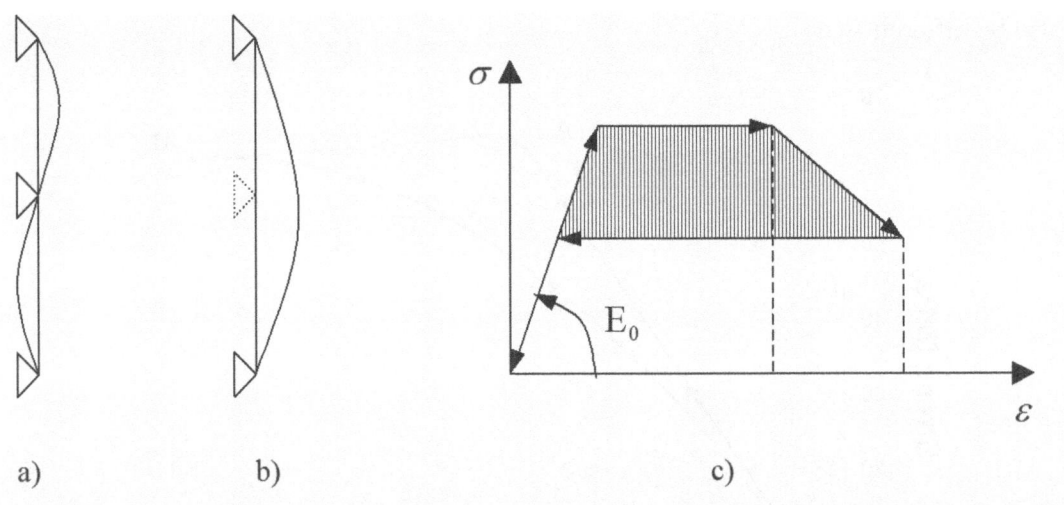

Figure 7.13 – Loading and unloading of a slender column: *a)* loading and buckling; *b)* middle support damaged and unloading; *c)* dissipated energy.

As a corollary, it can be stated that *the phenomenon of elastic instability in composites with long fibers dissipates energy* like *an evolutionary process of classic-elastic instability.* In both cases, the dissipation is caused by changes in the boundary conditions of the problem.

It should be remarked that the term *elastic* refers to the absence of plastic phenomena in the components of the composite. However, the whole process can be regarded as *inelastic,* as it dissipates energy.

Finally, the matrix degradation remains constant during the unloading process, since it is not possible to recover the original properties of the undamaged material. Therefore, the micro boundary conditions do not vary and should not vary in the loading process. Then, both phenomena are non-dissipative. When the reloading reaches again the unloading starting point (point C in Figure 7.11), the matrix exceeds again its damage threshold and thus the degradation processes are resumed (C-D curve in Figure 7.11). Consequently, the stiffness-loss due to buckling and the mechanical degradation of the fibers (if any) are also resumed.

7.6 Test example

An application example of the previously discussed formulation of the buckling model is presented in this section. The example numerically simulates a slender structure of composite with long fibers (reinforced concrete). The mechanical properties of the matrix correspond to the concrete ones having strength of 10 MPa. The fiber's properties correspond to steel having strength of 200 MPa (see Tables 7.3 and 7.4). Matrix and fibers participate with 40% and 60%, respectively. Figure 7.14 shows the finite element mesh. The fibers are arranged along the bar's longitudinal axis. A large-deformation plane-stress problem has been solved to consider the global instability of the bar.

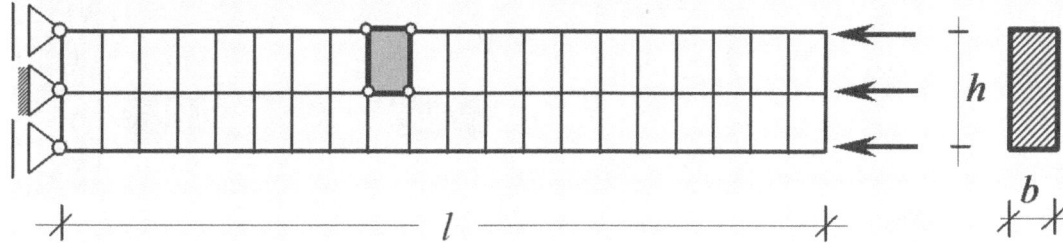

Figure 7.14 – Finite element mesh and boundary conditions for the numerical simulation.

The aim of the example is to compare the behaviors with and without local buckling. This will show the differences between the structure's global instability and the fiber's local instability.

Property	Value
Model type	Several: Elastic. Kachanov's isotropic damage
Young modulus (MPa)	$3.5 \cdot 10^4$
Compression strength (MPa)	10
Poisson's coefficient	0.20
Internal friction angle	30°
Crush compressive energy (N/m)	$2.0 \cdot 10^3$
Matrix volume (k_m)	0.40

Table 7.3 – Matrix mechanical properties (concrete).

Property	Value
Model type	Several: Elastic. Kachanov's isotropic damage and stiffness-loss due to buckling.
Longitudinal Young modulus (MPa)	$2.1 \cdot 10^5$
Transversal Young modulus (MPa)	$2.33 \cdot 10^4$
Compressive strength (MPa)	200
Poisson's coefficient	0.00
Crush compressive energy. (N/m)	$5.0 \cdot 10^3$
Fiber volume (k_f)	0.60

Table 7.4 – Fiber's mechanical properties (steel).

The *first* case is obtained through the simulation of global buckling and without local buckling, assuming elastic components. To reach the structural global instability, it is necessary to insert a perturbation into the system. This was carried out by a small transversal load at the right end of the mesh in Figure 7.14. The initial deformed shape is shown in Figure 7.15 (displacements in cm), where the global instability is evident. Figure 7.16 displays the reaction-lateral displacement curve at the middle points of each end of the structure. As shown in the figure, structural stiffening occurs beyond the buckling step (post-critical strength) for transversal displacements bigger than 40 cm.

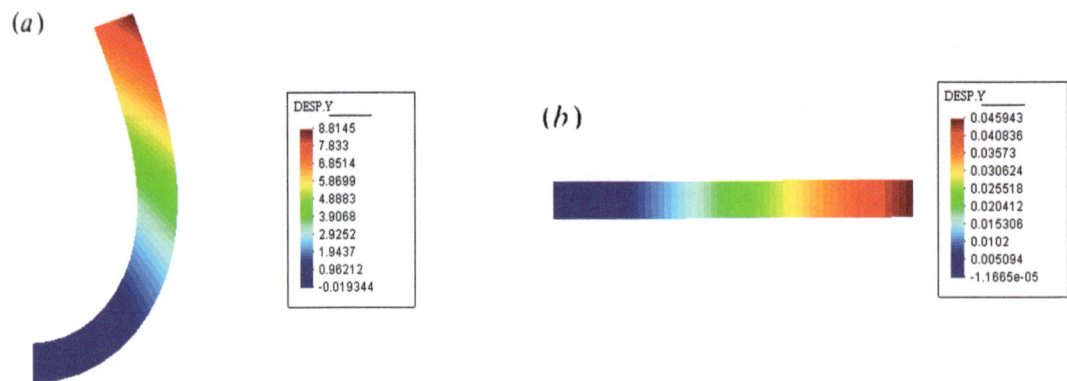

Figure 7.15 – Deformed shape: *a)* Matrix and fiber are perfectly elastic; *b)* Matrix simulation with a Kachanov's isotropic damage model and fiber simulation with an isotropic damage model having stiffness-loss due to buckling.

The second simulation models the matrix with a Kachanov's isotropic damage model and the fiber with a model having stiffness-loss due to buckling. The final deformed shape, drawn to the same scale and using the same units as in the previous case, is shown in Figure 7.15.b. In this case, the structure fails due to the micro buckling of fibers but any global instability occurs. In contrast, the structure displays a brittle failure when using a damage model. Similarly, the results in terms of reaction-lateral displacement at the same considered points in the previous case are shown in Figure 7.16. Due to the difference in scales observed in the figure, the first part of the figure is zoomed in Figure 7.17.

The results obtained in this second simulation are quite different to those obtained in the previous simulation. First, the displacements are much smaller than in the elastic case. Furthermore, these displacements quickly increase stress levels below those observed in the first simulation. Nevertheless, the most important qualitative difference can be observed in the final part of the figure where stiffening does not occur. Results in Figure 7.17 show the inability of the structure to withstand higher levels of strains, which would be equivalent to the structural failure in the laboratory.

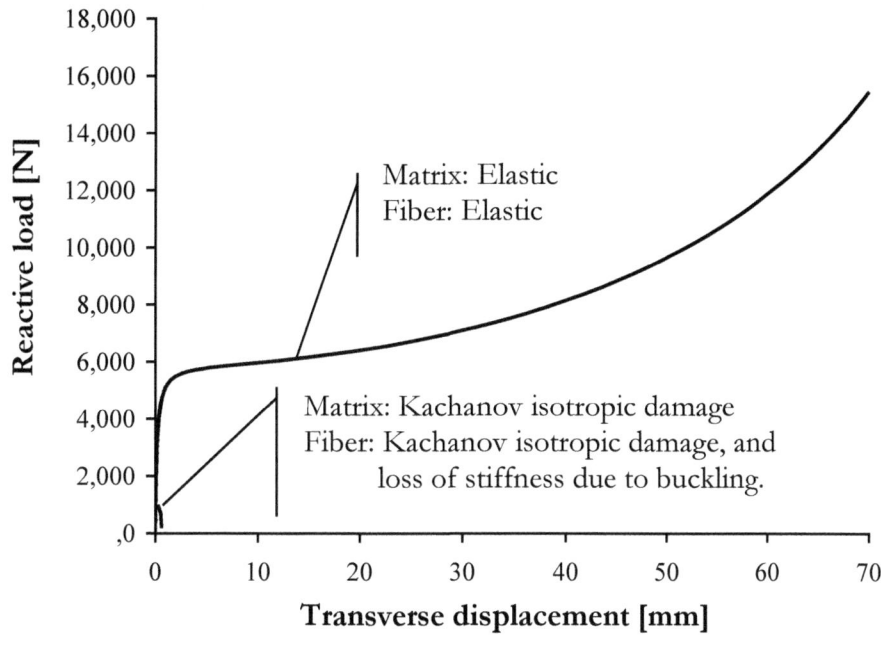

Figure 7.16 – Comparison of results obtained with different models.

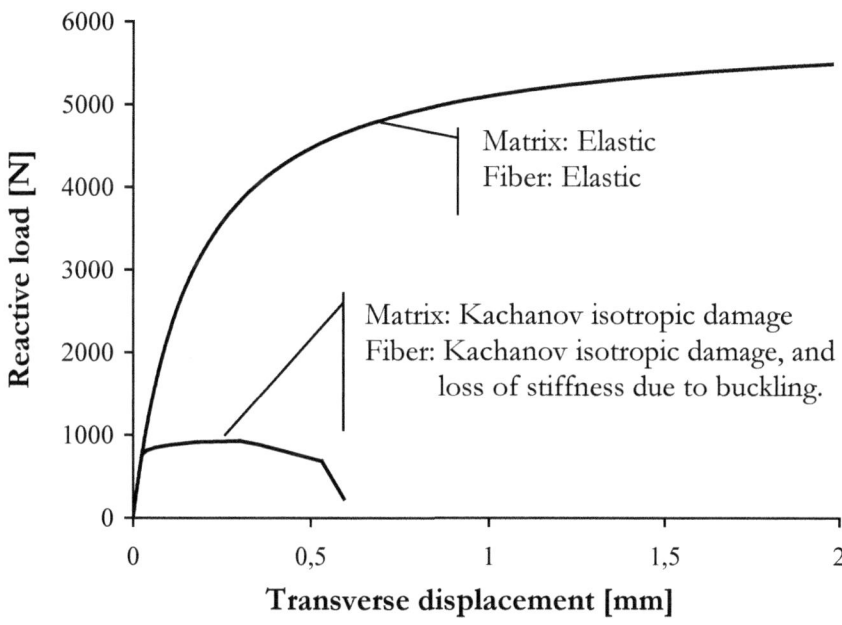

Figure 7.17 – Amplification of the left part of Figure 7.16.

Therefore, the differences between simulations are both quantitative and qualitative.

In the second case, fiber instability occurs at stress and displacement levels well below those arising in elastic instability. Furthermore, the post-critical strength obtained in the elastic model does not represent the real phenomenon. The damage model, by contrast, has no such post-critical strength, but a stiffness-loss due to the matrix degradation and fibers instability.

Therefore, it is concluded that the elastic model (perfect elasticity for each component) is unsuitable for simulating the long fiber buckling. Then, the damage model (isotropic damage for the matrix and stiffness-loss due to buckling for the fibers) must be used for these purposes.

Subject Index

Author Index

GPSR Compliance

The European Union's (EU) General Product Safety Regulation (GPSR) is a set of rules that requires consumer products to be safe and our obligations to ensure this.

If you have any concerns about our products, you can contact us on ProductSafety@springernature.com

In case Publisher is established outside the EU, the EU authorized representative is:

Springer Nature Customer Service Center GmbH
Europaplatz 3
69115 Heidelberg, Germany

Batch number: 09492640

Printed by Printforce, the Netherlands